A Diachronic Study of *Sus* and *Bos* Exploitation in Britain from the Early Mesolithic to the Late Neolithic

Sarah Viner-Daniels

BAR British Series 596
2014

Published in 2016 by
BAR Publishing, Oxford

BAR British Series 596

A Diachronic Study of Sus *and* Bos *Exploitation in Britain from the Early Mesolithic to the Late Neolithic*

ISBN 978 1 4073 1263 7

BAR Publishing is the trading name of British Archaeological Reports (Oxford) Ltd.
British Archaeological Reports was first incorporated in 1974 to publish the BAR
Series, International and British. In 1992 Hadrian Books Ltd became part of the BAR
group. This volume was originally published by Archaeopress in conjunction with
British Archaeological Reports (Oxford) Ltd / Hadrian Books Ltd, the Series principal
publisher, in 2014. This present volume is published by BAR Publishing, 2016.

Printed in England

BAR
PUBLISHING

BAR titles are available from:

BAR Publishing
122 Banbury Rd, Oxford, OX2 7BP, UK
EMAIL info@barpublishing.com
PHONE +44 (0)1865 310431
FAX +44 (0)1865 316916
www.barpublishing.com

Acknowledgements

This work could not have been completed without the help of a great many people. My supervisor, Umberto Albarella, provided hours of discussion and guidance for which I am extremely grateful. Mike Parker Pearson was supportive throughout my PhD studies and beyond. I am also grateful to both Peter Rowley-Conwy and Paul Pettitt for discussion of this research during my viva and the suggestions for improvement that they made.

Thanks also go to those that helped with access to the material on which this study is based and include the staff of the Natural History Museum, British Museum, Dorchester County Museum, the Alexander Keiller Museum, the Wiltshire Heritage Museum, the Corinium Museum, Newport Museum, West Berkshire Museum, Manchester Museum, Doncaster Museum and Wessex Archaeology. Particular thanks go to Peter Rowley-Conwy, Dale Serjeantson and Tony Legge for providing unpublished data and in prep. papers.

Nick Viner-Daniels read much of this volume, and extended limitless patience and support, especially during the final few months of writing up. Kate Harrell offered much welcome advice over the years during which we were office mates.

Contents

List of Figures

List of Tables

Chapter 1
Introduction – archaeological background and research questions

1.1 Introduction

The period from the Early Mesolithic to the Late Neolithic in Britain saw a substantial change in the way humans interacted with animals. The start of the Neolithic is heralded as the beginning of a new relationship between humans and their animals in which hunting was discarded in favour of bringing animals under human control. The purpose of this research is to investigate the changing human exploitation of *Sus* and *Bos* from a period in which it is assumed that hunting was the dominant economic practice (*i.e.* the Mesolithic), through the transition and into the period after animal husbandry is thought to have become widespread (the Late Neolithic). It aims to shed light on the approaches taken by prehistoric farmers to their animals and the motivations for the use of animals. Using zooarchaeological techniques a number of research questions will be addressed:

1) What was the situation in the periods before the introduction of animal husbandry in terms of a) the age and sex distribution of the animals deposited on archaeological sites and b) the biometrical character of the animals exploited?
2) What is the nature of the difference between the way animals were exploited in the Mesolithic and Early Neolithic?
3) What form did the introduction of animal husbandry take, *i.e.* what evidence is there for local domestication or an outside origin for cattle and pigs?
4) Are changes in practice evident from the Early Neolithic to the Late Neolithic?
5) Is there any evidence of introgression from wild animals into domestic populations?
6) What use was made of wild boar and aurochs after the Mesolithic period?
7) Can secondary product use be inferred from the zooarchaeological record of the Early and Late Neolithic periods?
8) What role did mobility play in the dominant husbandry regimes of the Neolithic?
9) What are the main differences between the treatment of different species in prehistoric Britain?

1.2 Outline of the volume

The remainder of this chapter will deal with the evidence of early animal domestication and how it has been identified in the zooarchaeological record, touching on the main intellectual standpoints that have been involved in its understanding. Background information about the animals being studied is essential to the understanding of their use by humans in both hunting and husbandry systems. The controversial subject of pre-Neolithic domesticated animal presence in some parts of Europe will be explored, as will the various arguments for and against local domestication events in Europe. A brief discussion of the archaeology of the study area during the two main periods of investigation will follow. Human diet in the periods concerned is of particular interest as is the evidence for wild and domestic animal exploitation as it is presently understood.

In order to begin to understand the process of change that transformed the human/animal relationship in the Neolithic period a number of important archaeological sites were selected for analysis. In addition to the geographical and chronological scope of the research, the sites chosen, along with the faunal remains recovered from them, will be discussed in detail in Chapter 2. Zooarchaeological and isotopic techniques were used to investigate the research questions established above, detailed discussion of which can be found in Chapter 2.

Results and analysis of the data will be presented in Chapter 3 for pigs, and separately in Chapter 4 for cattle. Broadly, these chapters will include analysis of the age, sex and biometry of the faunal remains studied. The results will be discussed in terms of diachronic and spatial variation. Parts of these two chapters were previously published (Viner 2011). Figures that have been published previously are identified in the captions.

Comparison of the results with the broader geographic context is an important undertaking and will be the concern of Chapter 5 in which the material studied will be compared with evidence from other British sites, and with zooarchaeological material from continental Europe (specifically Scandinavia, Iberia, France, Holland and Italy).

In Chapter 6 the results of isotope analysis will be presented and the possibility that cattle were herded over long distances in the Late Neolithic period will be discussed. The possibility of pig movement will also be briefly discussed. Some of the results included in this chapter have been published previously (Viner *et al.* 2010), those figures that have been published before are identified in the captions.

Chapter 7 will discuss the results, limitations and implications of the research to our understanding of animal husbandry in prehistoric Britain.

1.3 Background – *Sus scrofa*

The palaeontological record indicates the appearance of the first 'true pigs' (Suidae) during the Oligocene (Graves 1984, 482) between 23 and 34 million years ago. Uncertainty surrounds the date of the appearance of the genus *Sus*, but according to genetic evidence it was in existence by at least 5 million years ago (Mona *et al.* 2007, 757; Randi *et al.* 1996, 189). Eight living species are contained within the genus, six of which are endemic to south-east Asia. Thus a SE Asian origin is likely for the genus as a whole (Mona *et al.* 2007, 757). The *Sus* genus is very widely distributed at present, particularly *Sus scrofa* – the species with which this research is concerned. *Sus scrofa* was present from south-east Asia to Siberia, Japan and Taiwan in the east, as far as Portugal and Britain in the west, and in parts of northern Africa (Davis 2000; Mayer and Brisbin 2008; Oliver *et al.* 1993). It has also spread, with the help of humans, to many areas outside its natural distribution and is now common in parts of Australia, Argentina and the US. The wild boar is well attested from the beginning of the Mesolithic period in Britain (Maroo and Yalden 2000; Yalden 1999), but it may have been present significantly earlier and sporadically from the Cromerian interglacial (Yalden 1999). Wild boar are not considered native to Ireland, but were present by the Mesolithic period. Mesolithic hunters in Ireland made extensive use of wild boar perhaps due to the absence of other large mammal species in the country (McCormick 2004, 1). By the beginning of the Medieval period hunting of the rare wild boar was the preserve of the aristocracy, and documentary evidence suggests that the species was extinct by the end of the 13th century (Albarella 2010). Extant British populations are the result of accidental reintroductions during the 20th century (Goulding 2003; Goulding *et al.* 2008; Wilson 2003; Yalden 1999).

The presence of *Sus scrofa* in many parts of Europe, Asia and North Africa, in addition to the introduction of wild and feral populations to areas where it was not originally present, has increased the opportunities to study the species. Modern observations of Eurasian wild boar have revealed a great deal of variability in the behaviour, diet, phenology and appearance of wild boar under different conditions, variation that is to be expected given the substantially different areas that they inhabit. Of particular importance to this study is the observed variation in the size and shape of wild boar teeth and bones. Controversy surrounds the status of some *Sus scrofa* groups and it is unclear at present whether the number of subspecies has been inflated. Between 16 (Groves 1981; 2007, 22-3) and 22 subspecies have been identified within the species (Mayer and Brisbin 1991, 237-66; Mona *et al.* 2007, 757). However, based on cranial characteristics Genov (1999, 227-30) was able to separate *Sus scrofa* into only four groups: *Sus scrofa scrofa*; *Sus scrofa ussuricus*; *Sus scrofa cristatus*; and *Sus scrofa vittatus*. Irrespective of how many distinct subspecies are accepted, there is apparent variety between wild boar in different parts of their distribution. Wild boar increase in size from west to east, and from south to north according to a recent study by Albarella *et al.* (2009). Thus the smallest animals tend to occur in south west Europe and the largest in north east Asia. The variation in size of wild boar is of central importance and will be discussed in detail in the following chapters.

In studies of modern feral/wild animals, conception, and therefore farrowing period, are found to be variable and dependent on a number of factors, such as food availability (Fonseca *et al.* 2004, 62; Geisser and Reyer 2005, 267; Graves 1984; Maillard and Fournier 2004; Santos *et al.* 2006), temperature (Graves 1984, 487), photoperiod (Delcroix *et al.* 1990, 613) and rainfall (Fernández-Llario and Mateos-Quesada 2005, 244). Some wild boar populations exhibit highly synchronised births with farrowing periods being restricted to just a few months. One such example comes from Doñana National Park in Spain where births occurred almost exclusively between February and April (Fernández-Llario and Carranza 2000). The authors of the study blamed drought conditions in the park for the highly synchronous birth pattern. Such synchronicity is not exhibited by all modern groups. Wild boar in the Alentejo region of southern Portugal (Fonseca 2004, 62), western Iberia (Santos *et al.* 2006, 207), western France (Mauget cited in Fonseca *et al.* 2004, 62) and Switzerland (Moretti 1995) produce litters year round. Feral populations in the United States can also farrow at any time of year (Graves 1984), and wild boar in Punjab give birth between April and November, with a non-birthing period of 3-4 months (Ahmad *et al.* 1995, 166-67). This potential for variation in farrowing period, even over short distances (*e.g.* year round births in southern Portugal and highly seasonal births in northern and central Portugal, Fonseca *et al.* 2004), and its link to a number of factors, make it difficult to securely identify a season of birth for wild boar and domestic pigs in prehistory. Wild boar are multiparous, females can reproduce from around 6-10 months of age (Fonseca *et al.* 2004) and under some circumstances will produce more than one litter per year (Graves 1984). Modern litters contain between 1 and 9 piglets, a figure that is variable with latitude in European populations (Fonseca *et al.* 2004).

Wild boar sounders usually consist of females and their most recent piglets, with males joining groups most frequently during mating periods (Graves 1984, 483; Oliver *et al.* 1993, 114). Other groups of subadult animals are also sometimes formed (Oliver *et al.* 1993, 114). Group size usually remains small (up to 20 individuals).
Wild boar are highly adaptable and can flourish in a variety of habitats including the temperate conditions persistent in northern Europe and Britain during prehistory. They are omnivorous, but studies have shown that vegetable (*ca.* 90%) matter often makes up a large proportion of the diet of animals in the wild (Goulding *et al.* 2008; Oliver *et al.* 1993; Schley and Roper 2003, 44). In some areas wild boar or feral pigs are also known to consume earthworms (Baubet *et al.* 2004), molluscs, crabs, fish (Oliver *et al.* 1993), as well as small or young mammals, carrion and the eggs and chicks of ground-nesting birds (Goulding *et al.* 2008; Schley and Roper 2003; Wilcox and Van Vuren 2009). Mast has been cited as a particularly important dietary inclusion (Goulding *et al.* 2008; Hamilton *et al.* 2009; Schley and Roper 2003, 45; Yalden 1999), and can influence the timing of reproduction in areas where wild boar rely on its supply (Graves 1984). The eclectic diet of *Sus scrofa* has implications for the human/pig relationship, since domestic pigs can thrive on the detritus from human food consumption and can even be trained to subsist on human faecal matter (Nemeth 1998). Such a propensity to consume the waste from human habitation gives the pig a unique place amongst the main domesticated mammals.
The effect of rooting behaviour of wild boar, domestic and feral pigs has been the focus of much research with often contradictory conclusions drawn. Vegetation

growth is encouraged by rooting because of soil mixing, and certain tree species have been shown to sprout in response to wild boar/feral pig rooting (Singer *et al.* 1984, 467-8), in some cases vegetation diversity was increased by wild boar rooting (Arrington *et al.* 1999). Negative effects cited include soil erosion, leaching of nutrients from soil and loss of vegetation cover and diversity (Bratton 1975; Howe and Bratton 1975; Massei and Genov 2004). When available, wild boar will eat crop plants. In some areas the diet comprises almost exclusively crops (Herraro *et al.* 2006) and there is evidence to suggest that damage to farmland can be caused by rooting behaviour (Wilson 2004). Wild boars are likely to have significantly affected the composition of woodland in prehistory, and may have been partially responsible for maintaining woodland clearings utilised by hunters. It is also likely that if crops were grown during the Neolithic period, wild boar would have eaten them when possible.

1.4 Background – *Bos primigenius*

A long history of controversy surrounds the status of wild cattle in Europe. Throughout the 19th and for much of the 20th century debate raged regarding the number of wild *Bos* species that had been present and the relationship of these to modern domesticated cattle (see for example van Vuure 2005, 83-8). All European domestic cattle are the descendants of the extinct aurochs (*Bos primigenius*). The aurochs developed out of *Bos acutifrons*, commonly considered to be the first representative of the *Bos* genus (Zeuner 1963a, 203; van Vuure 2005, 35). *Bos acutifrons* inhabited an area of the Siwalik (foothills) of the Himalayas in northern India, and it is from here that *Bos primigenius* is known to have spread during the first part of the Pleistocene 1.8 million years ago. It is present in the Pleistocene record of parts of Europe (*Bos primigenius primigenius*), Asia (*Bos primigenius namadicus*) and Africa (*Bos primigenius africanus*). This population expansion reached what is now Britain by the middle of the 13th millennium BP (Smith 1992, 63). Aurochs remains have been found throughout England, Scotland and Wales (Yalden 1999). There is no evidence to suggest that they were ever present in Ireland (Yalden and Kitchener 2008).

With such a broad distribution, variation would undoubtedly have occurred across the aurochs geographical range. Zeuner (1963b, 12) identifies a north-south variation in coat type and colouration, and there is a great possibility that numerous and varied sub-groups existed (Clutton-Brock 1999, 83). The potential exists to identify spatial variation within the skeletal morphology of the aurochs, however no significant attempt has yet been made to understand the diversity within the species over large geographical areas. The often fragmentary nature of osteological data makes such an undertaking difficult (Clutton-Brock 1999, 83).

Aurochsen became extinct in Britain during the Bronze Age. Two late dated specimens are known, one from Charterhouse Warren Farm, Mendip, and dates to 3245BP (Cotton *et al.* 2006; Yalden and Kitchener 2008), *i.e.* 1620-1430 cal BC (Cotton *et al.* 2006, 159), and another from Porlock Weir, Somerset with a date of 1740-1450 cal BC (Cotton *et al.* 2006, 159). Some Roman specimens have been reported, for example at Vindolanda and Segontium but their identification is uncertain (Yalden 1999; Yalden and Kitchener 2008). Aurochs were present in central Europe for some time after their extinction in Britain, and, according to documentary evidence, the final aurochs died in the Jaktorowska forest in Poland in 1627 (van Vuure 2005; Yalden 1999, Zeuner 1963a).

Extinction of the aurochs has meant that it is impossible to study their behaviour through the observation of extant populations. The behaviour of other large ungulates, as well as observations made of the modern descendants of aurochsen have allowed some inferences of the likely behaviour of the species to be made. Although there is a tendency to view members of the *Bos* and *Bison* families as herd animals (for example Clutton-Brock 1999, 88; Grigson 1989, 78) it is most likely that the aurochs males were largely solitary animals, and that females (and young animals) formed small groups (Legge 2010; Smith 1992, 63).

Historical sources suggest that the main period of birth in aurochs was in May, and infrequently in September, with the latter often resulting in the death of calves due to winter conditions (van Vuure 2005, 269). Numerous studies of modern feral populations (Bertaux and Micol 1992; Hall and Moore 1986; 1987) and the related extant European bison (*Bison bonasus*) (Krasiński and Raczyński 1967) have found that reproduction of animals not under human control is seasonal, usually with a single period of birth. In common with other large mammals there is evidence to suggest that seasonal reproduction is related to food abundance (Bertaux and Micol 1992, 273).

Bos primigenius was a grazer and browser (Clutton-Brock 1999, 83; Legge 2010; Zeuner 1963b, 12). Variability between the diets of modern *Bovini* has been identified and depends greatly on availability of food (van Vuure 2005). In addition to grazing, mast may also have been an important source of food for the aurochs (Kyselý 2008, 10; van Vuure 2005, 217). During prehistory the open woodlands of temperate Europe would have been influenced by the activity of large mammals, such as the aurochs, that would have maintained open areas and prevented regeneration of woodland in some places (Bogaard 2004, 14; Maroo and Yalden 2000, 247; Piggott 1981, 15; Vera 2000). Historical accounts identify the aurochs as a forest dweller (Yalden 1999; Yalden and Kitchener 2008), and it seems likely that they would have required at least partial cover during the winter (Smith 1992, 63). But the assumption that woodland was the preferred habitat of aurochs populations in Europe has been challenged and it has been argued that woodlands identified as the main habitat of aurochs were refugia in which populations could survive when driven from other areas by human activity (Hall 2008). In addition, scrubland and open areas were suitable habitat for aurochs (Clutton-Brock 1999; Legge and Rowley-Conwy 1988; van Vuure 2005), and increasingly river valleys and damp areas have been highlighted as particularly hospitable to aurochs populations (Hall 2008; Zeuner 1963b, 12). Recently, work by Lynch *et al.* (2008) used nitrogen isotope analysis on aurochs bones in an attempt to reconstruct the habitat of the species in prehistory. No overlap existed between the preferred habitat of the aurochs and the area utilised by domestic cattle (Lynch *et al.* 2008; Noe-Nygaard *et al.* 2005). Differences in the isotopic composition of aurochs and cattle remains may be the result of the 'canopy effect'. This effect - a depletion of ^{13}C - is thought to be a result of photosynthetic recycling of CO_2 in soil that is then passed along the food chain to animals (Van der Merwe and Medina 1991).

1.5 Animal domestication and its identification in the archaeological record.

What is domestication?

Initial evidence of the lifestyle that we now call Neolithic comes from western Asia. Beginning with sheep and goats, followed by pigs and cattle (Moore 2003) animal domestication developed and spread from the Near East. The exact meaning of domestication has been hotly debated (Hesse 1982) and no consensus has been reached. Most researchers now agree that wild and domestic are two extremes of a spectrum which incorporates innumerable variations of the human-animal relationship (Vigne *et al.* 2005). Diverse definitions of domestication have been proposed that are of varying use to its zooarchaeological study. Meadow (1984) defined domestication as:

> a process in which humans have shifted their attention from the *dead* animal to securing and selectively maintaining the most important product of the *living* animal – its offspring.

Ducos (1978, 54) argued that any definition of domestication must not include *a priori* the causes and mechanisms of the process, and must be broad enough to incorporate all possible scenarios that could be deemed domestication. Consequently he claimed that "domestication can be said to exist when living animals are integrated as objects into the socioeconomic organization of the human group" (Ducos 1978, 54). In addition, Bökönyi (1969) highlighted the importance of intentionality in distinguishing between domestication and other types of 'interference'.

Many of these definitions have merit, but their applicability to the investigation of animal husbandry from the zooarchaeological record is variable. For the purposes of this study it is necessary to use a definition of domestication that can be investigated zooarchaeologically, and those that emphasise the social and ideological nature of domestication are inevitably of a more limited applicability than those based on a more biological approach. The argument of Ducos (1989) that it is unsound to assume domesticated animals must be different to wild ones in aspects of the skeleton is correct. However, zooarchaeological identification of domestication is restricted to those scenarios that have produced a recognisable effect on the animals being investigated. Many of the close human animal relationships that have or do exist are visible in the archaeological record (Jarman and Wilkinson 1972). The methods used to identify domestication from the zooarchaeological record will be discussed further below. It must be acknowledged that other human/animal relationships might have existed that are not identifiable in the zooarchaeological record, and that therefore fall outside the scope of this research.

The reason for the initial domestication of plants and animals has been as hotly debated as the question of what domestication is. The motivation for animal and plant domestication is not directly relevant to the study of its spread in northwest Europe and will be dealt with only briefly here. Numerous possibilities have been suggested that can be broadly separated into two groups; those that see some sort of change in the 'resource balance' (Thorpe 1996) in the Near East as the main causal factor; and those that invoke social change within human societies as the root cause of domestication. An increase in population that led to resource stress (Cohen 1977; Davis 2005), climatic deterioration that led to resource stress (Sherratt 1997) and conversely increased climatic stability that allowed increased sedentism (Bellwood 2005) have all been identified as important factors in the earliest development of agriculture. In terms of social development, the desire to produce surplus for feasting events (Bender 1978), and the need to maintain social control (Hodder 1990), are among many other possible scenarios that have been proposed.

The identification of domestication

Domestication events have taken place at different times and places, but for the purposes of this research the domestication of plants and animals in the Near East from around 10,000 years ago is of most interest. Sheep/goats appear to have been the first of the main domesticates to have come under human control around 7500BC and cattle some time later around 6000BC (Thorpe 1996) with pigs joining the spectrum a little later. Approaches to the identification of animal domestication from the zooarchaeological record have focused on four main strands of evidence:

1) the appearance of taxa outside their natural ranges (Grigson 1989; Jarman and Wilkinson 1972, 91-2; Meadow 1989).
2) an increase in frequency of 'prodomestic' species (*i.e.* those that would eventually become domesticated; Grigson 1989; Meadow 1989).
3) population structure (*i.e.* age and/or sex proportions) at archaeological sites that differs from that expected in a wild population (Grigson 1989; Hesse 1982; Jarman and Wilkinson 1972, 92-5; Legge 1996; Meadow 1989).
4) morphological changes in the animals themselves, most commonly a reduction in size

(Grigson 1989; Jarman and Wilkinson 1972, 84-91; Meadow 1989).

This section will discuss each of these approaches and provide examples of their use in identifying the domestic status of animals in the Near East and Europe.

Perhaps the easiest criterion to assess zooarchaeologically is the appearance of a species in an area outside its normal range. Wild sheep and goats were confined to a comparatively small part of the Near East and their appearance outside the boundaries of their natural range has been used as an indication of their domestication or at least control by humans. In the Near East goat remains were recovered from Jericho (southern Levant), which is some distance to the south of the *Capra aegragus* range, and offers evidence of early domestication of this species (Yalden 1999). Similarly sheep and goats found on archaeological sites during the Neolithic in Europe must have been domesticated as they were not present in the continent prior to the Neolithic (Whittle 1996; Wijngaarden-Bakker 1974; Yalden 1999) and appear in southeast Europe around 7000 BC (Whittle 1996). Aurochsen were not present in postglacial Ireland (Wijngaarden-Bakker 1974; Woodman *et al.* 1997), thus the first appearance of cattle bones in the zooarchaeological record of the country is a good indication of human agency. Wild boar and aurochs were both present within Britain immediately prior to the inception of the Neolithic making this method of little use to the present study.

Changes in the intensity of animal exploitation or significant changes in the population structure, particularly of those animals that were later domesticated has also been used to suggest that a group of animals had come under the control of humans. Bar-Yosef (1998) identified the beginning of sheep and goat husbandry at sites in Israel based on the relative frequency with which the bones of these animals were found on archaeological sites. In the PPNA (Pre Pottery Neolithic A) sheep and goats played a minor role in the economy of the region, while very large numbers of gazelle (*Gazella gazella*) bones on archaeological sites indicate the importance of this animal as a source of food. In the subsequent PPNB (Pre Pottery Neolithic B) period the pattern is reversed, and this, it is argued, reflects the appearance of domesticated sheep/goats and a corresponding reduction in gazelle exploitation (Bar-Yosef 1998, fig. 3:197). A shift of the same kind is also known to have occurred in other parts of the Near East during the PPNB (Bar-Yosef and Meadow 1995; Davis 1991, 382). At the Syrian site of Abu Hureyra the change was again from a dominance of gazelle to the use of *Ovis/Capra* (Legge 1975; Legge and Rowley-Conwy 1987), while at Mureybet XII (also in north eastern Syria) cattle became common while gazelle remains were found less frequently (Bar-Yosef and Meadow 1995). Intensification in the use of wild pigs in northern Europe (among other evidence) was cited as evidence for a close relationship between pigs and humans prior to the introduction of domesticates (Zvelebil 1995), but in fact the evidence for this intensification is limited (Rowley-Conwy *et al.* forthcoming).The argument from specialisation is problematic in general, due to the possibility that a concentration on one species might be an artefact of its availability rather than human control (*e.g.* Uerpmann 1987, 138).

Ageing of cattle based on tooth wear was used to support the prospect of incipient cattle domestication at Mureybet during the PPNA, but biometrical investigation of the assemblage provided no supporting evidence (Grigson 1989, 86). Early goat domestication was identified at Ganj Dareh, northwest Iran (Hesse 1982). Studies of faunal remains from Hagoshrim, Israel (Haber and Dayan 2004; Haber *et al.* 2005) incorporated both kill-off patterns and biometrical information in an interpretation that identified complex variation between the use of sheep/goats, cattle and pigs. Sheep/goats at the site were fully domesticated (in this case isolated from wild types) by the PPNC (Pre Pottery Neolithic C). Cattle kill-off patterns suggested continuity in herding practices throughout the site use but biometrical analysis indicated a reduction in size through time. In contrast, changes in the size of pig remains along with changing kill-off patterns (*i.e.* a tendency towards the killing of younger animals) allowed the identification of wild boar in addition to fully domesticated pigs at the end of the PN (Pottery Neolithic). Citing small sample size in a crucial transitional layer the authors could not securely identify *in situ* domestication or the introduction of pigs in a fully domesticated form (Haber *et al.* 2004). Using an increase in young animals at the expense of adult animals has been challenged frequently (*e.g.* Hesse 1982).

Using one of these lines of evidence is problematic (Meadow 1989) and the most secure identifications of domestication integrate one or more of these approaches. Some of the earliest evidence of pig domestication comes from Cayönü Tepesi in southern Anatolia (Turkey). This site has been intensively investigated since the 1960's and the evidence of pig domestication seems to fall in the middle part of the PPNB (Ervynck *et al.* 2001). At Cayönü Tepesi changes in the kill-off patterns within the pig population, change in the biometrical character of the animals and fluctuation in the occurrence of Linear Enamel Hypoplasia (LEH) contributed to the interpretation of gradual domestication of pigs (Ervynck *et al.* 2001, 70) despite the findings of earlier studies to the contrary (Hongo and Meadow 1998).

Finally, morphological changes, most significantly a change in size (a decrease in most species, but in rare cases an overall increase - for instance in guinea pigs, Davis 1987, 140) have been important in identifying domestication. The causal factors in the reduction of size in domesticates are not known but numerous suggestions have been made. One reason that has been proposed is that smaller animals were deliberately or unconsciously selected by humans in order to encourage docility. Jarman and Wilkinson (1972) oppose this view arguing that from modern comparisons small animals/breeds are no less ferocious than larger ones. Even if this is the case, and smaller animals are not more docile than larger ones, smaller individuals can perhaps be more easily physically restrained than large ones. Thus human selection for small animals should not be rejected as a possible factor in size reduction. Jarman and Wilkinson (1972, 86) argue that 'economic considerations often impose a high selective value on reducing the size of animals or maintaining body size' and suggest that a larger number of smaller animals would have been advantageous to herders as they could more easily survive periods of resource stress by making use of scant and widely distributed food supplies.

Clutton-Brock describes the "harsh conditions of primitive domestication" as unlikely to favour the successful breeding of some animals (1978), and malnutrition has been invoked to explain the size reduction of animals under domestication (Teichert 1993, 236).

Isolation probably occurred soon after the first stages in animal domestication (Higgs and Jarman 1972). Insular dwarfism is usually associated with island populations that have been isolated and results in significant size reduction in the affected populations. Similar conditions of isolation might have been created with the advent of human control of animal populations and could have contributed to the observed reduction in the size of domesticates (Grigson 1969). However, keeping wild and domestic stock separately would have been difficult (Higgs and Jarman 1972), and mixing might not have been discouraged (Jarman and Wilkinson 1972).

All of these methods can be applied to the investigation of early animal husbandry in prehistoric Britain. The most useful are changes in population structure and size of animals that provide information about the wild/domestic status of animals and also have the potential to provide specific information about the type of exploitation employed. Age/sex of cattle and pig populations, along with physical changes in the animals themselves from the Mesolithic to the Neolithic are to be investigated, and the methods used for this purpose will be presented in Chapter 2.

1.6 Evidence for introduced domestic types in pre-Neolithic Europe

The beginning of the Neolithic saw the appearance of domesticated animals in Europe on a large scale. Prior to this, the Late Mesolithic populations may have had the opportunity to acquire domesticated animals.

Evidence for pre-Neolithic 'domestic' cattle and pigs, has been claimed at a number of northern and western European sites. Degerbøl (1963, 73) suggested that cattle bones from the Ertebølle site of Dyrholmen (Stage I) came from domesticated animals, but later made a reassessment and confirmed that the bones came from wild animals (Degerbøl and Fredskild 1970). A further claim based on the small size of bones was made at the site of Rosenhof, northern Germany, but an investigation utilising mitochondrial DNA concluded that all of the animals from Late Mesolithic contexts were female aurochsen (Scheu *et al.* 2008, 1263). However, a recent study of pig mitochondrial and nuclear DNA from Ertebølle sites in northern Europe presented new evidence that domesticated pigs were present in Late Mesolithic contexts in northern Germany (Krause-Kyora 2013).

Despite repeated claims, there is still sufficient evidence that Mesolithic populations acquired domesticated animals on a large scale in southern Scandinavia (Rowley-Conwy 1995, 2003; Welinder 1998, 169). Parallels are frequently drawn between Britain and southern Scandinavia because of the general similarity of the situation in the two areas: both resisted the introduction of the Neolithic for a long period before finally succumbing (Thomas 1997). At present, there is no compelling evidence for pre-Neolithic domestic animals in either area.

The evidence is more convincing further to the east (Domanska 1998, 131). In Poland, domestic cattle and pig bones were recovered from the Late Mesolithic site of Dęby 29 (Lasota-Moskalewska 1998). Small numbers of domesticated animal bones are present at pre-Neolithic sites in the eastern Baltic (Timofeev 1998, 232). Conversely claims of pre-Neolithic domestic sheep/goats at sites in the Pyrenees, where a form of transhumance has been suggested (Geddes 1983; 1985), are problematic due to mixing of contexts (Rowley-Conwy 1995) and the bones are likely to be later intrusions. At Ferriters Cove on the southwest coast of Ireland substantial evidence of domesticates in pre-Neolithic contexts has been suggested (McCarthy 1999; Woodman *et al.* 1999). Cattle and sheep/goats bones were found in association with Late Mesolithic worked stone, and the absence of wild forms of these two species means they must have been imported to the country. Pig remains from the site were interpreted as belonging to local wild boar because their Mesolithic context, but no biometrical support for this assertion was gathered (McCarthy 1999) and these animals might also be introduced domesticates. Ferriters Cove is one example among many of the introduction of domesticated animals to Ireland during the Mesolithic. Wijngaarden-Bakker (1990) reviewed the available data from Late Mesolithic Ireland and found that all faunal assemblages from coastal sites (ten sites at the time the paper was written) included the remains of domesticated animals. Irish Late Mesolithic people were able, and prepared, to acquire domestic animals from their Neolithic neighbours.

No examples like those above have yet been found within the present study area. However, an enigmatic find at Lydstep Haven, Wales provides a glimpse of the possible interaction between Mesolithic and Neolithic in Britain. Dating to 4350-3940 cal BC (5,300± 100 BP OxA-1412), the remains of a possibly domestic pig were excavated with Mesolithic microliths imbedded in its neck (Chatterton 2006; Lynch *et al.* 2000). Hunting of escaped domestic animals by Mesolithic populations is likely to have occurred, and apparently took place in prehistoric Britain. Davidson (1989) argued that this kind of hunting would have been possible in Iberia and southern France, and presumably it could have taken place in other parts of Europe too. Although domesticated animals were present in Late Mesolithic contexts in some parts of Europe, the available evidence indicates that the acquisition of domesticated animals by Mesolithic communities was not ubiquitous, and Britain and southern Scandinavia stand out as areas for which there is little evidence of domesticated animals in the Mesolithic.

1.7 Local domestication versus introduction in Europe

Linked to the presence of domesticated animals in pre-Neolithic contexts is the possibility of local domestication events in Britain. Whether Europe saw local domestication events in the Mesolithic/Early Neolithic is linked to, but not synonymous with the issues of acculturation and demic diffusion that have concerned Neolithic archaeology for decades. Animal domestication in Europe was a possibility given the presence and importance of *Sus scrofa* and *Bos primigenius*. Bökönyi was perhaps the earliest and most fervent proponent of local domestication of animals in the European Neolithic. He argued that local domestication was evident in central Europe because of the presence of 'intermediate' groups in the zooarchaeological record (Bökönyi 1969). According to Bökönyi cattle remains from Berettyószentmárton in Hungary are comprised of large wild, small domesticates and intermediate forms. He argued that such transitional forms also occur in pig remains but provided no example in support (Bökönyi 1969). Although Bökönyi's logic is sound (*i.e.* if we agree that domestic animals are of reduced size compared to their wild counterparts, then we can expect some evidence of animals that fall between the two extremes) the data presented by him does not adequately support the conclusion that domestication took place in Europe.

The possibility of local domestication of the aurochs in northern and western Europe has received a great deal of attention, but as yet there is no firm evidence to support such a hypothesis (Rowley-Conwy 2003). Mitochondrial DNA studies have produced somewhat mixed results. A contribution by local aurochsen to domestic cattle groups has been firmly refuted (for instance Schen *et al.* 2008, Edwards *et al.* 2007) and a Near Eastern origin for domesticated cattle in Europe sustained by many researchers (Edwards *et al.* 2007; Troy *et al.* 2001). Beja-Pereira *et al.* (2006, 8115) provides an alternative view in which introgression from European aurochs was made possible in a system of free-ranging cattle herds. Attempts to shift the focus of DNA studies from matrilineal to paternal lineages in the study of European cattle origins have highlighted the possibility of repeated hybridization between introduced *Bos taurus* females and local *Bos primigenius* males (Götherström *et al.* 2005). This, however, is irrelevant to the issue of whether domestic cattle were imported or locally domesticated. In either scenario they could have eventually hybridised with local wild forms, particularly in free range husbandry conditions.

Research into the genetic make-up of pig populations has also been undertaken, and suggests that the situation is more complex than a simple introduction and spread from the Near East. Broad scale analysis of pig and wild boar mtDNA indicates that there were a number of independent domestication events in parts of Europe and Asia (Giuffra *et al.* 2000; Larson *et al.* 2005). Specifically regarding the progress of Neolithisation, Larson *et al.* (2007) have argued that indigenous and introduced animals both contributed to the appearance of domestic pigs in Europe.

That wild forms contributed to the genetic make-up of modern domesticates is unsurprising, but changes our understanding of the Neolithic relatively little. It is unlikely that, even had they wanted to, Neolithic herders would have been 100% successful in maintaining separation between their domestic and local wild animals. In addition, some deliberate small scale mixing might be envisaged. Documentary evidence suggests that aurochs were tamed/domesticated episodically in historic times; Virgil records that after an outbreak of cattle disease in northern Italy local people replaced stock with aurochs (Bökönyi 1969). It is not surprising that herders would have made use of aurochs when the opportunity or necessity arose. Mixing of domestic pigs and wild boar is currently regarded as undesirable by breeders of free range pigs, but there is abundant evidence that it occurred on a regular basis even in the recent past (Albarella *et al* 2007; Hadjikoumis 2010)

1.8 The Mesolithic period in Britain (10th -5th millennia BC)

After the hiatus in human occupation during the last glaciation, human repopulation in the Upper Palaeolithic of what is now the British Isles began around 13,000 years ago (Smith 1992, 2). The beginning of the Mesolithic period has been dated to around 10,000BP (Barton and Roberts 2004). The period is conventionally separated into two parts: the Early Mesolithic from *ca.* 10,000-8,500BP; and the Late Mesolithic from *ca.* 8,500-5,500BP (Barton and Roberts 2004). Excluding the temporally restricted lithic assemblages of the Initial Mesolithic, the British Early Mesolithic lithic repertoire is characterised by microliths along with end scrapers, burins, awls, axe heads and adzes (Barton and Roberts 2004).

It is assumed by most researchers that the Mesolithic inhabitants of the British Isles subsisted by hunting wild animals and gathering wild plant foods (McFadyen 2007). Such an assumption cannot be made without careful consideration,

> …man-animal and man-plant relationships similar to modern domestication will have tended to occur throughout the Pleistocene wherever this was the most profitable economic strategy in the prevailing circumstances."
>
> (Higgs and Jarman 1972, 12).

Despite Higgs and Jarman's assertion in the 1970's that human-animal relationships akin to domestication would have occurred whenever it was profitable, there is no evidence to suggest that anything other than hunting and gathering of wild resources took place anywhere in Britain prior to the beginning of the Neolithic. In coastal areas Mesolithic shell middens are abundant with the remains of molluscs that must have been an important dietary component (Kitchener *et al.* 2004). Marine mammals were also exploited and these, along with the molluscs found at coastal sites indicate a high reliance on wild resources during the Mesolithic period.

The use of stable carbon ($\delta^{13}C$) and nitrogen ($\delta^{15}N$) isotope analysis has allowed the determination of whether marine or terrestrial resources were the main constituent of the human diet during the Mesolithic in Britain and Europe. Relatively few studies of British human diet have been undertaken (compared with Scandinavia for instance; Barton and Roberts 2004) but some inferences can be made. Human bones from five British Mesolithic sites (including Thatcham III in Berkshire, see Chapter 2 for a description of the material from Thatcham sites used in this research) had $\delta^{13}C$ values of between -18.7 and -21.9 an indication of a diet with a largely terrestrial origin (Barton and Roberts 2004). In addition, the high $\delta^{15}N$ values from the same specimens are indicative of high meat consumption (Barton and Roberts 2004). Conversely, a high level of marine exploitation was found in Mesolithic humans from south western Britain (Schulting and Richards 2002).

Contrary to traditional perceptions of the Mesolithic that envisaged highly mobile groups comparable to examples of hunter-gatherers in the ethnographic record, there have been numerous attempts to highlight the variety of life-ways exhibited by Mesolithic people. As with much of the commonly used academic language there has been no attempt to standardize what is meant by 'mobile', and the word is often taken to have a meaning that is inherently understandable. Wickham-Jones (2005), however, highlights the vastly different understanding of mobility that is held by researchers in different areas of Europe. No single or simple pattern of subsistence, movement or habitation can be applied to Britain as a whole. In many parts of Europe Mesolithic foragers concentrated in coastal areas, river valleys and lake side environments (Price 2000), consistently with the importance of aquatic resources in the diet.

A lack of evidence is characteristic of the Mesolithic in Britain as a whole (Simmons 1996, 28), but generally the Early Mesolithic is better represented than the subsequent Late Mesolithic. Nevertheless, there are still some significant gaps in the record of the Early Mesolithic too. Reynier (1998, 176) identified three areas where there is little or no evidence of Early Mesolithic activity; lowland central England; the northeast; and Kent in the southeast. Significant faunal assemblages are rare, but the largest are found in the Vale of Pickering, Yorkshire, and the Kennet Valley, Berkshire.

The development that marks the transition to the Late Mesolithic is the appearance of broad blade microliths in place of narrow blade microliths (Lynch *et al.* 2000, 23). In Britain this transition took place around 8700BP (Lynch *et al.* 2000, 23). The Late Mesolithic period is characteristically ephemeral. A paucity of zooarchaeological assemblages is mirrored in the archaeological record in general. Evidence of habitation activity is rare, and sites are often identified only as scatters of flint working debris. Further, those animal bone assemblages that have been excavated are usually very small (some of the largest being <100 specimens). These are regularly of poor condition, can be highly fragmented and are often of little use for biometrical analysis. As a result very few Late Mesolithic assemblages have been available for inclusion in the study. The suggestion that there might have been intensification of resource exploitation in the Late Mesolithic of Britain (Thomas 1997) remains unsubstantiated.

A great deal of uncertainty surrounds the final stages of the Mesolithic and the beginning of the Neolithic in Britain. ^{14}C dating is limited by the lack of organic remains on many sites, particularly in the north of the country. Some researchers argue that there was chronological overlap between the Mesolithic and the Neolithic. Williams (1989) for instance highlighted the overlap between AMS dates from sites with Mesolithic and sites with Neolithic characteristics, but also acknowledged a number of problems with some of the dates used to identify the overlap (in particular the use of unreliable dating material used to achieve some of the dates). Numerous lines of evidence that could potentially be used to investigate the transitional period itself exhibit marked periods of discontinuity. The latest Mesolithic stone assemblage comes from Stratford's Yard, Buckinghamshire and has been dated to 5890±100 BP (Stainton reported in Barton and Roberts 2004), yet the earliest dates for the Neolithic in southern Britain are around 5200BP (Woodman 2000). Schulting and Richards (2002, 1019) identified a 3000 year gap in the ^{14}C record of human remains beginning *ca.* 7000BP, a gap that was only slightly mitigated by the find of human remains at Staythorpe, Nottinghamshire, that were dated to 5740-5620 cal BC (6790±40 BP BETA-144016) (Davies *et al.* reported in Conneller 2006, 142). Likewise, Chamberlain (1996, 951) highlighted a two millennia gap in the chronological record of human burials within caves in Britain. In the zooarchaeological record, a similar absence of dated faunal assemblages has been identified in some parts of the country (Albarella and Pirnie in prep).

1.9 The Neolithic period in Britain

An invasion by Neolithic people from continental Europe still has some proponents in British archaeology (*e.g.* Sheridan 2000; Tresset 2000; 2003; Whittle 1977), but it is also possible that Mesolithic people were substantially involved in the transition to the new Neolithic practices (*e.g.* Darvill 1987; Dennell 1983; Thomas 1996; 1997;

2008). Britain, like southern Scandinavia, resisted the spread of the Neolithic for an extended period before its ultimate adoption. Palaeoclimatic evidence indicates a period of environmental change at the time, leading to speculation that changing conditions prompted the adoption of agricultural practices (Bonsall *et al.* 2002; Rowley-Conwy 1984). In contrast, Price (2003) argues for social, economic and religious change, driven by indigenous people in northern Europe, influenced little by climatic variation. In a similar vein, Sherratt (1995) argued that conversion of indigenous people can explain the transition from Mesolithic to Neolithic. Whatever the motivation and process by which Neolithic practices were introduced into Britain, they had both economic and social impact.

The Neolithic period in Britain is conventionally separated into the Early Neolithic (4000-3200 cal BC), Middle Neolithic (3200-2800 cal BC) and the Late Neolithic (2800-2300 cal BC) (Parker Pearson *et al.* 2005). The earliest part of the Neolithic is poorly represented with no significant monument construction or settlements having been excavated at the very start of the period. The character of the Early Neolithic has also been debated and there is no consensus as to the Early Neolithic way of life. Tool assemblages contained leaf-shaped arrowheads and end scrapers and are consistent over much of the Early Neolithic distribution (Woodman 2000, 248). Ceramic styles also tend to be homogenous, as are the various monument types (Woodman 2000, 248).

Cattle dominates the earliest Neolithic faunal assemblages, a pattern shared with the Early Neolithic in other parts of Europe (central Europe, Bogucki 1989; Ireland, Lynch 1981). Milk was certainly used at least from the Early Neolithic in Britain (Copley *et al.* 2003; 2007; Legge 1981) and cereal cultivation was also an important economic component (Bogaard and Jones 2007). Wild mammals were rarely exploited in the British Early Neolithic (as is the case for many parts of Europe; Bogucki 1989), but seeds of raspberries and blackberries, sloes, crab apple and/or crab pear as well as hazelnut shell indicate the continued harvest of wild plant food during the both the Early and Late Neolithic (Robinson 2000 Table 8.1, 88). Along with the dominance of domestic cattle at many Early Neolithic sites, isotope analysis of human remains from Britain (and other parts of northern Europe) indicate that the significant use of marine resources in the Mesolithic was abandoned during the Neolithic (Richards and Hedges 1999; Schulting and Richards2002, but see Milner *et al.* 2004). By some, this abrupt change has been interpreted as a food taboo - further evidence of the ideological change from the Mesolithic to the Neolithic (Thomas 2003).

1.10 Summary

This research aims to use zooarchaeological analysis of assemblages from Mesolithic/Neolithic sites to answer a number of specific questions about early animal husbandry.

Definitions of domestication and the traditional methods of identifying it were assessed (Section 1.5). Of the methods discussed, two will be particularly useful to this investigation: population structure and skeletal morphology.

Unlike some other parts of Europe, there is very limited evidence for the presence of domestic animals in Britain prior to the beginning of the Neolithic. This, however, might also be a consequence of the dearth of substantial faunal assemblages in the Late Mesolithic period. Local domestication events took place in some parts of Europe (especially in pigs), but there is as yet no evidence from Britain.

The next chapter will present information about the faunal assemblages chosen for analysis, and discuss the methods used to address the research questions posed in this chapter.

Chapter 2 Materials and Methods

2.1 Introduction

To answer the research questions outlined in Chapter 1 a number of sites with substantial zooarchaeological collections were chosen for reanalysis. This chapter will begin by defining the chronological and geographical boundaries of the investigation, and in particular will provide background information about each of the chosen archaeological sites. Each site will be discussed briefly beginning with those that date to the Early Mesolithic. The level of detail afforded the discussion of each site is dependent on the scale of previous research. Sites that have been the focus of many publications can clearly be discussed in greater detail than those for which published information is limited. This concern applies particularly to some of the smaller Mesolithic assemblages. Where possible general background information about the excavation history, major findings of previous research, dating, relationship to other sites, and zooarchaeological information will be included. Following the site by site introduction, the second part of the chapter will be concerned with presenting the various methods employed to record information from archaeological skeletal material.

2.2 Materials

Introduction to the Mesolithic sites

In archaeological terminology this study begins with the Early Mesolithic period. When considering prehistoric husbandry practices the Early Mesolithic is somewhat removed from the main period of interest, but it was chosen in this case to provide a substantial set of comparative material, especially in light of the dearth of later Mesolithic faunal remains (discussed below).

Faunal assemblages from the Mesolithic period, with a number of conspicuous exceptions, are small and exhibit poor preservation. Two areas have been particularly important for the present study, the Vale of Pickering in North Yorkshire, and the Kennet Valley in Berkshire. A vast amount of research has been conducted in the Vale of Pickering, beginning with the work of Moore who identified a number of Mesolithic sites in the Vale (Mellars 1998). The Kennet Valley is likewise a source of unusually well preserved faunal assemblages. Sites at Thatcham near Newbury are particularly rich, but Marsh Benham, Faraday Road and Wawcott are also important. These final three sites are of particular interest as, contrary to the expected pattern of Early Mesolithic sites in Britain they were dominated by wild boar rather than red deer. Figure 2.1 gives the location of the Mesolithic sites investigated.

Thatcham sites

A series of sites at Thatcham, near Newbury, Berkshire were excavated between the 1920's and 1989 (Carter 2001; Churchill 1962; Healy *et al.* 1992; Wymer 1958; 1959; 1962). The earliest Mesolithic activity at Thatcham is characterised by the presence of Star Carr type microliths, which were replaced later in the chronology of the site by Deepcar microlithic technology (Chisham 2006). In addition to lithic evidence, a number of hearths were excavated at Thatcham III (Barton and Roberts 2004), but unfortunately no faunal remains were associated with the features. The oldest radiocarbon date from the any of the Thatcham sites is 10,900-9,700 cal BC (Q-659, Chisham 2006) (Table 2.1). As part of this research, faunal remains from the Thatcham sites were reanalysed using the protocol outlined in Chapter 2 to record all of the appropriate specimens.

Most of the faunal remains were concentrated in just two sites, Thatcham II and Thatcham V, and were excavated in 1958-61 (Carter 2001). By Mesolithic standards the bones from these two sites provide a relatively large zooarchaeological assemblage. King (1962) highlighted the similarity of the assemblage from Thatcham II and V to that found at Star Carr, Yorkshire. Red and roe deer were the most abundant taxa, and wild boar was also present in substantial quantities (Carter 2001). Aurochs remains were comparatively scarce; those that were found were highly fragmented (King 1962) and provided no age, sex or biometrical information. They were therefore not included in this study.

The diet of the Mesolithic inhabitants of Thatcham has been investigated through the application of carbon and nitrogen isotope analysis. Human and canid remains from Thatcham III and Thatcham IV were analysed, and it was discovered that they shared a similar diet that was largely based on terrestrial resources (Barton and Roberts 2004; Schulting and Richards 2000).

Wawcott sites

Numerous Mesolithic sites were identified and investigated by F. R. Froom in the Kennet Valley during the 1960's (Froom 1976). Flint tools and debitage characterise the many of the Wawcott sites, in addition to a number of structures dating to the Early Mesolithic period (Froom 1972; 1976). The chronology of this cluster of Mesolithic sites has been difficult to establish. A number of radiocarbon dates exist (see Table 2.1), many of which point to the Late Mesolithic utilisation of the area. The lithics from some sites are clearly of earlier Mesolithic date, and in one case might be of Upper Palaeolithic date (Chisham 2006). The situation is made more complex by the damage caused by modern ploughing at a number of sites that has destroyed stratigraphic information. Three of the Wawcott sites produced faunal remains that were available for further study. Wawcott XV is the earliest and is broadly contemporary with Star Carr and Thatcham (Carter 1976). Site IV is also of Early Mesolithic date, while Wawcott XXIII is slightly later in the relative chronology of the site cluster. Animal bones from sites XV, IV and XXIII were originally analysed by Carter (1976) and were reanalysed as part of this study. Even combined the amount of faunal material was very small. *Sus scrofa* and *Bos primigenius* dominated the assemblage, and a smaller contribution was made by *Alces alces* and *Cervus elaphus* (Carter 1976). All of the faunal material from the

Wawcott sites was reanalysed following the protocol outlined in Chapter 2.

Marsh Benham

Marsh Benham, Berkshire, was investigated by F.R. Froom in the 1920's, and further work was conducted by the Newbury Museum Archaeological Group in the 1960's. Another period of field work was carried out at the site by Roger Jacobi in 1972 (Reynier in press, cited in Chisham 2006). As at Thatcham, the lithic assemblages are characterised by Star Carr types in the earliest part of the Mesolithic and were followed by Deepcar type lithics slightly later (Reynier in press, cited in Chisham 2006). Further evidence of Early Mesolithic use of the site comes from a single radiocarbon date of 8270-7760 cal BC (OxA-5195, Chisham 2006), and indicates that Marsh Benham is of similar date to other important Early Mesolithic sites in the area namely Wawcott, Thatcham and Faraday Road (see below), as well as with Star Carr further to the north.

Faunal remains from Marsh Benham were well preserved and the assemblage included a substantial number of *Sus scrofa* bones. The remains were reanalysed using the recording protocol established as part of this research.

Faraday Road

Excavations at the site of Faraday Road, Berkshire, revealed Early Mesolithic deposits with abundant faunal remains (Ellis *et al.* 2003). Lithics belonged to the Deepcar tradition, and are therefore of Early Mesolithic date (Reynier 2000). ^{14}C dates, one of which was on wild boar bone, suggested occupation of the site between 9000 and 8500BP (R-24999/1 8510±60 cal BC, R-24999/2 9418±60 cal BC, Ellis *et al.* 2003). Faraday Road is therefore contemporary with Marsh Benham, the sites at Thatcham and others in the area (*cf.* Ellis *et al.* 2003). A. Powell and C. Ingrem originally analysed the faunal material from the site. Faraday Road is of particular importance to this investigation as the animal bone assemblage from the site was dominated by wild boar remains (Ellis *et al.* 2003; Ingrem and Powell 1999), and was reanalysed as part of this research.

Victoria Park

Victoria Park is located in the Kennet Valley, Berkshire, and is in close proximity to the sites discussed above, but there is relatively little published information about its form. It has not been directly dated, but lithic analysis conducted by Peake (cited in Wymer 1958) suggests that the site was comparable with nearby Thatcham. Therefore, it is considered here as an Early Mesolithic assemblage. The faunal assemblage was relatively small, however, all of the *Sus* and *Bos* remains were analysed as part of this research and resulted in an assemblage of 113 specimens.

Star Carr and Seamer Carr

Star Carr is one of many Mesolithic sites identified by John Moore in the Vale of Pickering, North Yorkshire (Clark 1954; Mellars 1998; Schadla-Hall 1990). Initial investigation of the site was undertaken by Clark between 1949 and 1950 (Clark 1954). Along with Thatcham, Star Carr provides some of the earliest evidence of Mesolithic activity in Britain (Barton and Roberts 2004) with the earliest dates being around 10,700 cal BC (Day and Mellars 1994, and Table 2.1).

Activity at Star Carr is evidenced by stone tools and debitage, platforms constructed from birch branches, occupational debris and faunal remains including abundant worked bone and antler (Clark 1954, 2). Organic remains were extremely well preserved in some areas of the site, and, as a result, the quality of information regarding animal exploitation is unparalleled in the UK. The faunal remains have been studied repeatedly; first by Fraser and King (1954), and subsequently by Legge and Rowley-Conwy (1988; 1989), Rowley-Conwy (1998), Noe-Nygaard (1983), Clutton-Brock and Noe-Nygaard (1990) and Carter (1997; 1998). Red deer, along with elk and aurochs occurred frequently within the Star Carr assemblage and Star Carr represents the largest collection of Mesolithic aurochs remains anywhere in Britain. It was therefore an essential inclusion in the study despite the detailed zooarchaeological work that had already been carried out. In addition to the red deer, elk and aurochs remains other species found at Star Carr include roe deer (*Capreolus capreolus*), domestic dog (*Canis familiaris*), brown bear (*Ursus arctos*), badger (*Meles meles*), beaver (*Castor fiber*), (Legge and Rowley-Conwy 1988; Noe-Nygaard 1983; Rowley-Conwy 1998), and a variety of wild birds (Clark 1954, 16).

Seamer Carr is located a short distance to the north of Star Carr and was investigated by Schadla-Hall in 1977-84 (Schadla-Hall 1990). Preservation at Seamer Carr was poorer than at Star Carr and the faunal assemblage comparatively limited. Analysis by Uchiyama *et al.* (2000) identified an abundance of red deer (*Cervus elaphus*), along with aurochs, dog and horse remains. Four radiocarbon dates are available and indicate activity between 9870-8720 cal BC and 7580-6700 cal BC (Schadla-Hall 1990).

Material from both Star Carr and Seamer Carr was reanalysed during the course of this research. The reanalysis did not incorporate all of the Star Carr/Seamer Carr faunal remains and was limited to the bones held in the Natural History Museum, London.

Figure 2.1 The location of Mesolithic sites investigated as part of the research.

Goldcliff East and Goldcliff West
Two main areas of activity were identified by the investigators at the sites on the banks of the Severn Estuary: Goldcliff West, from which few animal bones were recovered and which is a largely unstratified, mixed deposit; and Goldcliff East that produced a collection of Late Mesolithic faunal remains. Excavations took place in 1992 and 1993 and the bones are curated in Newport museum. Spatial distribution of bones and lithics at Goldcliff East suggest that different areas of the site were used for the butchery of specific taxa (Bell *et al.* 2000). The Goldcliff sites are of unique importance to this study as they are the only Late Mesolithic sites that have been included. Goldcliff East has been dated 5400-4000 cal BC (Bell *et al.* 2000).

A very small amount of material was also included in the analysis that was collected prior to the main investigation (Coard 2000). All of bones were studied initially by Noddle and later by Coard (2000). They were reanalysed for the purposes of this study in 2008 and resulted in a sample of 29 specimens. Red deer was the dominant taxa in terms of number of identifiable specimens (MNI = 4), but wild boar was also of importance (MNI = 3). No aurochs remains were recovered from either Goldcliff East or Goldcliff West (Coard 2000).

Introduction to the Neolithic sites
As the period during which animal husbandry was adopted the Neolithic forms an important part of the investigation. This importance is reflected in the greater emphasis that has been placed on data from sites of this period. Although the transitional period is central, a broader chronological approach was taken in order to gain an understanding of the changes in husbandry through time. Therefore, in addition to the Early Neolithic assemblages the Late Neolithic is also represented.

One of the problems encountered when studying the husbandry practices of the Neolithic period is the paucity of evidence from different site types. Many of the more significantly sized animal bone assemblages come from sites that have been interpreted as being ritual in character, and very few 'domestic' or habitation assemblages are available for the periods in question. Bradley (1998), discussing the monuments of northwest Europe states that:

> They played no part in everyday affairs and their prominent place in the prehistoric landscape contrasts sharply with what little is known about settlements and houses at this time.

Even the evidence for early agriculture
is very limited indeed.

(Bradley 1998, 3).

As a result we have for some time been forced to admit that the interpretations based on faunal assemblages from the Neolithic period might reflect ritual deposition rather than economic practice (although researchers should be careful not to artificially dichotomize the ritual and economic spheres). The selection of sites for this study undoubtedly reflects this bias against habitation sites, and four of the six Early Neolithic sites are monuments (a causewayed enclosure, two long barrows and a chambered tomb). Of the two remaining Early Neolithic sites one consisted of a large pit containing animal bones and is of unknown function, and the other is a rare habitation site.

The Late Neolithic, at least in terms of this research, is more balanced in that all of the remains come from Durrington Walls, a site with evidence of both ritual and domestic activity.

Hambledon Hill

Excavation of an extensive causewayed enclosure at Hambledon Hill, Dorset, took place between 1974and 1986 under the direction of Roger Mercer (Mercer 1980; Mercer and Healey 2009). A dating programme (that has included the analysis of 158 samples) suggests that Early Neolithic activity began around 3690-3640 cal BC and lasted for between 310 and 370 years in total, ending around 3340-3300 cal BC (Bayliss *et al.* 2009, 409).

Detailed analysis of the faunal remains was conducted by Legge (2009) and found an assemblage that was dominated by *Bos taurus* remains, but also contained significant numbers of *Sus* and caprine bones (Legge 2009, 537). The faunal remains came from very diverse contexts including the ditches of the main enclosure, pits and internal features and outworks such as the Stepleton enclosure. Reanalysis of the remains undertaken as part of this research included all material that could be securely attributed to the Neolithic period and resulted in more than 1500 records.

The results of carbon and nitrogen isotope analysis of human and animal remains from Hambledon Hill was reported by Richards (2009, 524-7). According to the results of $\delta^{13}C$ analysis the diet of 51 humans found at Hambledon Hill was entirely terrestrial in origin. $\delta^{13}C$ results from other mammals (*Bos* sp., *Sus* sp., *Canis* sp., *Cervus elaphus* and sheep/goat) also had terrestrial signatures. In terms of $\delta^{15}N$ the results were more varied, and a tentative sexual difference in meat consumption was suggested by the investigator (Richards 2009, 524). Neolithic pigs from Hambledon Hill had a diet that consisted almost exclusively of terrestrial plant foods.

Runnymede Bridge

The excavations at Runnymede Bridge, Berkshire, in 1978 and between 1984 and 1989 uncovered a large area of Early Neolithic and Bronze Age activity (Needham 1991; Needham and Spence 1996). The site provides evidence of domestic activity, and is the only non-ritual Early Neolithic assemblage included in this study. The vast faunal assemblage has been studied by Done (1980; 1991) and Serjeantson (1991; 1996). Despite the assertion by the original investigators that Runnymede Bridge should be considered Middle Neolithic, ^{14}C dates suggest activity from the Early Neolithic period with significant overlap with the use of Hambledon Hill (Table 2.1), and that range from 3970-3520 cal BC to 3302-2580 cal BC (Needham 1991). Therefore, for the purpose of this investigation Runnymede Bridge will be considered an Early Neolithic site.

Cattle dominated the faunal assemblage in all areas of the site, closely followed by the remains of pig (Serjeantson 1996). Serjeantson highlighted the importance of pigs at Runnymede Bridge when compared to other Early Neolithic sites. On the basis of their prominence Serjeantson suggested that perhaps pigs were consumed more frequently at domestic sites and cattle more frequently at ritual sites. This is an interesting observation, but unfortunately cannot be tested without the excavation of more faunal assemblages from Early Neolithic habitation sites.

The zooarchaeological assemblage from Runnymede Bridge is substantial in its size and due to time constraints a sample of the total assemblage was chosen for reanalysis. The bones investigated here came from areas 4, 6, 16, 19, 20, 22 and area 24 and provided almost one thousand records

Hazleton North

Hazleton North, Gloucestershire, was excavated between 1979 and 1982 (Saville 1990). The monument is a long cairn of the Cotswold-Severn type of monuments that are distributed in southern Britain from Gloucestershire and south Wales in the west, Somerset in the south and to Oxfordshire in the east (Thomas 1988).

The dating of the Hazleton North long cairn has been exceptionally thorough. More than forty ^{14}C dates have been gathered, and they place the pre-monument Neolithic phase of the site at 3900-3800 cal BC, and use of the monument for burial activity ceased in the 3620's cal BC (Meadows *et al.* 2007, 61). The first activity at Hazleton North began before the earliest Neolithic dates at Hambledon Hill, but the monument was probably partially contemporaneous with Hambledon Hill.

Faunal remains were originally studied by Levitan (1990). Those included here come from the pre-monument Neolithic levels, from the period of the monuments construction and the period during which it was used (see Saville 1990 or Meadows *et al.* 2007 for a detailed description of the monument sequence). Hazleton North is dominated by cattle remains, with pigs less well represented (Levitan 1990; Thomas 1988). During reanalysis of the assemblage as part of this research 21 bones were recorded. In addition, Thomas (1988) has identified the differential deposition of cattle and pigs at Hazleton North and other Cotswold-Severn type monuments. Cattle remains predominated in the chambers of Cotswold-Severn type tombs, whereas pig remains were more numerous in the blocking material and in forecourt areas of the monuments (Thomas 1988, 550). The faunal remains suggest cattle, pigs and other terrestrial mammals were of significance to the humans that were active at Hazleton North. $\delta^{13}C$ and $\delta^{15}N$ of human remains support this evidence and suggest that the diet was largely terrestrial for all of the people sampled (Meadows *et al.* 2007).

Coneybury Anomaly

Coneybury Anomaly, Wiltshire, is the earliest Early Neolithic site to be included in the study having been dated to 3980-3708 cal BC. It is the site of a large Early Neolithic pit that contained the remains of both wild and domestic animals that were studied by Maltby (1990). The early date of the feature has been questioned (Cleal 2004) and the possibility that it could be residual has not been fully addressed. Cattle (*Bos taurus*) and roe deer (*Capreolus capreolus*) were most prevalent within the zooarchaeological assemblage, and pig (*Sus scrofa*), red deer (*Cervus elaphus*), beaver (*Castor fiber*) and brown trout (*Salmo trutta*) were all present in the deposit (Maltby 1990). *Bos* and *Sus* remains were reanalysed as part of this research and resulted in a sample of 108 specimens.

Durrington Walls

Durrington Walls is one of the largest examples of a Late Neolithic henge monument in Britain. Excavations at Durrington Walls began in 1966 with the work of Wainwright (Wainwright and Longworth 1971), and more recent work has been carried out as part of the Stonehenge Riverside Project (Parker Pearson *et al.* 2006; 2007). A number of house structures, and a substantial midden that predate the construction of the henge ditch and bank produced huge quantities of bone. Ceramics from the site are mostly of the Late Neolithic Grooved ware type (Cleal and MacSween 1999). ^{14}C dating has narrowed the period of occupation at Durrington Walls during the Neolithic to between 2515 and 2460 cal BC at 95.4% probability (Peter Marshall pers. comm.).

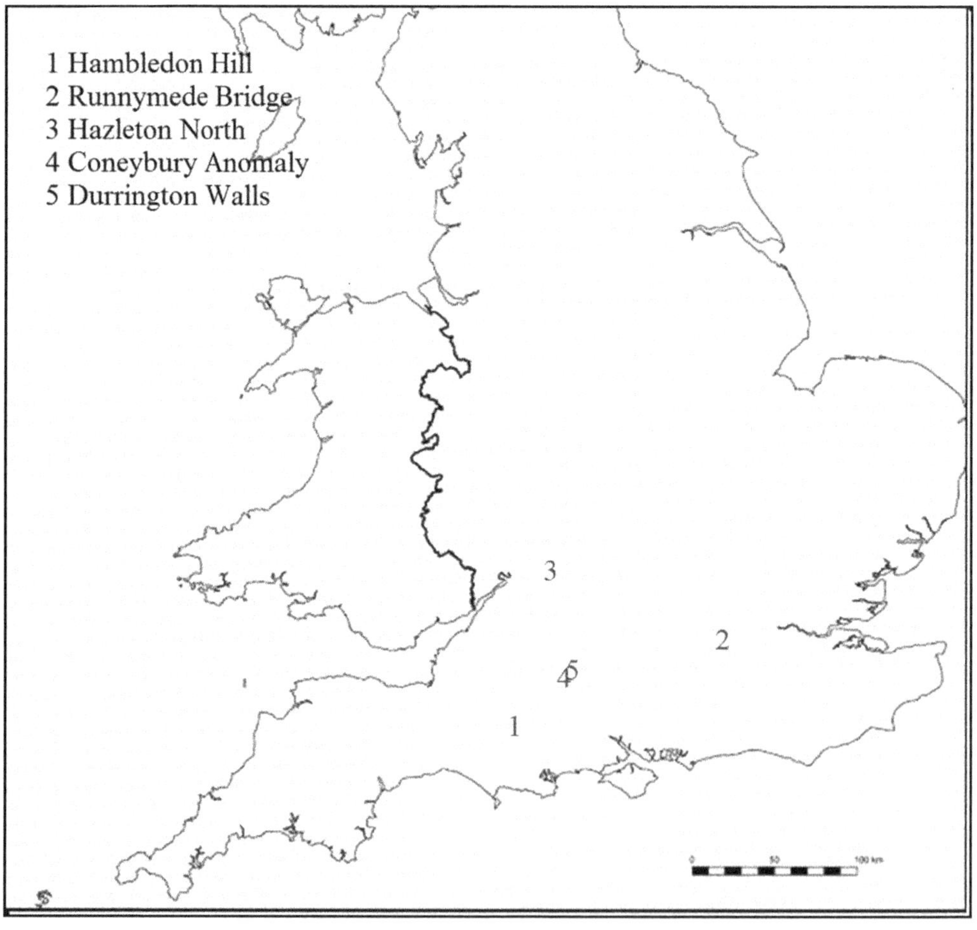

Figure 2.2 The location of Early and Late Neolithic sites investigated as part of this research

Animal bones from the original excavation were studied initially by R.A. Harcourt (1971), the same material was reanalysed by Richards and Thomas (1984). Faunal remains from the more recent Stonehenge Riverside Project excavations were analysed by Albarella and Serjeantson (2002), Albarella and Payne (2005). Material from Harcourt's investigation of the site that is held in Salisbury Museum was reanalysed as part of this research and new data from the Stonehenge Riverside Project was gathered by the author and U. Albarella. Pig remains were much more abundant than any other species at the site, and cattle were second in quantity. Sheep/goat and wild animal remains were infrequently found within the assemblage (Albarella and Serjeantson 2002).

Dates of the Mesolithic and Neolithic zooarchaeological assemblages

A significant problem in understanding the change from the Mesolithic to the Neolithic in Britain is the lack of evidence from the period at which the actual transition took place. This hiatus in evidence is not exclusive to the zooarchaeological record (see Chapter 1). The separation between the two main periods of interest has had an effect on this study. Table 2.1 presents some of the ^{14}C dates that have been published for the sites investigated. Of these, the latest Mesolithic site is Goldcliff (Bell *et al.* 2000) and the earliest Neolithic is Coneybury Anomaly (Richards 1990). Between the dates from the sites is a chronological gap of more than 1000 years. This gap has been highlighted elsewhere (Albarella and Pirnie in prep). Such a tremendous period for which there is little zooarchaeological information available has significant implications for any changes that are observed at the beginning of the Neolithic. The interruption will need to be carefully considered.

2.3 Methods

Introduction

The broad aim of the research is to investigate the earliest evidence for the husbandry of cattle and pigs in Britain. Any investigation of husbandry necessarily involves an attempt to reconstruct the structure of livestock populations as this generally reflects the different strategies employed by early farmers. Therefore, there is a need to understand the proportion of the sexes, the age at which animals were slaughtered, and the relative size and shape of the animals. These characteristics can be related to dietary regimes, to the purpose of the husbandry regimes and the conditions under which animals were kept. In Britain, the contribution made by local wild populations of aurochs and wild boar is also of relevance. Mobility, an important aspect of husbandry practices in many stock herding societies, will be explored in the Late Neolithic period by applying strontium isotope analysis to cattle remains. Parallel to these concerns is a need for efficiency in the collection of data. This is especially the case given the limited timescale under which some of the data collection has had to be conducted.

The present research is concerned with the early husbandry of two of the most important domesticates in prehistoric Europe, cattle (*Bos taurus*) and pig (*Sus domesticus*). The Latin nomenclature for domestic animals used in this volume will follow the recommendations of Gentry *et al.* (2004). An understanding of the domestication process (especially if domesticates are identified based on their variation from wild progenitors) cannot be reached without an analysis of the wild counterparts; the aurochs (*Bos primigenius*) and the Eurasian wild boar (*Sus scrofa*), which have therefore also been investigated as part of this research. The different behaviours, physiology and anatomy of these two species are likely to have led to the development of different approaches to their husbandry. This difference in the way that animals were treated would have resulted in different influences on the animals themselves and in turn led to variation in the physical responses to domestication. The animals have a different skeletal morphology, for example the number and type of bones varies between the cattle and pig lower leg. The size and robustness of bones is variable between the two species, and will therefore influence how susceptible cattle and pig bones are to taphonomic processes. All of these variations between the two, and the need to conduct separate data analysis, have made it necessary to establish two distinct recording protocols. Although broadly similar, there are differences in the approaches taken to the two animals, with different bones, parts of bones, and measurements recorded depending on taxa. The recording protocols are available in full in Appendices I and II. They include a comprehensive list of the skeletal elements and measurements included. Many of the aspects of the recording protocol that need to be discussed in the following passages are relevant to both species, and where possible they will be discussed together. Following the traditional zooarchaeological methodology, the approach taken to strontium isotope analysis will be presented.

The research will incorporate the study of previously un-investigated faunal assemblages, the re-analysis of some others, and the inclusion of a number of comparative studies in Chapter 5. The incorporation of numerous data sets makes coherency between the analysis of each particularly important. The information collected from the various faunal assemblages was input into two Microsoft Access databases that were constructed for the purpose of data collection.

The importance of ageing archaeological material

Understanding the age at death of the animals recovered from archaeological sites is important for any zooarchaeological investigation and is of particular relevance to the present study. Ageing techniques can be utilised in studies of early domestication by comparing population structures from groups that are assumed to be wild with those in which control by humans is suspected. Changes in the age profile of domestic species is considered to be one of the first indications that tighter control was being exerted on animal populations by human ones. In groups of animals that are domesticated, the structure of the population will be affected by the motivations of the humans that control it. Legge (1996) argues that a high number of juvenile bones along with large numbers of female animals at an archaeological site is an indication that domestication has taken place. Attempts to identify domestication of animals based on changes in population structure have been employed regularly in the Near East where the close temporal proximity to initial domestication events makes the assessment of size change in animals problematic (Hesse 1982; Zeder and Hesse 2000, 2254). Examples of this approach were discussed in Chapter 1. The reliability of arguing for husbandry or taming based on changes in population structure compared to what is considered normal wild population structure has been challenged (for instance Legge, 1996, 179). Of particular importance is the issue of whether a 'typical' wild population can be envisaged, especially at those sites for which evidence of wild exploitation is limited (Grigson 1989). In an attempt to resolve this issue the age and sex information from the Neolithic sites investigated will be considered in comparison with the pattern from the Mesolithic.

Detailed information about more specific husbandry practice can be obtained through an investigation of population structure. Differing mortality profiles observed in groups of domesticated animals have variously been interpreted as reflecting systems that concentrate on meat, or, for some species, secondary products such as milk and wool. Payne (1973) proposed a number of model kill-off patterns for sheep/goat economies. These models represent the expected kill-off patterns in systems that rely on the production of meat, milk or wool. Although archaeological material cannot be expected to conform strictly to the proposed models, they do provide a useful starting point for comparison. Cattle are capable of providing secondary products to humans, perhaps the most widely used of which is milk. Debate surrounding the likelihood of dairying in prehistoric Britain continues (see for instance Copley *et al.* 2003; Craig *et al.* 2005; Legge 2002; McCormick 1992), but numerous studies of cattle population structure have argued for the presence of a dairying economy. In particular, the work of Legge (1981), suggests that the presence of large numbers of female cattle at Neolithic Causewayed Camps is evidence of dairying during the Early Neolithic, and similar evidence has been proposed for Bronze Age Grimes Graves (Legge 2002).

Any attempts to use a population structure approach to find evidence for specific husbandry practices within pig keeping is more limited, predominantly by the fact that pigs provide no secondary products that are visible archaeologically. They are usually kept for the sole purpose of meat production and do not provide either hair or dairy products in the way that cattle and sheep/goats do. However there have been cases where differences in the practice of pig keeping have been identified by using such methods (*e.g.* Albarella *et al.* 2006).

Consideration of the age of zooarchaeological specimens is crucial when using the kind of biometrical methods that will be discussed in the following passages. The bones of young animals will continue to grow until fused (and to an extent after fusion), and are therefore not comparable with those from adult animals that are used in biometrical investigations. An inability to isolate juvenile bones from adult ones has the potential to lead to errors when analysing collected data. As an example, the inclusion of a number of young cattle bones in a sample might create a population in which animals are misleadingly small, or alternatively one in which more than one group of animals is identified. It is essential then that the age of each specimen is approximated so that only adult bones are used for biometrical analysis. Age also affects tooth size, though, while bones tend to get larger with age, teeth become smaller due to wear. Although wear mainly affects the height of molariform teeth, length and, in extreme cases, width can also be affected (for pigs see Albarella and Payne 2005; Payne and Bull 1988). It is therefore important that tooth size is analysed by taking into account tooth wear stages.

Finally, the identification of non-adult bones in an assemblage can be a useful indication of the level of preservation and the possible extent of bias within an assemblage. Young animal bones are usually much more susceptible to damage than those of the adult skeleton. A high proportion of young bones in a zooarchaeological assemblage are sometimes an indication that preservation conditions were favourable, and perhaps that taphonomic factors did not have a substantial impact on the composition of the assemblage. Such conclusions must only be made when the plethora of other potential bias causing factors have also been considered.

In addition, it is important to use information about changes in animal population structure in conjunction with other lines of evidence (discussed below).

Ageing methods used in this study

The method of estimating age used in this research is based on the state of epiphyseal fusion in conjunction with the eruption and wear of mandibular teeth. Most of the bones included in the recording protocol were chosen specifically to maximise the amount of information that could be collected per specimen, including ageing information. Consequently, most of the selected skeletal elements have a recordable area of fusion. Bones were recorded as 'f' when the epiphysis and diaphysis was fully fused and 'g' when partially fused, whereas 'ud' was used for an unfused diaphysis, 'ue' for an unfused epiphysis and 'ux' when the bone was unfused but both parts were present. A final category, 'h', was used in cases where it was unclear whether a bone was entirely fused or fusing; this usually applied to broken specimens. The astragalus is an exception to the rule that the elements used had an area of fusion. The element can be difficult to age because it has no area of fusion, therefore during data collection the astragalus was characterised as normal, light or porous. Those that were normal or light were measured, whereas those that were classified as porous were not measured. The level of lightness or porosity is inevitably subjective, but the characterisations provided a rough guide to the age of an animal and by recording them the problem of accidentally including young animals in the analysis of astragalus size was avoided.

The use of epiphyseal fusion in attempts to reconstruct animal population structure must be conducted with consideration of a number of caveats. Most researchers interested in epiphyseal fusion in domestic animals recognise that the sequence of fusion within the skeleton follows the same order of succession for animals within the same species, but it is accepted that the exact timing of fusion is variable. The rate of fusion is known to vary under the influence of a number of factors. These factors include individual variation, nutrition, environment, sex of the animal and castration (Davis 1996; Reitz and Wing 1999, 75). In addition, many of the sequences that provide an approximate age for the fusion of specific bones are (necessarily) based on studies of modern animals. Silver's (1969) sequences are commonly used (including in this study) but there is no way to be sure that these ages are comparable with the age of fusion in prehistoric animals. Many factors may have affected the age at which bones fused in early domestic types and wild animals, therefore, this study used a more comparative system in which bones were separated into categories depending on their placement within the fusion sequence. The specific category to which individual bones were assigned is provided in Tables 2.2 and 2.3. Age categories for cattle were established from O'Connor (2003) and Silver (1969).

In pigs the Earliest category only contains the scapula and pelvis. It was decided that these two elements should be placed in a separate category from the other bones based on the information provided by Bull and Payne (1982) (which includes data from Lesebre) that suggests these bones fuse when an animal is approaching 1 year. In addition the scapula and pelvis are the only bones that were fused by 7-11 months of age in animals from Kizilcahamam (Bull and Payne 1982).

The Early category consists of the distal humerus, proximal radius, phalanx 1 and 2. According to Silver's research there is a gap between the timing of fusion in the first and second phalanx. Despite this gap the first and second phalanges have been included in the same group here based largely on the data presented by Bull and Payne (1982). These categories were established based the work of Silver (1969), Bull and Payne (1982) and O'Connor (2003).

Further information about the population structure of zooarchaeological assemblages can be gathered by employing measurements of parts of the skeleton that are thought to be very age dependent, such as the width of the scapula neck SLC (*sensu* von den Driesch 1976) (Albarella and Payne 2005). One of the main potentials of this measurement is for the identification of seasonal slaughter patterns, and it has been used for instance at the Mesolithic site of Star Carr to suggest seasonal wild boar exploitation (Rowley-Conwy 2001).

Teeth must be approached in a different way to the ageing applied to postcranial elements. The eruption and wear of teeth (both deciduous and permanent) can provide more extensive information about the age of an animal than the fusion of postcranial bones. Epiphyseal fusion provides information about the development of an animal until it reaches maturity (*i.e.* when its bones are fully fused). This means that once adulthood is reached no further information about advancing age can be gathered from the state of epiphyseal fusion. In contrast, tooth wear can continue to provide information throughout an animal's life. For the purpose of this investigation wear on the occlusal surface of lower molars, 4th premolar and deciduous premolar was recorded. According to Grant (1982, 94) these teeth provide the most reliable indication of age. The wear stages established by Grant (1982) were employed for both cattle and pig dentition.

In common with the fusion of postcranial elements, the eruption of an animal's teeth may be closely linked to a number of factors that might complicate our understanding of past animal population structure. Sex and castration (Moran and O'Connor 1994), as well as nutrition (Grant 1978, 103) have all been suggested as possible reasons for differential tooth eruption times. Moreover, it is possible that variable environmental or nutritional influences will have an impact only on specific teeth within the mouth (Grant 1978, 103). It cannot be assumed that these variation causing factors affected all aspects of the dentition uniformly. The sequence of eruption, however, should remain unchanged over time. Variation in tooth wear can be linked to nutrition, but is more likely to be affected by the type of food consumed. Those animals that consume very coarse or abrasive food will sustain a much greater level of wear over a shorter space of time than their counterparts eating less abrasive food (Moran and O'Connor 1994, 270). This assumption has been challenged by Grant (1978, 105) who found no difference in the level of wear on the teeth of sheep from Scotland that consumed a diet with substantial sand inclusions, when compared with Roman animals from the south of England. Variable tooth wear is an important consideration, but is not likely to have impacted greatly on the animals investigated during this study. Many of the sites used in the present research are located in areas of similar environmental conditions and therefore the animals are likely to have been exposed to similar inclusions in their diet.

As with epiphyseal fusion the variability inherent in the eruption of teeth and the rate at which they wear makes it unproductive to try and attribute very specific ages to events within the sequence. Therefore, this analysis will again employ a method in which mandibles are grouped following the system outlined by O'Connor (2003). O'Connor identified a number of categories based on the wear stages of Grant (1982), ranging from Neonatal to Elderly. In some cases it will only be possible to attribute mandibles to one of six broad categories, but for some mandibles it will also be possible to identify a more specific period within the categories. In the case of the latter, mandibles will be attributed to one of ten sub-categories: Neonatal (N); Juvenile (J); Immature (I1 and I2 in pigs); Subadult (SA1 and SA2); Adult (A1, A2 and A3); and Elderly (E). The categories are shown in Table 2.3 (pigs) and Table 2.4 (cattle) including the stages from Grant (1982) that they correspond to (the tables have been produced following the information in O'Connor 2003).

Rationale for employing sexing techniques

Many species of mammal exhibit marked sexual dimorphism (Reitz and Wing 1999, 83), a difference in size between female and male members of the species. This is the case for aurochs (Degerbøl and Fredskild 1970, 25, 51; Grigson 1969, 280; van Vuure 2005, 108), domestic cattle (Degerbøl and Fredskild 1970, 141; van Vuure 2005, 106), wild boar (Albarella *et al.* 2009, 107) and domestic pigs. All of these, to varying extent, show a trend towards females being smaller than their male counterparts. There is also some evidence to suggest that the level of sexual dimorphism may decrease with domestication (van Vuure 2005, 109) and could provide an indication of changing exploitation practices. A potential difficulty that must be borne in mind when analysing sex in archaeological populations is that even in prehistory castration may have occurred. Clutton-Brock argues that castration would have been a common practice from the beginning of domestication (Clutton-Brock 1999, 37), but it may be very difficult to identify in zooarchaeological assemblages. Castrated animals have characteristics that can be different from both females and entire males, but these will vary according to the age of castration (Davis 2000, 384). The presence of castrates in an archaeological assemblage may lead to the identification of three - sex determined - size groups, but it is more likely to obscure patterns of sex variation. Whatever the case, such potential variation must be considered in the interpretation of data collected for the purpose of sex differentiation.

The ratio of female to male animals in an animal bone assemblage is important and can elucidate hunting concerns as well as husbandry practices. In addition, a major concern when using biometrical analysis is whether the variation seen is the result of wild/domestic variation, age variation, sex variation or other factors. Therefore, a number of methods that should allow the identification of sex ratios within the pig and cattle populations were included in the recording protocol. Sexual dimorphism in both wild boar and domestic pigs is manifested in part by the smaller size of the females and by some morphological differences. Biometrical analysis of skeletal elements should allow distinctions to be identified between female and male, but caution was necessary to avoid mistaking different sex groups for different groups of animals. The dentition of female and male *Sus* varies, and the most striking difference can be seen in the mandibular and maxillary canine teeth. In females these teeth are comparatively small, while males have large distinctive tusks. There are also clear morphological differences between female and male canines. Canine teeth were recorded as female 'f' or male 'm', and where possible the alveolus of the tooth will be recorded as 'af' or 'am'. In some cases sex determination was difficult, for example when canines were deciduous or damaged specimens. When this occurred, the presence 'p', was inserted into the database. Permanent lower canines still embedded in the jaw cannot be sexed unless a sufficient proportion of the tooth has erupted.

Sexual dimorphism in cattle is pronounced but there are no obvious morphological differences that can be employed to distinguish between the bones of females and those of males. This study therefore relies most heavily on identifying the differential size of females and males. One of the most commonly used methods to distinguish between female, male and castrated cattle is through the measurement of the metapodials, particularly the breadth of the distal end. This method has been employed by a number of researchers working on cattle from various time periods and different geographical areas. For example, Degerbøl and Fredskild highlighted the potential of distinguishing between female and male aurochs by investigating the size and shape of their metapodials (Degerbøl and Fredskild 1970, 51), and Albarella discussed the use of these bones in detail (Albarella 1997). A further example of sexual dimorphism in cattle metapodials is that the metacarpal is more sexually dimorphic than the metatarsal, due to the greater weight supported on the front legs of bulls compared with that of cows (Bartosiewicz 1987, 48). If a sufficient number of cattle metapodial measurements are available these will be analysed in an attempt to detect sex groups. Another bone that is potentially useful in detecting sex in cattle is the pubis (Grigson 1982), but this is unfortunately not commonly found in a well preserved form within archaeological assemblages and as such was not sexed during the analysis.

Rationale for employing biometrical analysis

The inclusion of at least a minimal amount of biometrical analysis has become commonplace in zooarchaeological investigations carried out by European researchers (Albarella 2002). Most modern reports of faunal assemblages produced in Britain contain a biometrical element, many of which to a greater or lesser extent, attempt to tie measurements into the wider zooarchaeological picture. However, reports are commonly limited to the inclusion of measurements that have not been chosen with consideration of the specific research questions of the investigation in mind. This can result in biometrical analysis that provides little more than a descriptive account and makes no contribution to the interpretation of the archaeological record of which it is a part. It has repeatedly been argued that the measurement of animal bones from archaeological sites is important to zooarchaeological investigation (Albarella 2002; Boessneck and von den Driesch 1978; von den Driesch 1976, 1). But these must be taken in a consistent way, and most importantly with close regard to the questions that investigators are attempting to answer. In order to efficiently exploit the potential of taking measurements from the assemblages under investigation, the aims of the research must be central to recording protocol decisions. This section will discuss the potential of biometry to provide information relevant to the present study, will provide some examples of previous zooarchaeological work that has produced positive results using similar methods, and explain why certain decisions have been made when choosing measurements.

Traditionally one of the most commonly utilised criteria to identify domestication in animals recovered from archaeological sites is a reduction in body size. Domesticated cattle (*Bos taurus*) are usually smaller than wild aurochsen (*Bos primigenius*) (Grigson 1969, 278; 1989), and the same pattern applies to domestic pigs and their wild counterparts. Size decrease may be partly due to the circumstances of domestication: the restriction of movement or confinement of the animals, poor quality or insufficient food, disease and parasites exacerbated by forced proximity of groups of animals, and the accidental or deliberate selection of smaller animals by humans could all have contributed to produce animals with smaller body size (Meadow 1989, 86). Biometrical analysis can be used to identify the presence of wild or domestic animals within an assemblage. It has been employed in numerous investigations that have concentrated on the initial stages of domestication in the Near East (for example the work of Ervynck *et al.* (2001) at Çayönü Tepesi). Similar methods have been regularly employed by researchers investigating the spread of domesticates through Europe (*e.g.* Albarella *et al.* 2006; Rowley-Conwy 2003; Tresset 2000; 2003) and the presence of wild animals in assemblages otherwise dominated by domesticates (Davis and Payne 1993, 15). Many more examples could be cited. The measurement of animal bones has the potential to contribute to the debate surrounding the earliest husbandry of cattle and pigs in Britain. Whether the earliest domesticates found in Britain were introduced as fully domesticated types, or some contribution was made by local populations, can be investigated through employing the biometrical method. In addition the possibility of continued hunting of wild animals in the Neolithic can be addressed.

Changes to the size and shape of animals have been linked to a number of other, non-human factors, and it is important that these are excluded before changes in the animal skeleton can be attributed to human influences. Davis (1981) conducted a study in which temperature was linked to the size of animals. He observed a trend that suggested that over large geographical areas the body size of some animals is inversely correlated with temperature. In other words, as temperature was seen to increase, body size decreased. Similarly, Albarella *et al.* (2009) have observed a west/east and a south/north cline in the body size of wild boar. The largest examples of this animal in the archaeological record can be found in north-eastern areas, and the smallest in the south west. This phenomenon is explained as being the result of temperature and continentality. Although these factors will create differences between animals from Britain and those from other parts of Europe, the relatively small size of the research area being investigated makes it unlikely that significant differences would exist between groups of animals in different parts of southern Britain. Most of the sites used in this study that date to the Neolithic period are at similar latitudes, and the temperature change associated with the south/north cline identified by Albarella *et al.* (2009) would have had little effect. Similarly the west/east cline is thought to be linked to continentality, a factor that is unlikely to have had a great effect on animals over the short distances of southern Britain. Confinement to islands has been seen to result in sometimes drastic reduction in the size of animals (Lister 1996; Lomolino 1985; Reitz and Wing 1999, 293), but again insularism has not had a great impact of the animals of the British mainland. Clearly there is no single factor that is responsible for the change in animal size. It is therefore essential that different variables are considered – in addition to the domestication process - as potential factors of size variability in an assemblage.

An important consideration when investigating domestication through size changes is that these changes do not necessarily affect the whole body or all skeletal elements to the same degree. A study of the changes through time of the domesticated pig skeleton in Poland suggested that skeletal elements responded to domestication in different ways. Such variation between elements resulted in decreased stature but also changes in the physical proportions of the animals (Lasota-Moskalewska *et al.* 1987, 51). As an example, teeth tend to show less pronounced variation than postcranial bones do. Even between different teeth the third permanent molar tends to be more variable than the first and second molars (Albarella and Payne 2005, 596; Payne and Bull 1988). A further example is the lachrymal bone of suids, which shows a marked and fast deviation under the influence of domestication, and has often been used to distinguish between different groups of wild boar and between wild and domestic forms (Clutton-Brock 1999, 91). Much research has concentrated on the measurement of aurochs and cattle skulls (*e.g.* Aaris-Sørenson 1999; Bartosiewicz 1999; Chaix and Arbogast 1999). Unfortunately, the skull of both *Bos* and *Sus* is somewhat fragile and complete examples rarely survive in the archaeological record. Therefore skull measurements have not been included in the recording protocol.

Many of the sites included in this investigation provided faunal assemblages with a small number of specimens. This is particularly the case for those sites dating to the Late Mesolithic. Assemblages from the Mesolithic in general tend to be small, and are often poorly preserved and highly fragmented. Some skeletal elements also provided very small total samples, for example cattle mandibular second molars recorded during the data collection period of this project only provided 22 examples that could be measured despite numerous assemblages from multiple periods having been investigated. As a result, in isolation this element could provide very little useful archaeological information. The problem of small sample size is pervasive in zooarchaeology (Albarella 2002, 52), a discipline that relies on the incorporation of large amounts of data to make interpretations on the population rather than the individual level. To deal with this issue and gain useful information from sites, measurements can be pooled together using a size index scaling technique. Multiple examples of such methods have been devised (see Meadow 1999 for a detailed discussion), but all allow the amalgamation of material to some extent, essentially combining measurements to create larger units that can then be compared. Common to all of these scaling techniques is the requirement of a standard measurement or set of measurements from a modern population (as in that used by Ducos, reported in Meadow 1999, and the modern Turkish wild boar used by Payne and Bull 1988), or similar mean measurements from archaeological examples. Comparison could also be made with single measurements from either archaeological or modern specimens, however this final option creates the problem of whether a female or male animal should be used, and whether the age of the individual chosen will affect results. For the purposes of this research the mean measurements from a number of archaeological and palaeontological specimens will be used. In order to apply size index scaling techniques to pig remains a standard has been established from the measurement of *Sus* remains from the Late Neolithic site of Durrington Walls, Wiltshire, southern Britain (Albarella and Payne 2002). No such large assemblage of cattle remains is readily available from Britain from which a cattle standard could be established. A new standard using palaeontological remains of aurochs from Ilford, Essex, was calculated (see Appendix III). Most aurochs skeletal elements are represented in this collection that has been dated to Marine Isotope Stage 7 (186-242kya) (Andy Currant, pers. comm.).

The importance of biometrical analysis and its potential for addressing the aims of the research have been outlined above. Measuring bones of cattle and pigs will form a central part of the analysis of faunal remains. It is essential that each element is measured as consistently as possible throughout the investigation. Measurements of the cattle skeleton are based on the recommendations provided by Davis (1992), and Von den Driesch (1976). Measurements of the pig skeleton are those defined by Von den Driesch (1976), Payne and Bull (1988) and Albarella and Payne (2005).

Dental pathologies

Linear enamel hypoplasia (LEH) is a pathological condition that is frequently observed on archaeological animal as well as human teeth. The root causes of this condition are much discussed and no firm consensus has yet been reached, but many investigators highlight the possibility of nutrition and disease as contributing factors in the formation of hypoplasia (Dobney and Ervynck 1998, 263). As a result the presence (or absence) of hypoplasia on the teeth of domesticated animals may provide valuable information for the study of early animal husbandry. If the condition can be linked to nutritional or other kinds of stress, or to infectious disease, conclusions about the conditions and quality of life of prehistoric animals might be inferred. LEH was recorded when it occurred on the buccal surface of permanent mandibular molars of both pigs and cattle. Single or multiple lines or linear hypoplasia were distinguished, but no measurements of the height of occurrence of LEH on the crown were taken. Measuring the height of LEH bands can be a very time consuming activity, which produces uncertain results that are not considered of central importance for this research.

A second condition was observed frequently in the archaeological material once data collection had been started and the recording protocol for pig teeth was amended to accommodate it. This phenomenon occurs as a distinctive linear line of caries on *Sus* molars and has been observed and reported on pig teeth from Late Neolithic Durrington Walls. Although the cause of this pathology is as yet unknown it has been observed frequently on domestic pig teeth within some assemblages, and relatively infrequently on wild pig teeth. It may therefore be a useful future indicator of pig living conditions and was recorded in the column previously reserved for LEH as 'c'. Any other pathological or unusual characteristics of specimens were recorded in the comments box of the database. Recording of pathological information, such as caries and LEH, can provide important information to the investigation of the earliest animal husbandry in Britain. Pathological conditions that could reflect confinement of animals appear within the lifetime of an individual animal, a much faster physical manifestation of the effects of human control than the size diminution often associated with domestication that can take some time to become apparent (Meadow 1989, 86-7).

Burnt Specimens

Depending on the temperature of exposure and other specific conditions, burning can result in varying degrees of bone modification. Most importantly, in some cases the result is warping or shrinkage (von den Driesch 1976, 3-4; Reitz and Wing 1999, 133) to the extent that measurements are no longer an accurate representation of the original dimensions of the bone. The effect that burning has on a particular element depends on an array of factors: the taxon; the age of the animal; the skeletal element affected; the condition of the bone when exposed to burning (for an overview see Lyman 1994, 384-92). The interplay of all of these factors must be taken into account when investigating assemblages that contain burnt material. Therefore, to avoid potential biases and inaccuracy of measurements, bones that are clearly heavily burnt or calcined were not included in the biometrical analysis. Those that were partially burnt or singed were assessed on an individual basis to determine whether they should be included. Any bones that showed signs of exposure to fire/heat had a description included in the comments box of the database. This allowed for their exclusion at a later point if necessary.

Table 2.1 Selected radiocarbon dates from sites included in the study, from various sources (order as discussed in the text). The available dates from Wawcott have been included here, but none of them refer to sites XV, IV and XXIII from which faunal remains are available. 154 ^{14}C dates have been published for Hambledon Hill, some of the earliest ones from different parts of the site are included here. A date from Runnymede Bridge Area 8 has been included here, although no bones from this part of the site were recorded.

Site	Dates BP	Dates cal BC	Lab code	Source
Thatcham	8,100±180	7,550-6,550	BM-65	Chisham 2006
	10,030±170	10,700-9,100	Q-658	
	10,365±120	10,900-9,700	Q-659	
	9,200±90	8,630-8,260	OxA-2848	
Wawcott I	5,260±130	4,350-3,750	BM-449	Chisham 2006
Wawcott III	6,120±134	4,700-5,400	BM-767	
Marsh Benham	8905±80	8270-7760	OxA-5195	Chisham 2006
Faraday Road	8510±60	7600-7480	R-24999/1	Ellis *et al.* 2003
	9418±60	8830-8550	R-24999/2	
Victoria Park	n/a	n/a	n/a	n/a
Star Carr	8940±90	8100-7840	OxA-4377	Dark 1998
	9385±115	8430-8090	OxA-4376	
	9390 ±70	8510-8320	OxA-3342	
	9420±70	8580-8340	OxA-3343	
	9360±70	8600-8370	OxA-3344	
	9580±70	8710 8460	OxA-3345	
	9560±70	8770-8520	OxA-3346	
	9680±70	8810-8590	OxA-3347	
	9700±70	8830-8630	OxA-3348	
	9640±70	8950-8700	OxA-3349	
	9500±70	8980-8770	OxA-3350	
	9630±100	9050-8830	OxA-3351	
Seamer Carr	9940±100	9860-9240	OxA-1030	Schadla-Hall 1990
	9790±180	9870-8720	BM-2350	
	8740±120	8240-7570	BM-1841-R	
	8210±150	7580-6700	HAR-6498	
Goldcliff East		5600-5200		Bell *et al.* 2000
Hambledon Hill				Bayliss *et al.* 2009
Main enclosure	5220±110	4330-3780	HAR-2370	
	4900±40	3770-3640	OxA-7029	
	4870±45	3710-3530	OxA-8851	
	4830±30	3660-3530	OxA-7099	
	4800±65	3660-3380	OxA-7024	
Stepleton enclosure	4855±60	3660-3380	OxA-7020	
	4772±19	3640-3520	UB-4151	
	4648±21	3500-3350	UB-4138	
Runnymede Bridge				Needham 1991
Area 4	4630±70	3621-3105	HAR-6132	
	4920±80	3950-3524	HAR-6128	
Area 6	4830±70	3780-3381	HAR-6130	
	4930±90	3970-3520	HAR-6131	
	4270±110	3302-2580	HAR-6136	
Area 8	4690±110	3770-3100	HAR-6133	
Hazleton North				Meadows *et al.*2007
Pre-cairn phase	5110±90	4230-3700	OxA-13374	
	4875±80	3900-3630	OxA-646	
Completed construction		3695-3650		
Coneybury Anomaly		3980-3708		Richards 1990
Durrington Walls		2515-2460		Marshall pers.com.
Alice, Uskmouth	4660±70	3640 - 3130	CAR-1096	Noddle 1989

Selection of skeletal elements and measurements
The recording method used a highly selective system that was based on previous work carried out by Davis (1992) and Albarella and Davis (1996). Skeletal elements were selected based on a number of factors:

1) potential to provide information on the wild or domestic status of an individual
2) potential to provide ageing information
3) potential to provide sex information
4) survivability
5) usefulness when trying to establish species frequency
6) usefulness when trying to determine the adverse effects of taphonomic processes.

The elements selected for recording were: the deciduous 4th premolar; the permanent 4th premolar; 1st, 2nd and 3rd permanent molars; atlas; scapula; humerus; radius; metapodials (excluding lateral metapodials in *Sus*); 1st and 2nd phalanges; pelvis; femur; tibia; calcaneum and astragalus.

Analysis of the material from the Late Neolithic site of Durrington Walls followed a more comprehensive recording system, as the investigation of the assemblage is ongoing and has additional objectives to those of this research. However, despite differences in the recording protocol, all the data gathered from the other assemblages used for this research were also collected for Durrington Walls.

Teeth
The durability of teeth renders them relatively resistant to attack from taphonomic processes. When compared to other elements of the cranial and postcranial skeleton they tend to survive well and are found in abundance in the archaeological record. Cattle and pig teeth are both included in the current analysis. With the exception of the *Sus* canine, sexual dimorphism does not significantly affect the size and shape of teeth of either taxa (Grigson 1982, 7-8; Payne and Bull 1988, 30). There is also little dependence on age (but see discussion above under 'Ageing'), as once formed the teeth will not continue to grow as an animal ages. As a result of these characteristics the teeth of both cattle and pigs can be very useful when attempting to distinguish between wild and domestic types. Shortening of the rostral region associated with domestication in pigs has also led specifically to a shortening of the molars (Mayer *et al.* 1998:40), particularly the third molar (Albarella 2002, 57; Payne and Bull 1988). Consequently, measurements of the third molar have regularly been used to investigate wild and domestic differences (*e.g.* Albarella *et al.* 2005). However, using the size reduction of the third molar as an indicator of domestication can be problematic as a consequence of its great intra-population variability (Payne and Bull 1988). The analysis of the third molar must therefore be integrated with that of other molariform teeth, particularly width measurements as these tend to be more numerous, less age dependent and less variable (Payne and Bull 1988). It has also been demonstrated that the measurements of the lower second molar are affected by the shortening of the jaw as a consequence of domestication (Mayer *et al.* 1988).

Atlas
This bone is easily identifiable both to skeletal element and its designation to taxa. It is the only element of the vertebral column to be included in the analysis. Its importance lies in its contribution to understanding assemblage composition. The presence or absence of parts of the vertebral column will give an idea of whether whole or partial skeletons were being deposited. It also exhibits a high level of sexual dimorphism (Grigson 1982, 9) and the measurements chosen (H and BFcr) may be useful in distinguishing between the sexes and between wild and domestic animals.

Scapula
The scapula can be easily attributed to taxa. According to Silver (1969, 285) this bone fuses at around 7-10 months in cattle and within the first year in pigs. This relatively early fusion leads to its high level of survival in the archaeological record. The width of the neck (measurement SLC as defined in Payne and Bull 1988, 42) is highly dependent on the age of an animal (Albarella and Payne 2005, 596-8), and some parts of the bone will continue to increase even once the bone is fully fused (Payne and Bull 1988, 30). This part of the bone is perhaps the most robust and can often be found within assemblages, even if other parts have been destroyed. Therefore the main benefit of taking this measurement is for its use in investigating the age of animals in the populations being studied.

Humerus
This bone has a variety of uses when investigating early animal husbandry. Different parts of the bone develop depending on different factors meaning that much information can be gained from a single bone. The distal humerus being an early fusing bone (Silver 1969, 285) survives better than the later fusing proximal end and will form the focus of the investigation of this bone. Two measurements, the HTC (height of the trochlea) and the BT (breadth of the trochlea) will be taken on all specimens. The HTC varies between wild and domestic animals (Payne and Bull 1988, 42), while in contrast the BT is more related to sexual dimorphism (Payne and Bull 1988). It provides the opportunity to compare the wild/domestic status with age variation within a single bone. The SD was also taken because of its use when investigating age in populations (Albarella and Payne 2005, 596). Another commonly taken measurement of the distal end – the greatest width (Bd *sensu* von den Driesch 1976) – was not taken due to its high level of age dependence (Payne and Bull 1988), and because the same information can be obtained through BT, which is located on the same axis but is less variable with age.

Radius
The proximal radius fuses early while the distal epiphysis remains unfused almost until an animal reaches adulthood (Silver 1969, 285). The development of the proximal end

is highly dependent on age, and as with the scapula may continue to increase in size even after fusion (Payne and Bull 1988, 30). No measurements on the proximal end were therefore taken, as they would replicate – but less effectively – the data already available through the scapula. Where possible the greatest length of the radius was measured. Though this measurement is not commonly available, the radius is the most likely of all long bones to be found complete and due to the scarcity of other height measurement, this could prove viable. Once fusion has occurred this measurement shows almost no change with age, as post-fusion size increase affects widths and depths far more than lengths.

Metapodials

The benefit of using the metapodials (the metatarsal and metacarpal) in biometrical analysis has been discussed above. The proximal ends of the metacarpal and metatarsal of cattle and pigs fuse before birth and are therefore of little biometrical use, due to their huge post-fusion increase. Distal ends fuse much later (between 16 months and 2 ½ years in cattle and between 2 and 2 ¼ years in pigs (Silver 1969, 285-6), but they tend to survive well in the archaeological record (Bartosiewicz 1987, 47). A combination of length and distal width measurements of these bones have been used to distinguish between female, male and castrated cattle. They are generally broader in bulls than cows, and longer in castrated animals than complete males (Grigson 1982, 10; Kyselý 2008, 16). A further aspect of sexual dimorphism that could be investigated by measuring metapodials is the difference between the front and rear legs. It has been observed that the front legs of cattle and pigs are more robust – relative to rear legs – in males than females (Bartosiewicz 1987, 48; Kyselý 2008, 16; Lasota-Moskalewska *et al.* 1987; Payne and Bull 1988, 30). Metapodial measurements can also be of use in distinguishing between wild and domestic cattle (Degerbøl and Fredskild 1970), different breeds of cattle (Albarella 1997, 45), and sex (Higham 1969, 139). For cattle, a series of measurements will be taken that will allow a number of variables to be investigated. The measurements taken on cattle metapodials were: the greatest length (GL); the breadth at fusion (B at F); the breadth of the distal end (BFd); width of the lateral condyle (b); width of the medial condyle (a) and the heights of each at 3 and 6. All of these measurements have been defined by Davis (1992). The length measurements of the pig central metapodials were taken when they were complete.

Pelvis

The measurement LAR in pigs (defined by Payne and Bull 1988, 42) and LA in cattle (after von den Driesch 1976, 82-83) was taken on the archaeological material. This measurement can most usefully be used to distinguish between wild and domestic animals (Payne and Bull 1988, 32).

Femur

Both the proximal and distal ends of the femur fuse late in the sequence of skeletal development (Silver 1969, 286) making them susceptible to destruction and a less common occurrence in the archaeological record than other elements. The only measurement to be taken on the femur was the DC (depth of the caput). This measurement varies little with age once fully fused and is not highly sexually dimorphic (Albarella and Payne 2005, 597).

Tibia

The tibia tends to be robust and survive well in the archaeological record. Two measurements, the breadth and depth of the distal end (Bd and Dd for cattle as defined in von den Driesch, 1976; and the BdP and Dd following Payne and Bull 1988, 42) were taken along with the greatest length of the bone whenever possible. The breadth of the distal end is particularly useful for distinguishing between wild and domestic animals as it is not greatly influenced by sexual dimorphism or the age of an animal (Payne and Bull 1986, 32: Albarella and Payne 2005, 595; Lasota-Moskalewska 1987, 67).

Calcaneus

Two measurements will be taken on the calcaneus. These are the greatest length (GL) (von den Driesch 1976, 90; Payne and Bull 1986, 42; Albarella and Payne 2005) and greatest depth (GD) (Albarella and Payne 2005). Measurement of the calcaneus will help to distinguish between wild and domestic animals (Lasota-Moskalewska 1987, 67). The calcaneum is arguably the least sexually dimorphic skeletal element (Kyselý 2008, 16) and is therefore a good element to distinguish between groups of animals without the confusion of different sex groupings arising.

Astragalus

For pigs, two measurements the GLl and Glm were taken (following Payne and Bull 1986, 42 and von den Driesch 1976, 89). Similarly for cattle, the GLl, GLm as well as the Bd will be measured (defined by von den Driesch 1976, 88). Although the length measurements are likely to provide the same kind of information (Albarella and Payne 2005, 597) it was decided that both the GLl and GLm would be taken. This will increase the number of measurements available for analysis and will also allow the construction of scatter plots that compare the two. The astragalus has no epiphyses but, being restricted in its potential growth by surrounding bones, shows little variability once care has been taken to exclude the lightest and most porous specimens. It is therefore a useful bone for the distinction of wild and domestic forms (Albarella and Payne 2005, 598; Payne and Bull 1988).

Phalanges

The first, second and third phalanges were recorded for both cattle and pigs. The relatively early fusion of the first and second phalanges makes them useful in assessing the contribution of very young animals to faunal assemblages.

2.4 Strontium isotope analysis

Introduction

Strontium isotope analysis has the potential to provide useful archaeological information to researchers interested in past animal husbandry, and will be used to investigate the movement of animals to Durrington Walls. The technique can be applied in a number of ways, but the most relevant to the present study is the method of comparing the $^{87}Sr/^{86}Sr$ of individuals (humans or animals) with the conditions prevalent in the locality of their burial (Evans *et al.* 2006, 265). Animals take in strontium from their environment through the food and water that they consume. The $^{87}Sr/^{86}Sr$ signature provided by analysis of skeletal material will reflect the underlying geology and soils of the area in which an animal was raised (Balasse and Ambrose 2002; Evans *et al.* 2006, 266; Hoppe *et al.* 2003, 21; Nehlich 2009; Sillen and Sealy 1995). More specifically, the ratio of ^{87}Sr to stable ^{86}Sr gives an indication of the age of underlying geology (as older rocks are richer in rubidium, from which ^{87}Sr is derived, than younger ones) (Montgomery *et al.* 2006), so that the highest $^{87}Sr/^{86}Sr$ ratios are derived from the oldest rocks (Bentley 2006). This tendency for $^{87}Sr/^{86}Sr$ in the skeleton to reflect geological conditions has been exploited by archaeologists interested in migration and movement. The $^{87}Sr/^{86}Sr$ signature is set within animal teeth during their development, and once the teeth are fully formed will not be significantly altered (Price *et al.* 1994; 2002; Schweissing and Grupe 2003). Variation between the $^{87}Sr/^{86}Sr$ signature of a human or animal and that expected at the location of deposition can indicate movement. $^{87}Sr/^{86}Sr$ ratios have been used to investigate past human migration, for instance in the central European Linearbandkeramik (Price *et al.* 1994), Neolithic Dorset (Montgomery *et al.* 2000), the Bronze Age in Wiltshire (Evans *et al.* 2006), and Roman Hampshire (Evans *et al.* 2006).

A number of complicating factors need to be addressed. Firstly, the composition found in animals that lived in an area is unlikely to be identical to that of the rocks from which the strontium originated (Price *et al.* 2002). Secondly, a great deal of variation can exist between soils, even if they originate from the same underlying rock, due to factors such as weathering, alluviation and loess (Price *et al.* 2006; Bentley 2006), and seaspray (Evans *et al.* 2010) resulting in the potential for variable $^{87}Sr/^{86}Sr$ in plants that grow on them (and therefore the strontium that enters the food chain. In some areas for instance, it has been found that an almost equal contribution to the $^{87}Sr/^{86}Sr$ signature of surface plants was made from underlying substrate as from rain water (Chadwick, cited in Price *et al.* 2002). A lack of homogeneity in source rocks can also affect the strontium that enters the food chain (Price *et al.* 2002). Therefore, $^{87}Sr/^{86}Sr$ information gathered from rocks themselves can only be used as a point from which comparisons are made, and it is important to consider the *biologically* available strontium in a research area. A map produced recently by Evans *et al.* (2010) illustrates the main variations in biospheric $^{87}Sr/^{86}Sr$ will be used extensively in this research.

There has been significant debate surrounding the potential contamination of archaeological specimens. Due to the porous structure of some skeletal material, specifically bone and tooth dentine, significant post-depositional alteration in bone/dentine chemistry can occur (Bentley 2006; Nehlich 2009; Nelson *et al.* 1986; Price *et al.* 2002). As a result of diagenesis strontium can enter skeletal material, with the result that analysis provides information about the burial environment rather than the geographical origin of an individual. Attempts to mitigate the problem of diagenetic alteration through the development of methods that remove intrusive strontium, while maintaining the signature of biological strontium have had mixed success (Hoppe *et al.* 2003; Price *et al.* 2001; Sealy *et al.* 1991; Sillen and Sealy 1995). No existing technique can remove all strontium of diagenetic origin (Bentley 2006; Hoppe *et al.* 2003). A less problematic approach to strontium isotope analysis is to avoid the use of postcranial material and dentine, and only use tooth enamel. The less porous structure of tooth enamel ensures that it is more resilient to diagenetic alteration than other skeletal material (Budd *et al.* 2000; Fricke and O'Neil 1996; Price *et al.* 1994; 2002) and it can be used without the necessity of techniques that isolate biological from contaminant strontium.

The main focus of strontium isotope analysis during this research was on cattle mobility, but in addition two pig teeth were also analysed. At the time that this research was undertaken strontium isotope analysis of pig teeth was considered problematic due to the porosity of *Sus* tooth enamel. Two teeth were analysed in an attempt to provide some information regarding the suitability of *Sus* enamel for strontium isotope analysis. Subsequent work by Madgwick *et al.* (2012) largely resolved the issue of porcine enamel use in strontium isotope analysis and showed that pig teeth can be used in studies using strontium isotope analysis.

Method

The following protocol for analysing $^{87}Sr/^{86}Sr$ was developed by Dr Jane Evans, NERC Isotope Geosciences Lab, Keyworth. For ease of preparation one cusp was initially separated from each cattle tooth, and the tooth surface thoroughly cleaned to remove calculus and dirt. Next, the dentine was separated from the enamel, and a small sample retained so that a comparison between dentine and enamel values could be made. Finally, the slices were removed from the remaining tooth enamel. For cattle teeth these samples were removed from the root/enamel margin, the approximate middle of the tooth, and where possible the apex (see Figure 2.3 for approximate sample locations). For pigs a single sample was removed as shown in Figure 2.4. The resulting samples were transferred to a clean (class 100, laminar flow) working area for further preparation. In a clean laboratory, the samples were first cleaned ultrasonically in high purity water to remove dust, rinsed twice, dried down in high purity acetone and then weighed into pre-cleaned Teflon beakers. The samples were mixed with ^{84}Sr tracer solution and dissolved in Teflon distilled 16M HNO3 (nitric acid). Strontium was collected using Dowex resin columns and was loaded onto a single Re Filament with TaF following the method of Birck (1986). The isotope composition and concentrations were determined by Thermal Ionisation Mass spectroscopy (TIMS) using a Thermo Triton multi-collector mass spectrometer. The international standard for $^{87}Sr/^{86}Sr$, NBS987, gave a value of 0.710284 Â± 10 (n=20, 2σ) for static analysis. All strontium ratios were corrected to a value for the standard of 0.710250. Blank values were in the region of 100pg.

Finally, a number of plant specimens were collected and analysed so that an accurate picture of the local Durrington Walls strontium signature could be established. These were analysed following the method outlined in Evans *et al.* (2009).

Table 2.2 The attribution of pig skeletal elements to age category, based on Bull and Payne (1982), O'Connor (2003), and Silver (1969).

Earliest (<1 year)	Early (1-2 years)	Intermediate (2-3 years)	Late (3-4 years)
Pelvis	Distal humerus	Distal metacarpal	Proximal humerus
Scapula	Proximal radius	Distal metapodials	Proximal femur
	Phalanx 1	Calcaneus	Proximal tibia
	Phalanx 2	Distal tibia	Distal radius
			Distal femur

Table 2.3 The attribution of cattle skeletal elements to age category, based on O'Connor (2003) and Silver (1969).

Earliest (<1 year)	Early (1-2 years)	Intermediate (2-3 years)	Late (3-4 years)
Pelvis	Distal humerus	Distal metacarpal	Proximal humerus
Scapula	Proximal radius	Distal metatarsal	Proximal femur
Phalanx 1		Distal tibia	Proximal tibia
Phalanx 2			Distal radius
			Distal femur
			Calcaneus

Table 2.4 Categories and sub-categories used in the analysis of *Sus* mandibular wear ageing (reproduced following O'Connor 2003, 160).

Category	Description	Grant 1982
Neonatal	dP_4 unerupted or erupting.	
Juvenile	dP_4 in wear, M_1 not in wear.	
Immature	M_1 in wear, M_2 not in wear.	
Immature 1	M_2 present in crypt.	
Immature 2	M_2 erupting to the occlusal plane.	
Subadult	M_2 in wear, M_3 not in wear.	
Subadult 1	M_3 in crypt.	
Subadult 2	M_3 erupting to the occlusal plane.	
Adult	M_3 in wear.	
Adult 1	M_3 with enamel attrition.	Stage 'a'.
Adult 2	M_3 with minor dentine exposure.	Stage 'b' to 'd'.
Adult 3	M_3 with dentine exposure merging on the mesial cusps.	Stage 'e' to 'h'.
Elderly	Three main zones of dentine exposure merging.	Stage 'j'.

Table 2.5 Categories and sub-categories used in the analysis of *Bos* mandibular wear ageing (reproduced following O'Connor 2003, 160).

Category	Description	Grant 1982
Neonatal	dP_4 unerupted or erupting.	
Juvenile	dP_4 in wear, M_1 not in wear.	
Immature	M_1 in wear, M_2 not in wear.	
Subadult	M_2 in wear, M_3 not in wear.	
Subadult 1	M_3 forming, to just erupting.	
Subadult 2	M_3 erupting to the occlusal plane.	
Adult	M_3 in wear.	
Adult 1	M_3 with minor dentine exposure.	Stage 'a' to 'b'.
Adult 2	M_3 with dentine exposure across central column.	Stage 'c' to 'd'.
Adult 3	M_3 with dentine on distal column.	Stage 'e' to 'h'.
Elderly	Dentine exposure to or beyond stage 'j'.	Stage 'j'.

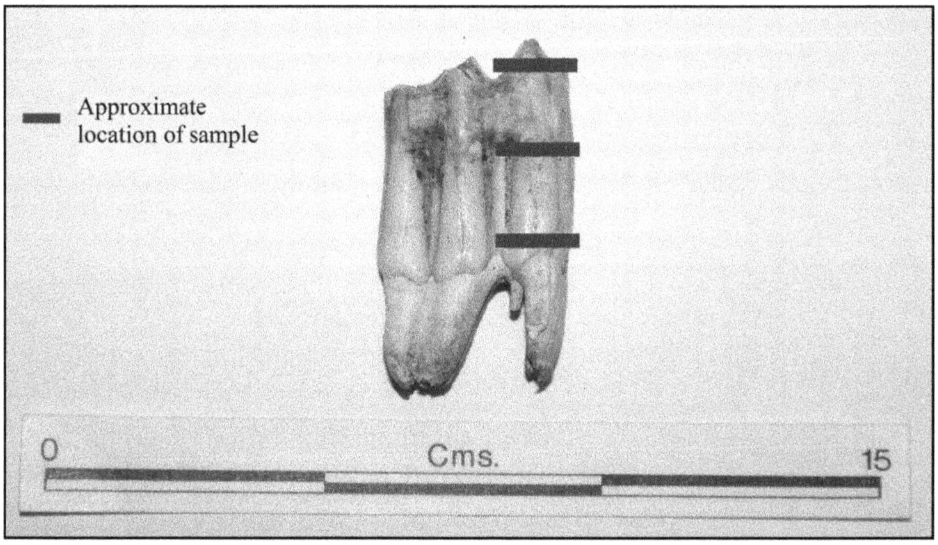

Figure 2.3 Approximate position of samples taken from cattle teeth for strontium isotope analysis.

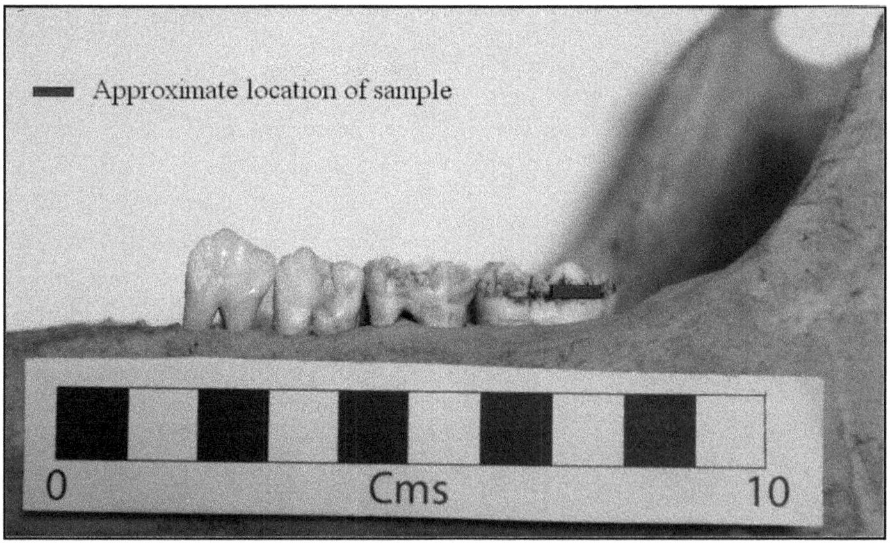

Figure 2.4 Close up of a pig jaw showing the approximate position of the samples taken for strontium isotope analysis.

2.5 Conclusion

Isolating specific characteristics of each skeletal element was not an easy process, and one which was made more complicated by the multiple strands of evidence that each bone has the potential to provide. The discussion in the previous section justified the selection of specific skeletal elements and measurements in terms of their potential to provide appropriate information. As with any investigation it has been impossible, due to the constraints of time and access to material, to collect all potential data from the faunal assemblages that were investigated. The preceding sections have provided justification for the exclusion of some lines of evidence, and in the development of the protocol it was important that each choice of method provided an appropriate amount and quality of relevant information. Other researchers may have chosen a different approach, but by using the methodology as it has been outlined here the aims that are particular to this research can be efficiently addressed.

The chapter has also highlighted some of the potential limitations of the various methods employed during data collection. The need to identify sources of confusion or complication is paramount and it is hoped that by choosing the methodology outlined here these will be limited as far as possible. The age and sex of specimens need to be recorded effectively if population structure is to be understood. Numerous external factors, such as environmental variation, could have influenced body size of the animals. Taphonomic processes affect different bones to varying extents, and excavation can create a further bias. The type of information that was collected by the recording protocol should help to shed light on these factors. When it is not possible to remove the sources of uncertainty in the data, the impact of problematic factors will need to be isolated so that emergent patterns can be seen to be the result of husbandry practices rather than the uncontrollable bias within a given assemblage.

Chapter 3
Age, sex and biometrical analysis of pig remains

3.1 Introduction

Changes in exploitation practices and husbandry through time, along with inter-site and inter-regional variation in the use of animals are central concerns of this research. Chapter 3 will concentrate on the findings that relate to pig exploitation in the study area, and will include details of the age, sex and biometrical characteristics of pig assemblages from all periods of study. General themes will be explored initially (in Section 3.2), with emphasis placed on highlighting overarching trends within evidence from the main periods of investigation. This somewhat coarse approach to the data will be balanced in the second part of the chapter by finer resolution analysis, which will focus mainly on inter-site comparisons of the Early Neolithic data (Section 3.3). Although the main focus of the comparison is the Early Neolithic, the inclusion of data from other time periods is necessary for a consideration of the evidence in its chronological and geographical context. The inherent differences between pigs and cattle make it necessary for the results of each to be considered separately in the first instance; therefore cattle will be discussed in Chapter 4. Some of the figures in this chapter were published previously (Viner 2011).

3.2 Variation in pig exploitation through time

Age

Age profiles provide important information about past husbandry practices. In this section the results of epiphyseal fusion will be discussed first, followed by the information provided by tooth eruption and wear stages. The method that was used to analyse the state of fusion of postcranial *Sus* bones was discussed in detail in Chapter 2. Briefly, it involves the organisation of bones into one of four categories depending on the age at which they fuse. As a very rough estimate those bones in the Earliest category are expected to fuse before 1 year of age, the Early bones between 1 and 2 years, the Intermediate bones between 2 and 3 years, and the Late bones between 3 and 4 years (see Table 2.2; Silver 1969).

Figure 3.1 provides a representation of the epiphyseal fusion data collected during the research. In the Mesolithic period between 15-30% of the bones from the first two categories were unfused. This indicates that a significant proportion of the animals were very young or young when killed. More Intermediate category bones (around 50%) were unfused, and an even greater proportion of Late category bones (70%). Overall, the epiphyseal fusion information suggests that Mesolithic human populations were not targeting animals of any particular age group in the wild boar population. They exploited both very young animals and adult members of the population. The limitations of the epiphyseal fusion method (*i.e.* that it can provide no indication of post fusion age) means that there is no way to tell whether this untargeted use included the exploitation of very old animals, although it does appear possible. One way to explore the relative ages of adult pigs is through the use of mandibular tooth wear, a method that will be employed in the following paragraphs.

The Mesolithic and Early Neolithic patterns appear similar, and the absence of a focus on animals of any particular age group is apparent in both time periods. Fusion data from the Early Neolithic suggests the presence of some very young individuals at the sites investigated. Around 30% of bones in the Early category were unfused. Many animals died between the second and third year, as about 70% of the specimens from the Intermediate category were unfused. Some Early Neolithic pigs undoubtedly exceeded three years of age (>80% of the Late category specimens were unfused), but these were a definite minority. This pattern is reminiscent of the Mesolithic plot, but there is some variation. In both periods animals of all ages are represented, from very young (Earliest and Early category) to adult (Late category), but in general there is a tendency in the Early Neolithic for a greater emphasis on the use of younger animals.

The Late Neolithic plot provided a more exaggerated pattern than that of the preceding periods (Figure 3.1). Young animals were well represented, with 40% of bones in the Earliest category and 60% of bones in the Early category found unfused. This is an indication that a large proportion of the animals died before reaching the age of two years. An even larger proportion of bones in the Intermediate category were unfused (*ca.*85%), which indicates that approximately another quarter of the population was slaughtered in the third year. A few more animals died in the fourth year, as more than 90% of the bones in the Late category were unfused, which means that less than 10% of the pigs made it to the fifth year. Pig exploitation in the Late Neolithic seems to have made more extensive use of young animals than in the Mesolithic and, to a lesser extent, the Early Neolithic. Of course, it must be considered that all the Late Neolithic data derive from Durrington Walls and may therefore reflect a site-specific scenario rather than a pattern to be applied to the whole of the British Late Neolithic.

Pig tooth eruption and wear can be used to support and complement the bone fusion evidence. Mandibles were separated into one of six stages (following O'Connor 1988): Neonatal, Juvenile, Immature, Subadult, Adult or Elderly (see Chapter 2 for a detailed discussion). The specimens that were included in the analysis can be seen in Appendix V. The sample of *Sus* mandibles from Mesolithic sites was too small to provide any meaningful results regarding the age of animals that were exploited during the period. In total only nine mandibles were complete enough to be securely attributed to an age stage. Of the nine mandibles that could be used four were Juvenile, one was Subadult and a further four were placed in the Adult category. No mandibles came from either Immature or Elderly wild boar. Despite the small sample size, the occurrence of animals from three of the age categories confirms the findings of epiphyseal fusion that animals of a variety of ages were exploited during the Mesolithic period in Britain.

The Early Neolithic is better represented by adequately complete mandibles than the Mesolithic (48 mandibles were complete enough to be included in the analysis, Figure 3.2). Immature mandibles were numerous (25%) but the greatest number of specimens came from Subadult pigs with 40% of all the specimens belonging to this category. A significant number of animals, more than 30% of the total sample, reached adulthood based on the state of their teeth. Although the emphasis is clearly on Subadult mandibles, evidence from mandibular wear stages does give the impression of animals of different ages being culled. As can be seen from Figure 3.2 no mandibles belonging to Elderly of Juvenile animals were found, and only one mandible was attributed to the Neonatal category. Broadly, these results concur with the epiphyseal fusion data discussed above. In both the Mesolithic and Early Neolithic animals at varying stages of maturity were utilised, but very young and very old animals were scarce or entirely absent.

Tooth wear data can be made more specific by introducing a greater number of age categories. Although introducing more categories can lead to each category having very few specimens the method will in some cases provide further interesting results (O'Connor 2003, 160-1). Figure 3.3 presents the mandibular wear data from Early Neolithic sites separated into ten categories from Neonatal to Elderly (following O'Connor 2003, 160). This more detailed approach provided three interesting pieces of evidence that are useful in clarifying the Early Neolithic exploitation of pigs (Figure 3.3). First, all of the Immature mandibles were attributed to the earlier part of the category (*i.e.* I1 rather than I2). This suggests that all of the animals with immature mandibles at Early Neolithic sites were killed around or soon after 1 year of age (O'Connor 2003). Secondly, the four categories from Sub-Adult 1 category to Adult 2 were all well and quite evenly represented, and this would seem to suggest that pigs were killed at a steady rate from the point that they reached the Sub-Adulthood. Finally, the large proportion of Adult 1 and Adult 2 pigs, and the absence of Adult 3 and Elderly pigs at sites of the Early Neolithic is suggestive of a system in which most animals were killed soon after reaching adulthood. The Adult category stretches from 2 years to 5-6 years in pigs (O'Connor 2003), and it appears that most of the pigs studied here died in the earlier part of this period.

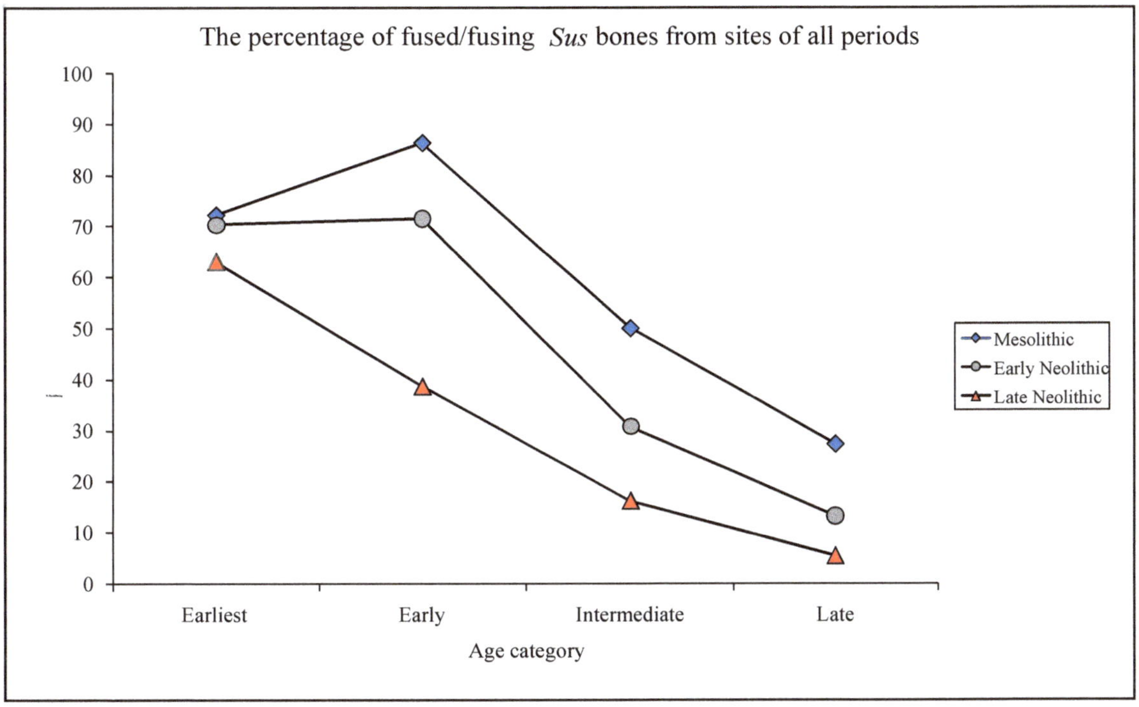

Figure 3.1 Epiphyseal fusion of *Sus* bones from the Mesolithic, Early Neolithic and Late Neolithic. Postcranial bones that were recorded as fused or fusing have been included (apart from the scapula) (Viner 2011, Figure 14.3)

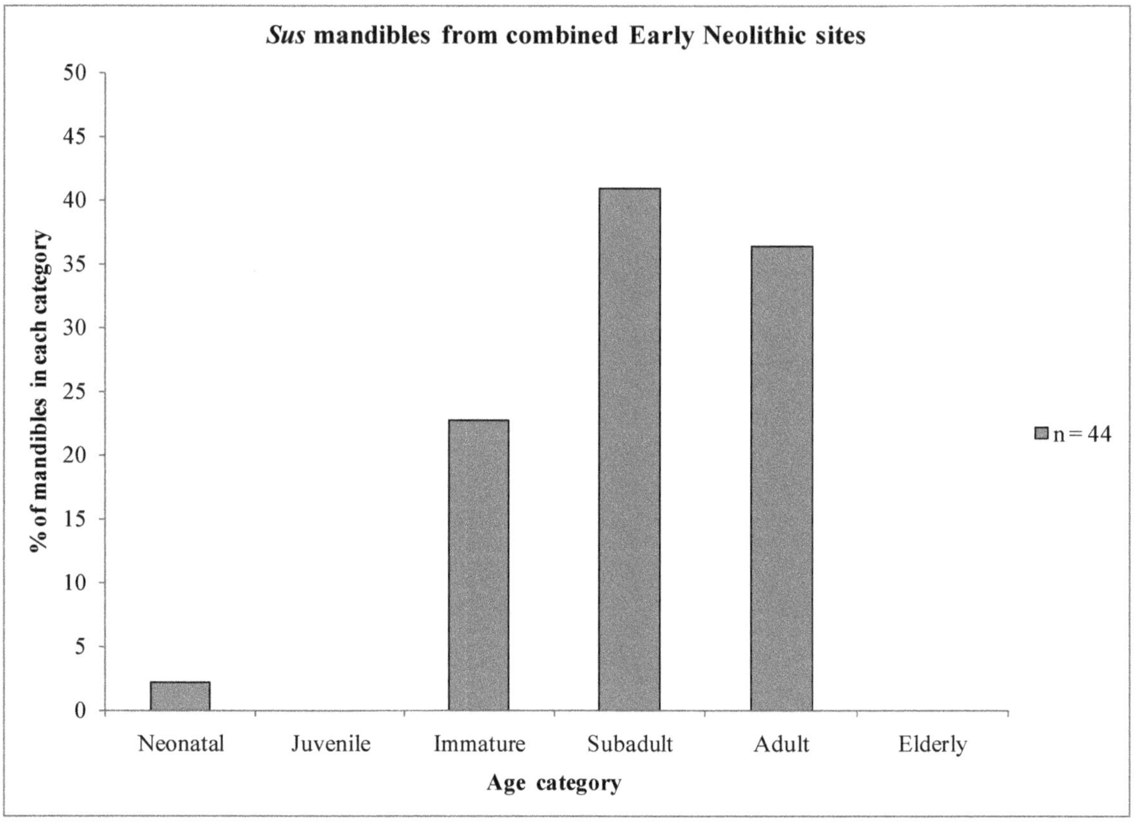

Figure 3.2 Attribution of *Sus* mandibles to age category from Early Neolithic sites. Categories are after O'Connor (1988).

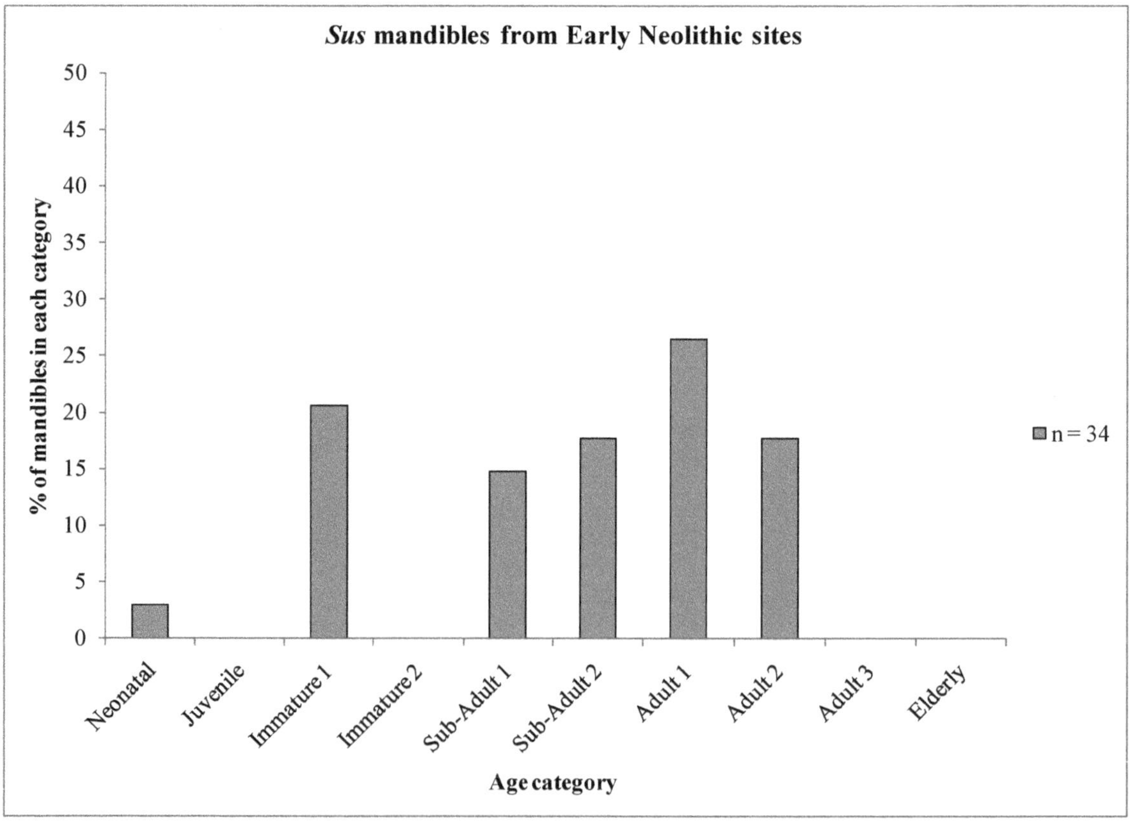

Figure 3.3 Attribution of *Sus* mandibles to age category from Early Neolithic sites. Categories are after O'Connor (1988).

The Late Neolithic mandibular tooth wear plot is in some ways similar to that of the Early Neolithic period (Figure 3.4). Immature, Subadult and Adult mandibles were all well represented with the Immature and Adult mandibles making the greatest contribution (around 35% of the sample each). Like the Early Neolithic, mandibles from young pigs were present, but in very low numbers. Similarly, no Elderly animals were identified from Durrington Walls.

With the introduction of more age categories (Figure 3.5) there is no clear period at which pigs were killed. A slight peak in the number of deaths occurred during the Immature 2 category indicating that more animals died towards the end of the period than the beginning. The peak in the number of deaths seen soon after 12 months of age in the Early Neolithic was not apparent within the Late Neolithic sample. In fact, with the exception of the rather limited peak during Immature 2, the number of pigs that died at each category was fairly constant. Mandibles were attributed to all three of the Adult categories (*i.e.* A1, A2 and A3), but the number of animals was less in Adult 3 than A1 or A2. This pattern is suggestive of a population in which animals began to die in large numbers sometime around one year of age (Immature 2), and killing progressed relatively steadily so that few animals reached advanced adulthood (Adult 3), and none became Elderly. The usual caveat, that the data for the Late Neolithic derive exclusively from one site – Durrington Walls – applies here.

The greatest difference between the Early Neolithic sample and the Late Neolithic sample is the relative abundance of Immature mandibles. In the Early Neolithic over 20% of the assemblage came from Immature animals (*i.e.* those just over one year of age; O'Connor 2003) compared to 35% in the Late Neolithic. In the Early Neolithic a larger proportion of the sample came from Sub-Adults and Adults. In both periods, very few animals were killed before around one year of age, and the specific time at which pigs were killed after one year varied by period, the Early Neolithic emphasis was occurred soon after one year but in the Late Neolithic a short delay is apparent. The slightly greater frequency of Immature animals in the Late Neolithic confirms the pattern identified in the fusion of epiphyses of a younger kill-off pattern in this period.

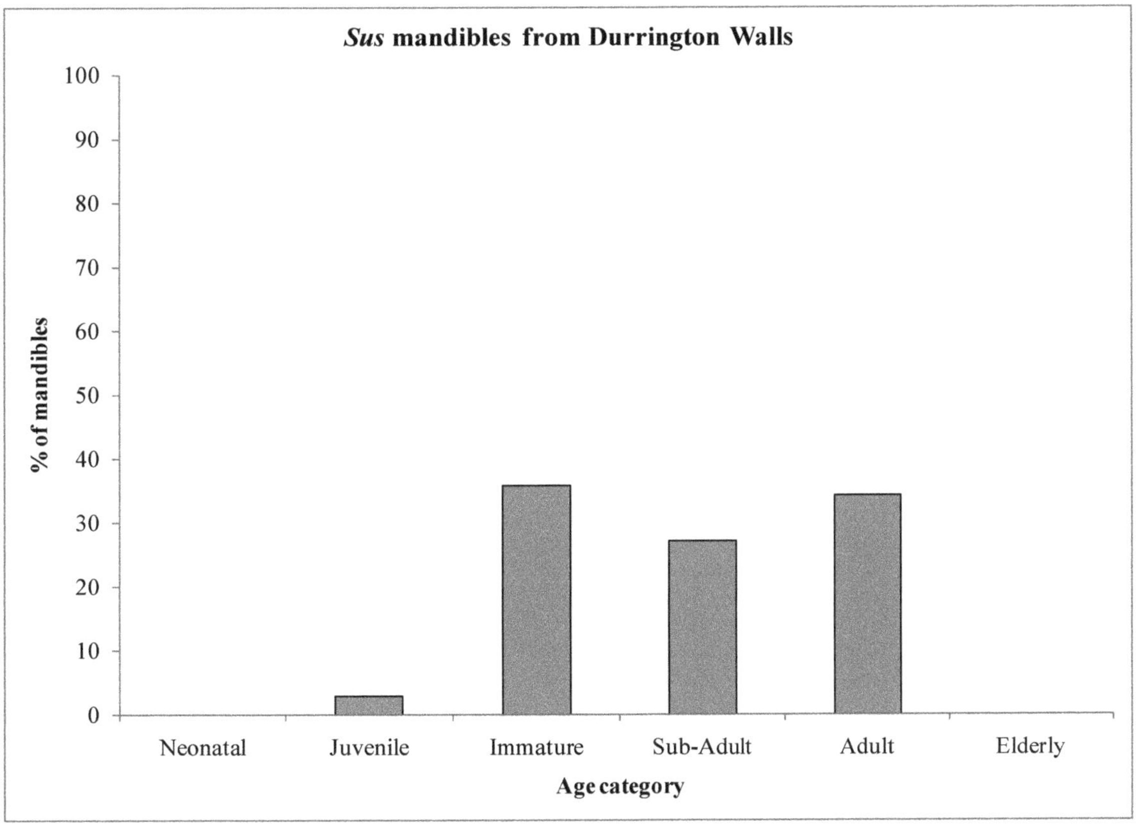

Figure 3.4 Attribution of *Sus* mandibles to age category from Late Neolithic sites. Categories are after O'Connor (1988).

Ageing by epiphyseal fusion and tooth wear stages of mandibles both have limitations. The epiphyseal fusion method is limited by the age at which the final bones fuse in the skeleton. For instance, in pigs, most of the commonly occurring bones in archaeological contexts are fused by 3.5 years (Silver 1969) and animals could potentially have survived a number of years after fusion was complete. In contrast, the method of ageing mandibles that was used above extends to 5-6 years of age, but is limited by the low frequency of suitably complete specimens. Adult animals were exploited in the Mesolithic, Early Neolithic and Late Neolithic, both ageing methods failed to identify any animals that would be considered elderly. Many loose third molars were recorded from all of the periods under discussion, but so far none have been used in the assessment of pig age. This is because the mandibular wear method employed above requires that at least two teeth be present in the jaw for a specimen to be included. Assessing the wear on loose third molars does not increase the period of life that is covered, but the ubiquity of the tooth on archaeological sites means that a large number of specimens were available for study. Bull and Payne (1982, 57, Table 2) suggest eruption occurs between 16 months in early maturing breeds and 20 months in late maturing modern breeds. Wild boar tooth eruption follows the same pattern as domestic pigs, but is arguably generally later in wild animals than in domestic types (Bull and Payne 1982, 55; Matschke 1967, 109). Matschke (1967, 109) found that third molars of wild boar in Tennessee, US, erupted at around 21-26 months. Even taking the earliest estimate, it is clear that third molar eruption occurs at an advanced point in the animal's life, and it should provide information of longer duration than epiphyseal fusion.

In the Mesolithic assemblage, the most abundant condition in which third molars were found was unworn (U on Figure 3.6). It is important to consider that, since eruption stages cannot be assessed on loose teeth, these will all be classified under 'U', a category that includes all stages from the formation of the tooth to the onset of wear. Unworn third molars must have belonged to Subadult animals (or younger) according to O'Connor's system, but without eruption information it is impossible to be more specific. Most of the animals represented were less than around two years of age at death. Figure 3.6 shows that the number of specimens of each wear stage decreased gradually from 'U' and there were no teeth that were recorded at a more advanced stage than 'c' in Grant's system (1982). According to O'Connor (2003) this would mean that most animals died before reaching stage Adult 3, and all are likely to have died soon after two years of age. When taken in conjunction with the other methods of ageing discussed above, wear on third molars reiterates that no animals of advanced age were present on Mesolithic sites.

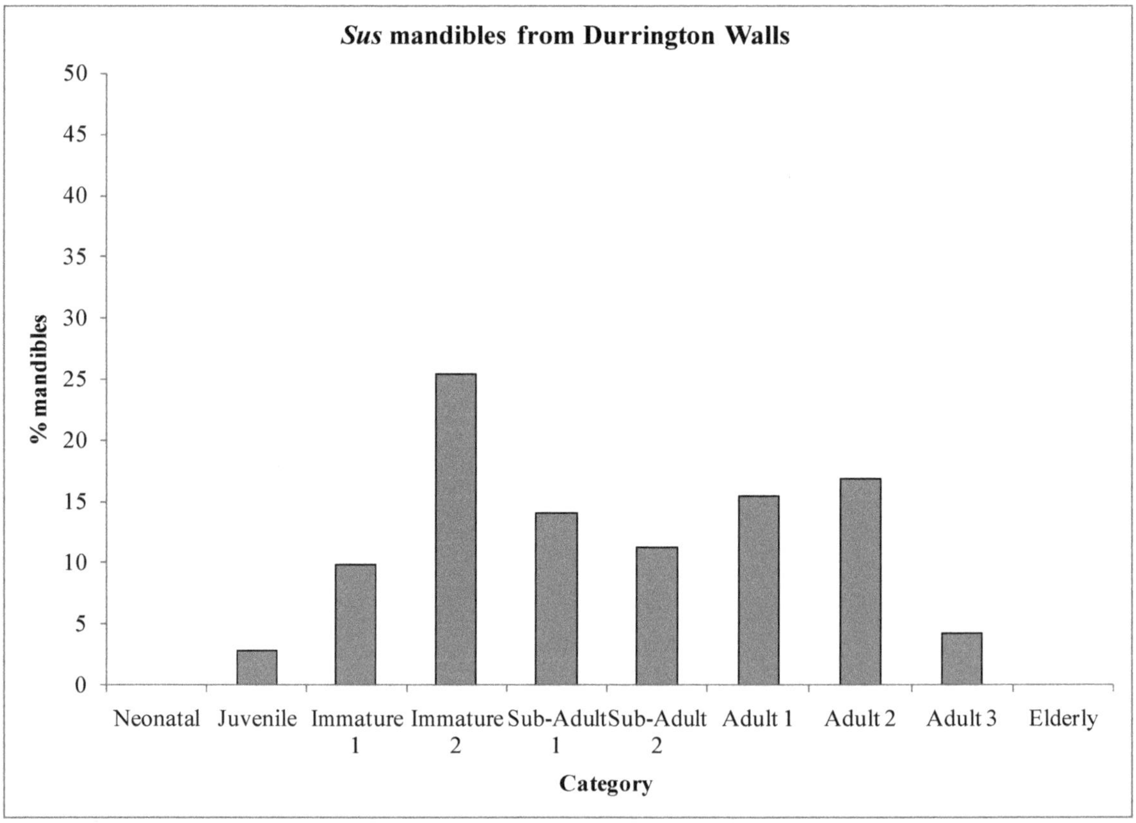

Figure 3.5 Attribution of *Sus* mandibles to age category from Late Neolithic sites. Categories are after O'Connor (1988).

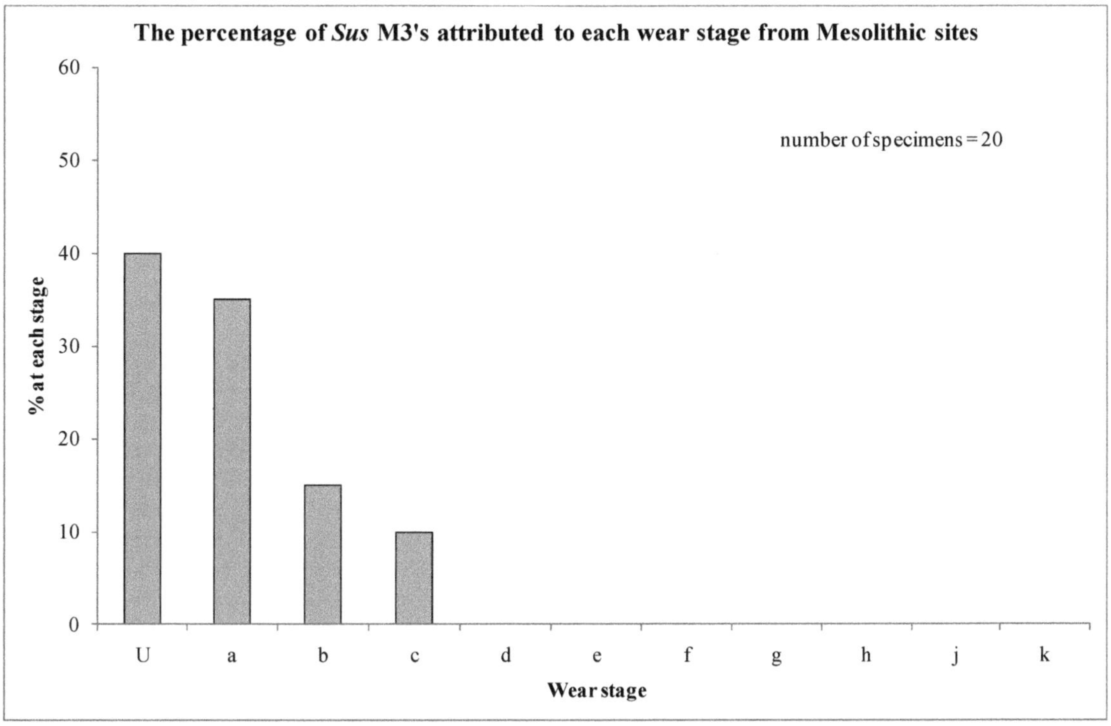

Figure 3.6 Wear stages of pig mandibular third molars from Mesolithic sites. Wear stage U includes all eruption stages as these cannot be recorded for loose teeth.

Figure 3.7 identifies that a large proportion of third molars from Early Neolithic sites were unworn (18 specimens, 58% of the sample). Again, those animals with unworn third molars must have been Subadult or younger when they died and are likely to have been killed before reaching two years of age. The loose third molars do not contradict the evidence from mandibular wear stages presented above that suggested a dominance of Subadult animals in the Early Neolithic assemblages. A number of more severely worn teeth (*i.e.* those at stages 'd' and 'e' *sensu* Grant 1982) indicate that some slightly older animals were present on Early Neolithic sites compared to the Mesolithic. In common with the Mesolithic, no specimens were found in a heavily worn state and therefore there is no evidence of older adult or elderly animals in the Early Neolithic. We must also consider that the larger sample from the Neolithic means that the probability of encountering less common wear stages is higher.

In the Late Neolithic unworn molars were most abundant (Figure 3.8), a pattern common to all of the periods explored, and one that suggests that Subadult animals predominated. This would appear to contradict the findings from mandibular wear (as Subadult and Adult mandibles were found in almost equal proportions; Figure 3.4), however, the Adult category includes all other teeth up to stage 'h' and overall they are equally well represented. As with the mandibular wear stages, individual third molars suggest that Adult animals were being killed continually and at a variety of ages, and in contrast with the earlier periods a number of teeth were recorded that exhibited heavy wear (two specimens were recorded as 'k', the final stage in Grant's system). These molars, although few in number, suggest that occasionally pigs of advanced age were utilised during the Late Neolithic at Durrington Walls.

Sex

The morphological and size differences between female and male pig canines allowed many of the specimens recorded during the study to be sexed. Figure 3.9 presents the relative abundance of female and male canine teeth from the Mesolithic, Early Neolithic and Late Neolithic. The diagram includes both loose teeth and those found in jaws, and maxillary and mandibular teeth combined.

Results from the Mesolithic indicate that both female and male wild boars were deposited on sites of the period. The sample is relatively small (37 specimens), and canines came from a number of sites from which information has been pooled. The proportion of teeth identified as male was greater than those identified as female; canines and jaws from male animals contributed around 65% of the total sexed sample, compared with 35% females. Similarly, in the Early Neolithic the relative abundance of canines from male pigs was greater than canines from female pigs (Figure 3.9). 60% of the sexed canines came from males and 40% from females. The similarity in the proportion of females and males observed between the Mesolithic and the Early Neolithic continued into the Late Neolithic period, during which time male pig canines continued to dominate the sample. Figure 3.9 indicates that 80% of the canines from the Late

Neolithic were from male pigs, and that far fewer female canines were recovered. According to these results, male animals were more commonly represented on the archaeological sites investigated.

In addition to the female and male canine teeth recorded from archaeological sites of all periods, a small proportion of the recorded canines could not be attributed to sex and are therefore considered under the category undesignated. Canines are most likely to be undesignated if they are deciduous, if they are permanent but in the early stages of eruption (as these specimens are often hidden within the jaw and cannot be easily identified), or if the specimen is damaged resulting in the loss of distinguishing morphological features. Therefore the analysis of *Sus* sex distribution presented here cannot consider animals in the early stages of life. The contribution made by undesignated canines was minimal at all of the sites investigated. Figure 3.10 presents a comparison of the sexed and unsexed canines by period.

In the Mesolithic period no undesignated canines were recorded, providing a substantial difference between the Mesolithic and the Early Neolithic sample. In the latter 15% of the total canine sample was unattributed. An increase in the proportion of undesignated canines at Early Neolithic sites when compared to Mesolithic ones could be a result of greater fragmentation at the later sites, or, more likely, a result of the greater number of younger animals deposited at Early Neolithic sites, a pattern also suggested by the ageing evidence.

A characteristic of the Late Neolithic sample is the comparatively large proportion of the assemblage that was made up of canines that could not be sexed. Undesignated canines made up 26% of the total sample. Again this can be linked to increasing utilisation of young animals during the Late Neolithic period and is supported by the evidence from epiphyseal fusion. To determine whether the pattern of male *Sus* exploitation that persisted from the Mesolithic to the Late Neolithic is real, potential bias against female canines must be identified. Robust male canines tend to survive preferentially in the archaeological record. Recovery bias can be investigated, to an extent, by comparing the number of loose canines with those found in jaws. The latter should reflect more accurately the proportion of males and females deposited on an archaeological site, and significant differences between the sex distribution of jaws compared to loose teeth might suggest that some of the smaller female teeth were missed during excavation. Both loose canines and sexed jaws from the Mesolithic period (Figure 3.11) suggest that males were more common than females, however the sample of sexed jaws is far too small. The overall sex ratio is mainly based on loose teeth, which are very likely to be affected by a recovery bias. Consequently the Mesolithic sex evidence cannot be regarded as reliable, and can only indicate that both female and male pigs were present.

Canines recovered from Early Neolithic sites (see Figure 3.12) indicate that both sexes were exploited during the period. The number of female and male sexed jaws is similar (12 and 11 specimens respectively), but male loose canines outnumber female ones, suggesting that perhaps recovery bias (*i.e.* preferential collection of large male canines) may have contributed to the prevalence of male specimens. It should, therefore, be questioned whether a reliance on male animals was really a characteristic of Early Neolithic husbandry practices. The more reliable evidence from jaws indicates that the two sexes were present in roughly equal numbers.

The Late Neolithic saw the most pronounced difference in the contribution made by female and male animals of the periods studied so far. Males greatly outnumbered females, and this pattern is reflected in both the number of loose teeth and the number of jaws attributed to each sex. The differences between the representation of females and males during the Late Neolithic do not appear to be the result of recovery bias, which is not surprising considering that the Durrington Walls material was fully sieved. As a result, the difference in the presence of female and male animals within the Late Neolithic sample can be interpreted as a real emphasis on the use of males, and a corresponding under reliance on the use of females.

The Mesolithic evidence can merely indicate that both male and female wild boar were hunted. In the Early Neolithic there is evidence to suggest that recovery bias affects the sex ratio. Despite the apparent predominance of males it is likely that in this period females and males were exploited in equal numbers. More pronounced differences were identified in the Late Neolithic which saw the most extreme expression of male *Sus* dominance in the study. During the Late Neolithic period males dominated the sample, and there was a substantial decrease in the proportion of female canines recorded from the earlier periods. The predominance of males may be explained by the fact that Durrington Walls has the characteristics of a typical consumer site, where the majority of pig remains dumped on site would derive from the slaughtering of young males. More females would be kept into adulthood for reproduction.

Further analysis, in this case using fused distal humeri (Figure 3.14), provided more evidence of the sex distribution of pigs at the sites investigated. Measurements of the distal humerus from all periods exhibited very similar distributions with a great deal of overlap. According to the results there was no great physical difference between the animals exploited by human populations in different periods. However, two different groups can be identified; a group of small animals in the bottom left and a group of larger animals in the upper right. One explanation for these groupings, given the sexually dimorphic nature of the breadth of the humerus (BT) (Albarella and Payne 2005; Payne and Bull 1988), is that they represent two sex groups.

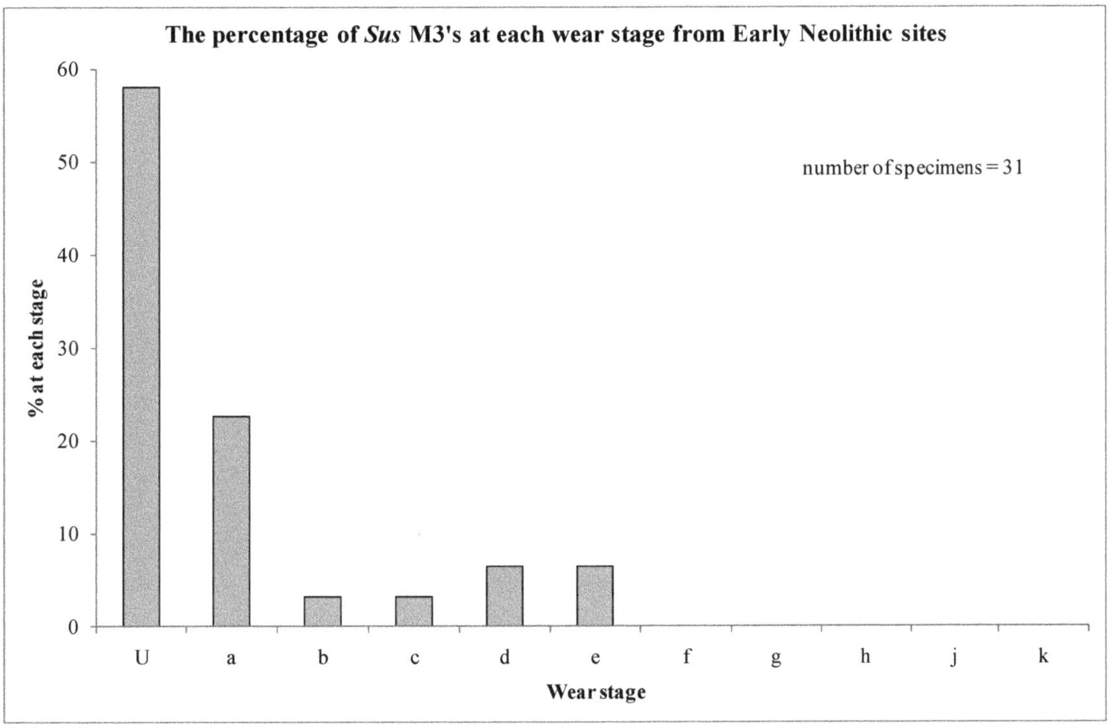

Figure 3.7 Wear stages of pig mandibular third molars from Early Neolithic sites. Wear stage U includes all eruption stages as these cannot be recorded for loose teeth.

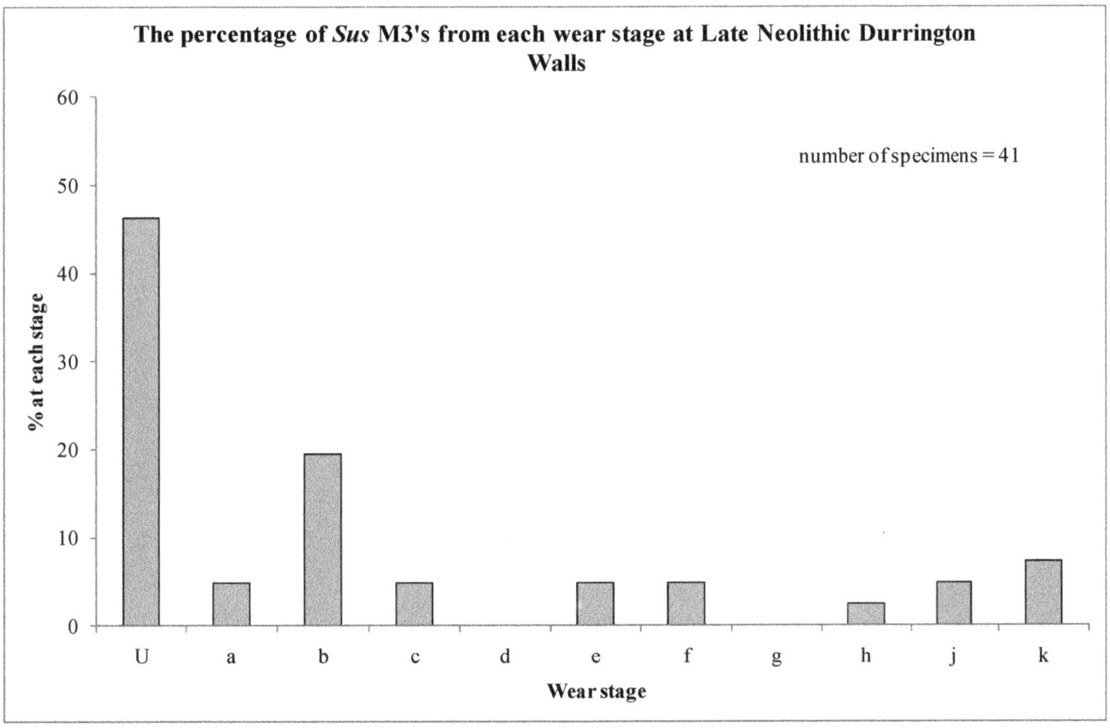

Figure 3.8 Wear stages of pig mandibular third molars from Late Neolithic Durrington Walls. Wear stage U includes all eruption stages as these cannot be recorded for loose teeth.

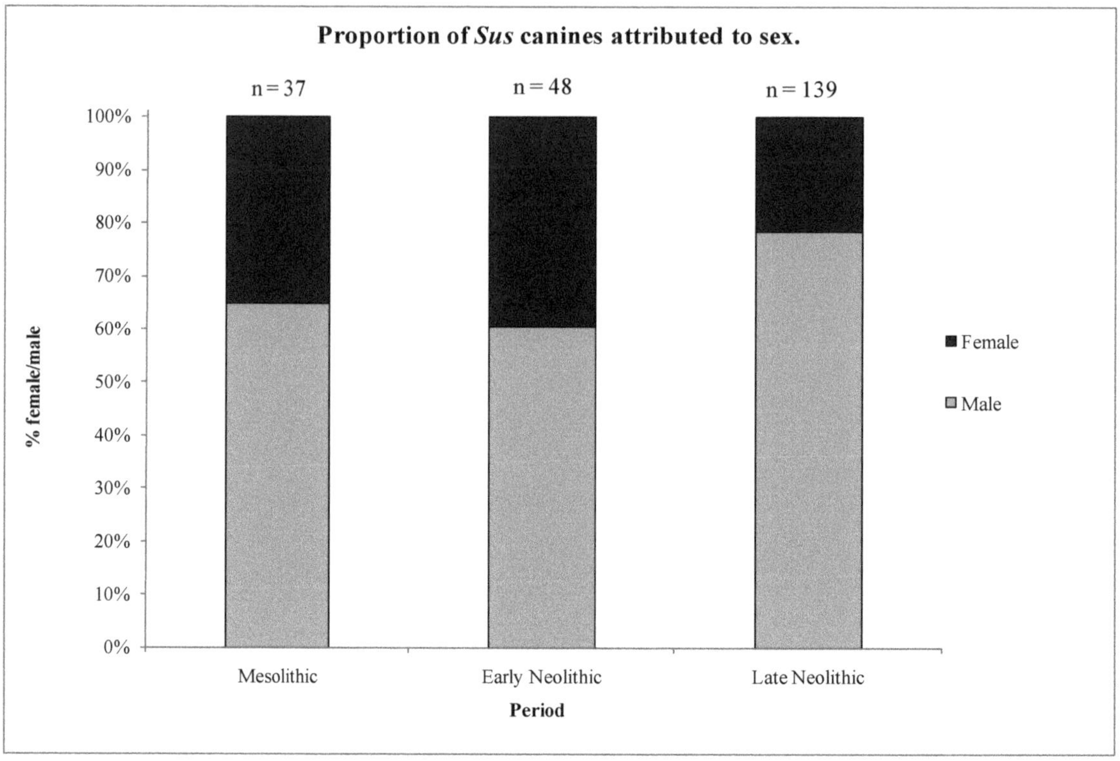

Figure 3.9 The proportion of *Sus* canines attributed to sex from the Mesolithic, Early Neolithic and Late Neolithic periods (Viner 2011, Figure 14.4).

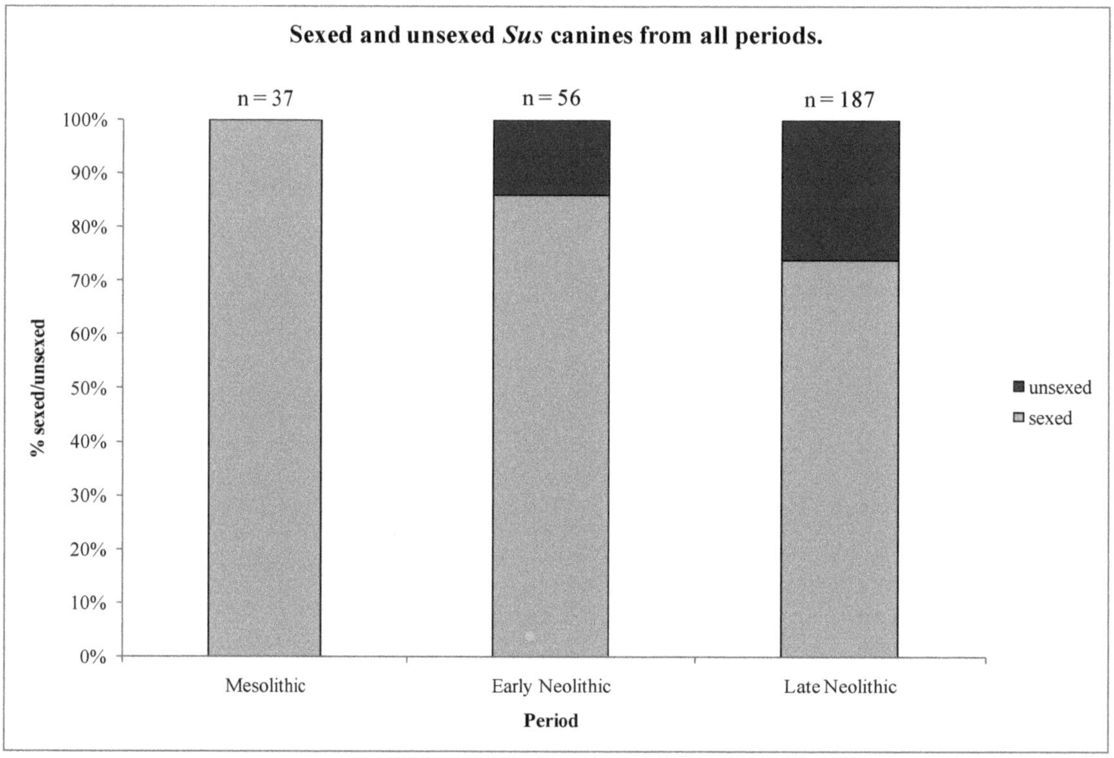

Figure 3.10 Sexed and unsexed (undesignated) *Sus* canines from the Mesolithic, Early Neolithic and Late Neolithic periods. Sexed jaws and loose teeth have been included from the maxillar and mandible.

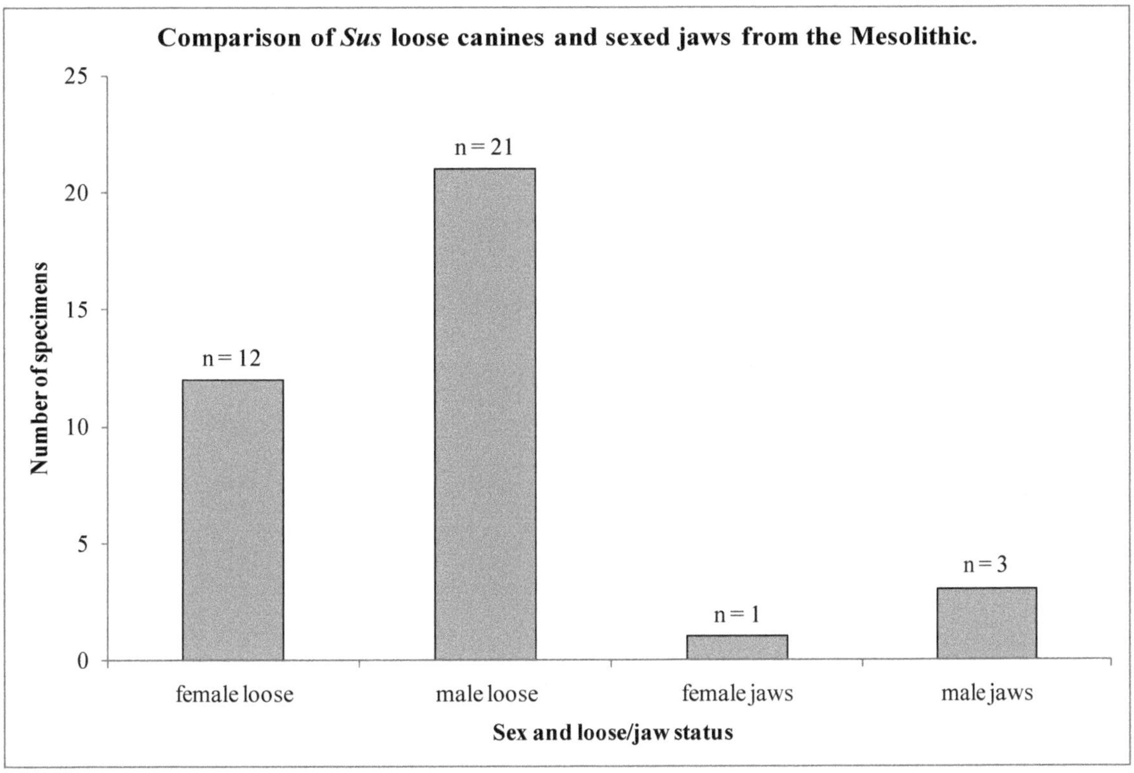

Figure 3.11 The number of *Sus* jaws and loose canines recorded from Mesolithic sites. Upper and lower specimens have been included.

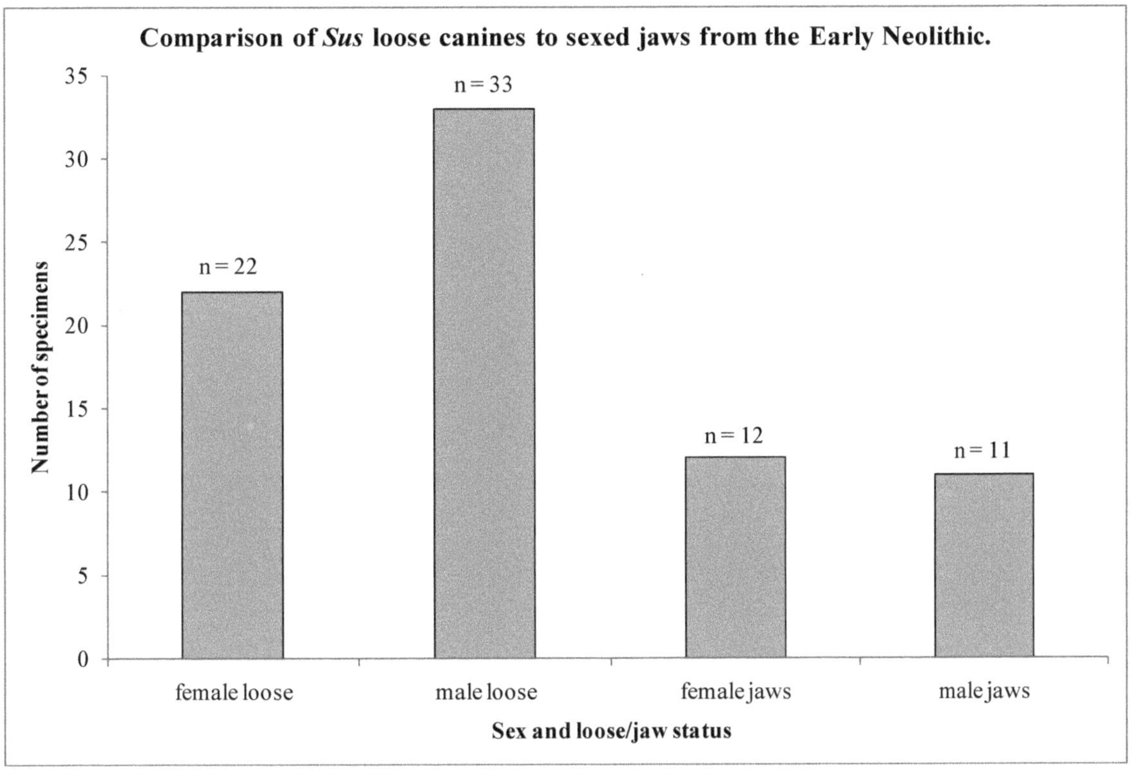

Figure 3.12 Comparison of *Sus* loose canines and sexed jaws from Early Neolithic sites. Both maxillary and mandibular specimens included.

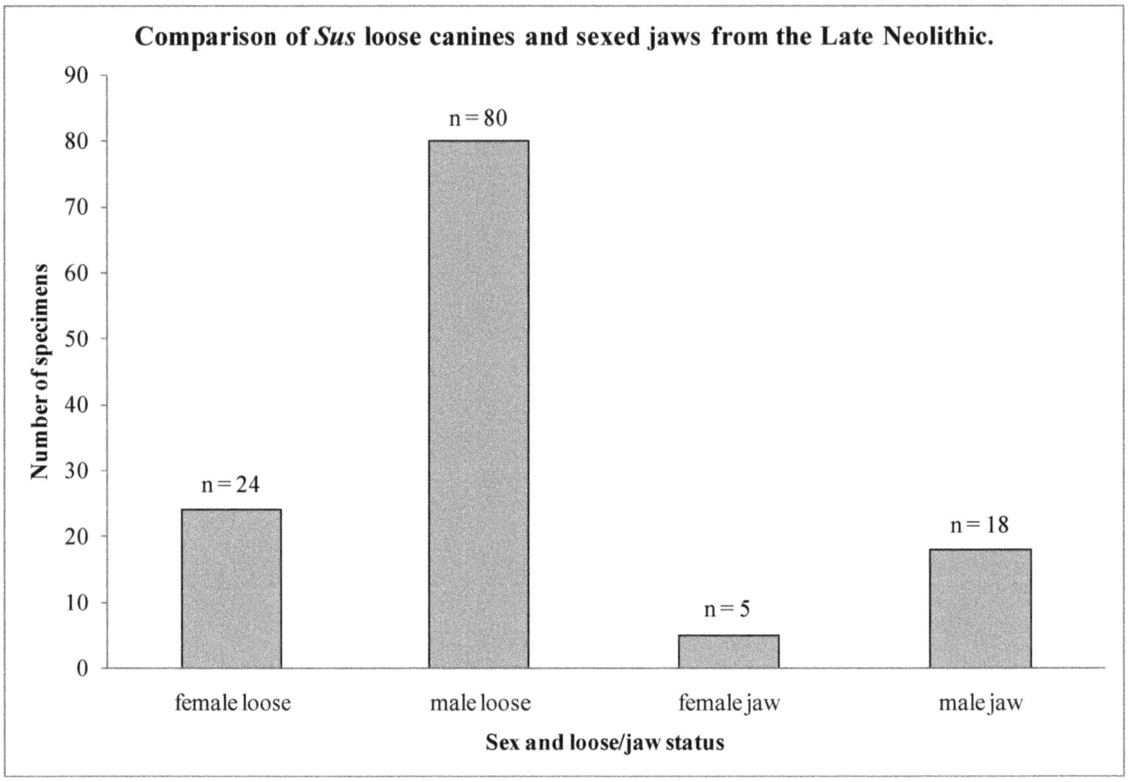

Figure 3.13 Comparison of *Sus* loose canines and sexed jaws from the Late Neolithic (Durrington Walls). Both maxillary and mandibular specimens have been included.

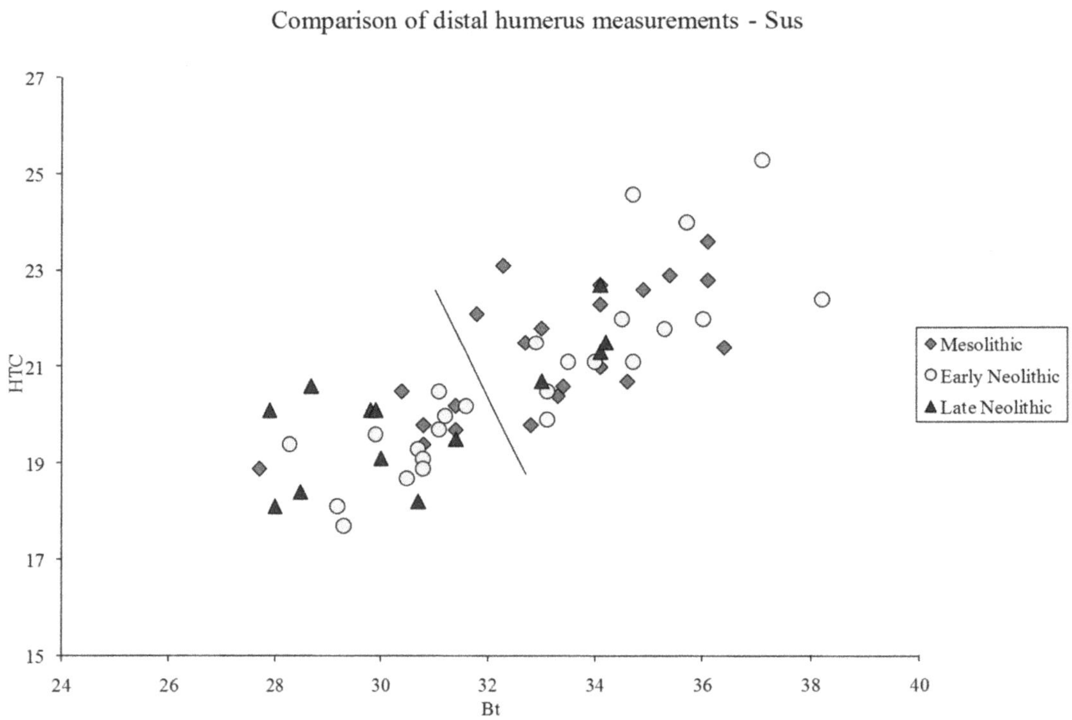

Figure 3.14 Comparison of the distal humerus measurements (Bt = breadth of the trochlea, HTC = height of the trochlea) from Mesolithic, Early Neolithic and Late Neolithic sites. Only fused specimens are included.

If the two groups in Figure 3.14 can be attributed to female and male animals, then the plots mirror what was found in sexed jaws with males (tentatively) outnumbering females in the Mesolithic, more equal sex distribution in the Early Neolithic, and a dominance of male pigs during the Late Neolithic. In Figure 3.14 only fully fused specimens have been included. The fused humerus measurements taken do not vary significantly post fusion and the possibility that the sex ratios would change if bones from younger animals were also plotted needs to be explored. This is especially important given the changing emphasis on young animals identified above, and the possibility that the perceived sex differences occurred because one sex was killed at a younger age. The sample of Mesolithic unfused distal humerus measurements was too small to provide productive results. In Figure 3.15 the breadth and height of the distal humerus trochlea has been plotted for specimens at different stages of fusion from the Early Neolithic. The fusing specimens are, as expected, all towards the smaller end of the plot. The majority of the fused specimens overlap in distribution with the unfused and fusing ones, but on average they are clearly larger. The sex groups suggested above are less likely to work with the unfused and fusing specimens and the main size differences are probably due to growth rather than sex differences.

In the Late Neolithic unfused specimens were more abundant than the Early Neolithic, and as a result form a much clearer group in Figure 3.16. These are distinctively smaller than the specimens in the other fusion categories and form a rather compact group. Their size is therefore likely to be mainly the consequence of incomplete growth rather than any sex bias. Interesting the fusing specimens overlap much more with the fused specimens than in the Early Neolithic. Considering that in this period there is a peak of animals killed around the year or age (when the distal humerus would be fus*ing*) it is likely that there would be mainly males, which would explain the rather large size of fusing humeri. This would fit well with the evidence obtained from the canines.

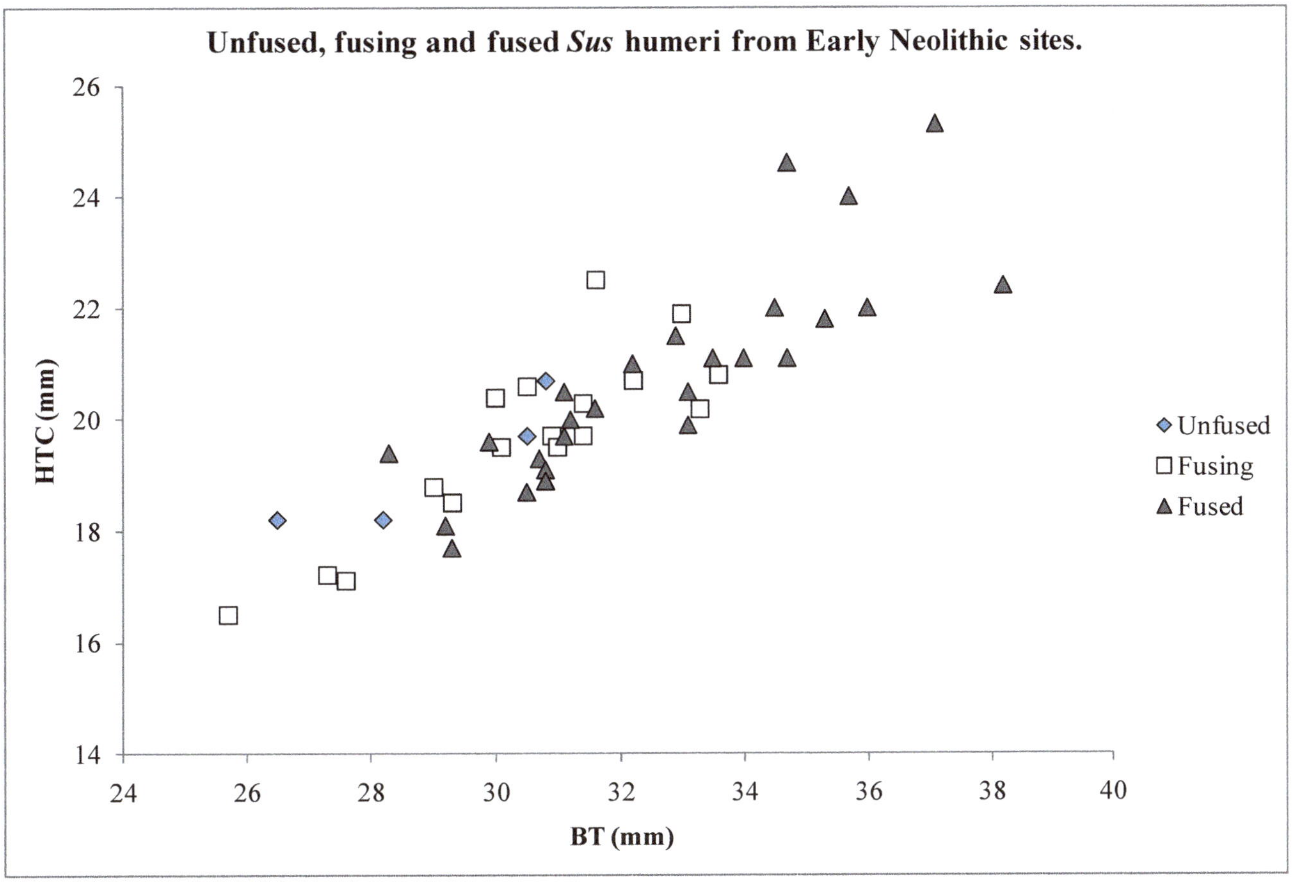

Figure 3.15 Unfused, fusing and fused *Sus* humerus measurements from Early Neolithic sites.

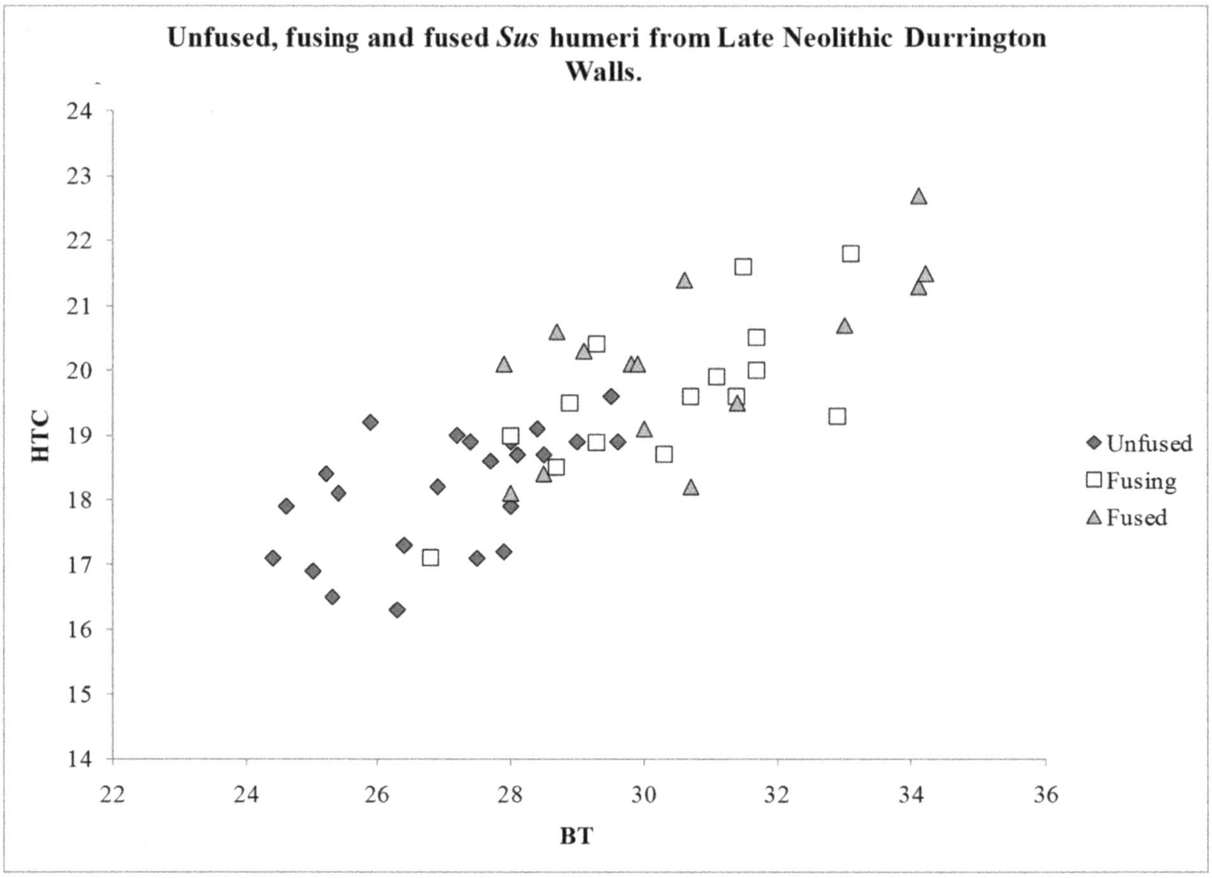

Figure 3.16 Unfused, fusing and fused *Sus* humerus measurements from Durrington Walls.

An alternative explanation of the groupings seen in the Neolithic is that two different groups of animals were being exploited. These could, for instance, be a large wild population and smaller domesticated types. The occurrence of the same pattern in the Mesolithic – when domestic animals are unlikely to have been present – makes this explanation rather unlikely.

Dental pathologies

This section will discuss the pathological conditions that were encountered on *Sus* teeth from the prehistoric sites investigated. The only pathologies that were routinely recorded were linear enamel hypoplasia (LEH) and dental caries (see Chapter 2).

Linear enamel hypoplasia were relatively uncommon during all of the periods studied. The condition has been linked to nutritional stress and disease by a number of researchers (*e.g.* Dobney and Ervynck 1998). No occurrences of LEH were recorded on Mesolithic specimens, and this is perhaps a result of the small number of specimens available from the period. Dating to the Early Neolithic, LEH on a mandibular first molar and second molars from Runnymede Bridge suggest the possibility that periods of stress occurred during the life of prehistoric pigs at the site. The presence of LEH on the mandibular first molar corresponds to a period of stress caused by birth and/or weaning in the first few months of an animal's life (Dobney *et al.* 2004).

Dental caries are occasionally found on archaeological pig teeth (an example can be seen in Figure 3.17). Caries were infrequent within most of the assemblages studied; just two Mesolithic examples were found (both specimens came from Thatcham, Berkshire, see Figure 3.18) and 12 Early Neolithic occurrences were recorded (six each from Hambledon Hill and Runnymede Bridge). In light of the infrequency of dental caries from sites of the Mesolithic and Early Neolithic, the abundance of dental caries on pig teeth from Durrington Walls is anomalous. Caries of the kind observed on pig teeth were probably caused by a component of the *Sus* diet that must have been high in sugar and could for instance have been caused by consumption of large quantities of apples, acorns or other fruit that occurred in prehistoric Britain.

The intermittent appearance of the condition on some teeth indicates that this was associated to seasonal events, such as the availability of a certain type of food at a particular time of the year. Animals from Thatcham are considered wild, and assumed to have been hunted by Mesolithic humans. If this is accepted (a detailed discussion of the Mesolithic hunting issue can be found in Chapter 1) then the dietary factors that cause tooth decay in pigs must have been available in pre-Neolithic Britain. A slightly higher incidence of both caries was encountered in the Early Neolithic (and is perhaps to be expected given the much larger Early Neolithic sample size). The proportion of teeth that showed this condition was very small at all sites of the Early Neolithic (less than

1% of the total). Although more caries were encountered when studying Early Neolithic than Mesolithic specimens, it is not necessarily the case that they were more common during this period.

Dental caries were observed by U. Albarella, S. Payne and D. Serjeantson in their re-analysis of the material from Durrington Walls. The condition consisted mainly of cervical caries, exhibited by one or more lines of caries that run horizontally across the tooth (as in Figure 3.17). This type of caries was most commonly found on molars, but also occurred on the fourth permanent premolar and deciduous premolars. The condition was recorded almost exclusively on maxillary teeth in all periods (Figure 3.18). This is a pattern that has been observed on the teeth of many mammals (including humans), and is thought to be a result of either the greater complexity of enamel folding in upper teeth or alternatively the increased distance from salivary glands in many mammal mouths.

A large increase in the presence of caries occurred, as mentioned, in the Late Neolithic (91 incidences). This might indicate a change in the diet of animals at Late Neolithic Durrington Walls to include a greater contribution of food that was detrimental to the integrity of the enamel. A change in diet as signified by increased dental caries could be an indication of greater control being exercised over pigs in the Late Neolithic. This control probably included feeding the animals on a restricted diet and controlling access to other kinds of food. It has been suggested that spent grain, a by-product of the ale making process that has high sugar content, could have been fed to pigs during the Neolithic (Dinelay 2006) and might account for the damage to pig teeth.

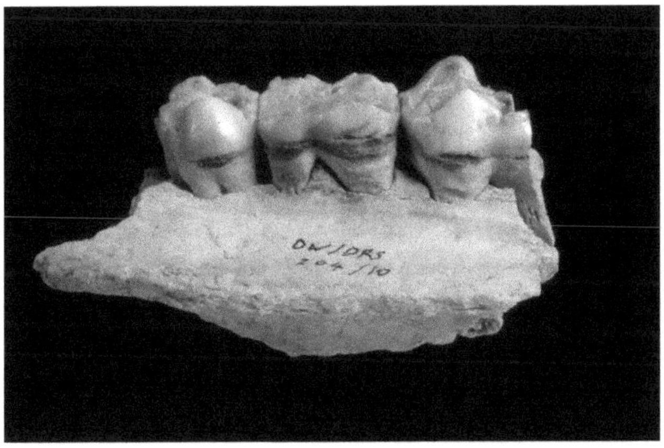

Figure 3.17 An example of dental caries on pig maxillary teeth from Durrington Walls. Photograph by U. Albarella

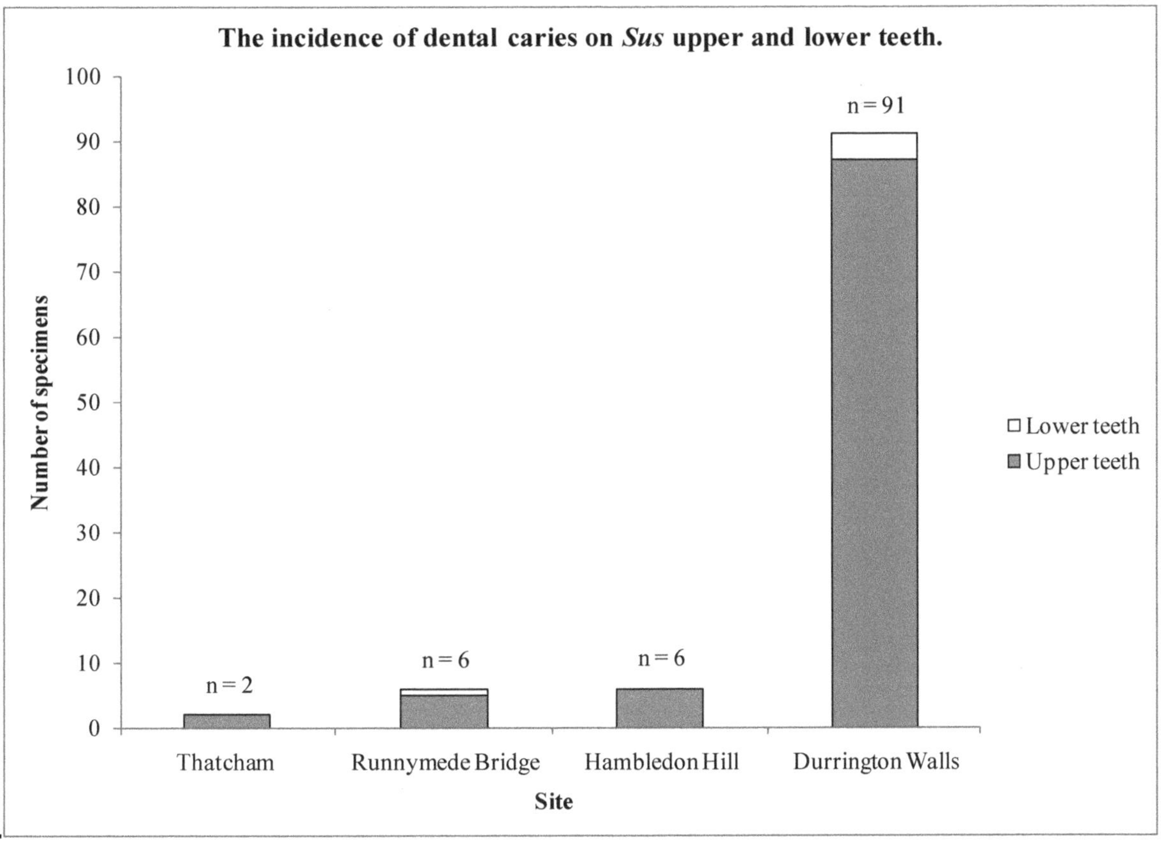

Figure 3.18 The number of teeth with caries from Mesolithic Thatcham, Early Neolithic Runnymede Bridge and Hambledon Hill, and Late Neolithic Durrington Walls. In jaws, each tooth with the condition has been included individually.

Biometrical analysis - Goldcliff in focus

As the only Late Mesolithic site in the study, the small assemblage from Goldcliff in the Severn Estuary holds a position of special importance to understanding the transition to animal husbandry in Britain. The assemblage contained a number of postcranial elements, along with mandibles and loose teeth (all mandibular) from animals assumed – on the basis of their chronology - to be wild boar.

Many of the postcranial elements from Goldcliff were unfused (and therefore must have come from young wild boar), making them unsuitable for biometrical analysis. Two astragali were complete enough to provide two length measurements (the GLl, greatest lateral length, and the GLm, greatest medial length). These specimens have been plotted, along with those available from the Early Mesolithic, in Figure 3.19. Both of the specimens from. Goldcliff are at the smaller end of the Early Mesolithic distribution. Although the sample is too small to be the basis for firm conclusions, the small size of bones from Goldcliff could point to a decrease in the size of wild boar during the Late Mesolithic. Teeth from Goldcliff, which were more abundant, do not support this hypothesis (Figure 3.20). Despite the greater quantity of teeth only 22 measurements are available for comparison. All of the measurements fall within the distribution of measurements from Early Mesolithic wild boar, and are generally towards the larger end of the plot.

Since the differences between wild boar remains from Goldcliff and those from the Early Mesolithic are very limited, the biometrical analysis presented in the following sections will be presented for all Mesolithic sites (including Goldcliff) in combination.

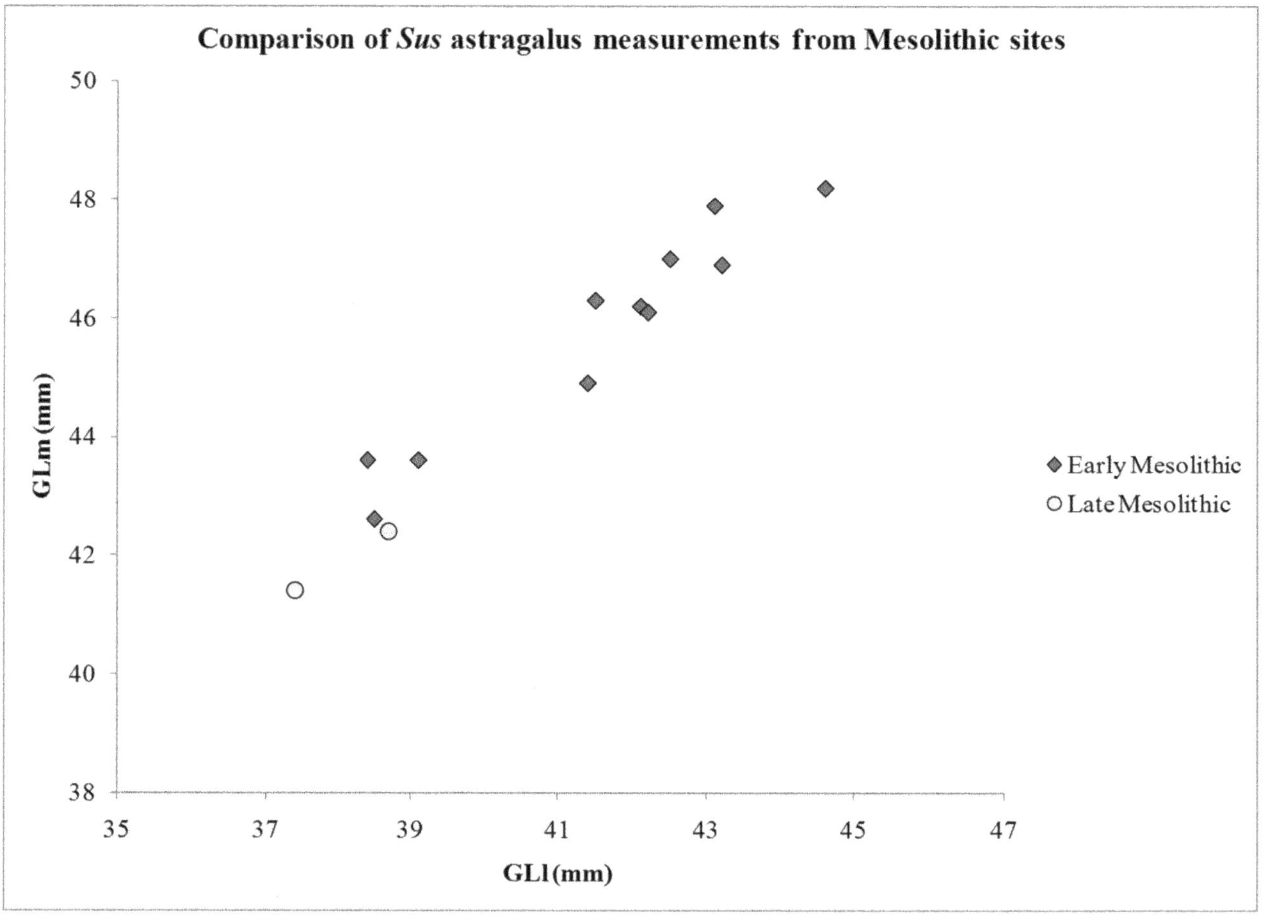

Figure 3.19 *Sus* astragalus measurements from Mesolithic sites (GLl = greatest lateral length, GLm = greatest medial length after von den Driesch 1970).

Biometrical Analysis – from the Mesolithic to the Late Neolithic

One of the aims of biometrical analysis of pig remains was to identify changes in the physical characteristics of the animals and how these could be associated with the beginning of the domestication process and husbandry practices in general. Previous studies have shown that the distal tibia is particularly suitable for comparing the size of different pig populations as they are not dependent on sex and age (Albarella and Payne 2005, 595; Lasota-Moskalewska 1987, 67; Payne and Bull 1986, 32). The distal tibia breadth and width measurements have been plotted in Figure 3.21. Mesolithic, Early Neolithic and Late Neolithic distal tibia measurements overlap suggesting that no substantial differences occurred in the size of the animals. In particular, Mesolithic and Early Neolithic tibia measurements are indistinguishable (and there is no statistically significant difference between the groups, see Table 3.1). If interpretations were based on this measurement alone there would be little evidence of change between the Mesolithic and the Neolithic, and if accepted in isolation there is no basis for the conclusion that different patterns of human exploitation were operating on animals of the two periods. Late Neolithic points congregate towards the bottom left of the plot, overlapping only with the smallest animals from the Mesolithic and Early Neolithic sample. Decreased diversity of animals during the Late Neolithic can be attributed to the fact that this period is represented by only one site but it may also be the result of varying husbandry practices during this period. It is perhaps an indication that increased control of pigs or more intensive husbandry was being applied during the Late Neolithic. A comparison between Durrington Walls and other Late Neolithic assemblages is necessary to determine whether this is a real pattern, and will be undertaken in Chapter 6.

Such uniformity between animals of different periods is not a characteristic of the entire dataset and other skeletal elements have provided evidence of variation in pigs through time. Measurements of the third permanent molar are often used in attempts to distinguish between wild and domestic populations (see for instance Dobney *et al.* 2007; Ervynck *et al.* 2001; Rowley-Conwy and Dobney 2007) due to the speed with which this tooth changes when domestication is imposed on pigs. Mandibular third molar measurements from the Mesolithic and Neolithic samples were compared and it became clear that there were differences between the populations in terms of their size (Figures 3.22, 3.23, 3.24). Mesolithic animals provided the largest measurements, as expected, as these samples are assumed to have originated from wild animals. Early Neolithic specimens overlapped substantially with those from the preceding Mesolithic period, but showed a tendency towards the smaller end of the distribution. This size reduction was clear in both the length and width of the third molar and the difference between the groups is statistically significant (Table 3.1). One very large outlier is located in the upper right of all three plots and can certainly be identified as belonging to a wild boar. This is an indication that some wild boar were still being utilised during the post-Mesolithic periods, although, it appears, in relatively small numbers. Analysis of other skeletal elements will be necessary to identify the extent of wild boar use in post-Mesolithic periods. Finally, there was some overlap between the distribution of Late Neolithic specimens and those of other periods, but the Late Neolithic specimens tended towards the lower left part of all three plots, indicating generally smaller toothed animals than in the Mesolithic and Early Neolithic. A number of interpretations of the third molar data need to be explored. Differences in the distribution of Mesolithic and Neolithic specimens are possibly the result of distinct populations of *Sus* being exploited, an interpretation that supports the hypothesis that domesticated animals were introduced at the beginning of the Neolithic. A further interpretation is that the same type of animals were used (*e.g.* they are all wild boar), but the concentration on smaller toothed animals in the Neolithic is evidence of increased exploitation of female animals. However, the third molar is not highly sexually dimorphic (Grigson 1982, 7-8; Payne and Bull 1988, 30) and it is unlikely that differences between female and male animals would provide the disparate pattern observed. Finally, it has been shown that the third molar could be influenced by environmental factors (Albarella and Payne 2005) and inter-population variation (Payne and Bull 1988). Therefore, to establish whether the differences are due to environmental factors or to a genuine genetic difference between populations we must interpret them in the context of other lines of evidence.

Due to the absence of morphological features that allow the accurate differentiation of first and second molars when they are loose, all isolated examples of these teeth were recorded as M1/2. The first and second molars can often be separated on the basis of their size (the second molar is larger than the first), but it is unwise to rely on observation of the relative size of loose specimens during recording. A more accurate method involves comparing the width measurements of teeth that could safely be identified as first or second molars (*i.e.* those found in jaws) with measurements from loose unidentified specimens. Figure 3.25 presents the pooled Mesolithic measurements, Figure 3.26 and 3.25 the main Early Neolithic sites of Hambledon Hill and Runnymede Bridge, and Figure 3.26 Late Neolithic Durrington Walls. In all periods investigated the separation between known lower first molars and known lower second molars was very clear and when loose teeth were compared to the measurements of known specimens these, too, fell neatly into two different groups. All loose mandibular teeth could therefore be assigned either first or second molar status. The attributed status allowed the inclusion of loose specimens in further biometrical analysis including scatter plots and log ratio comparisons. The results of the separation reiterate the greater plasticity that is inherent in the second molar when compared with the first molar (Albarella and Payne 2005). The first molar points form a much tighter group in the lower left of the diagrams, while in the upper right second molars have a wider distribution and the group exhibits more internal variation.

Isolated maxillary teeth were treated in the same way as mandibular teeth and the results can be seen in Figures 3.29, 3.30 and 3.31. On the whole there is separation between upper first and upper second molars from Mesolithic sites (Figure 3.29). Two specimens fall between the main first and second molar distributions, and it is not possible to determine which group these two teeth belong to. Similarly, in the Early Neolithic plot, the known first and second molar groups are for the most part distinct but there is a significant area of overlap. Because of this overlap it is not possible to consistently distinguish between isolated maxillary first and second molars, therefore, only those found in jaws will be included in the biometrical analysis presented below.

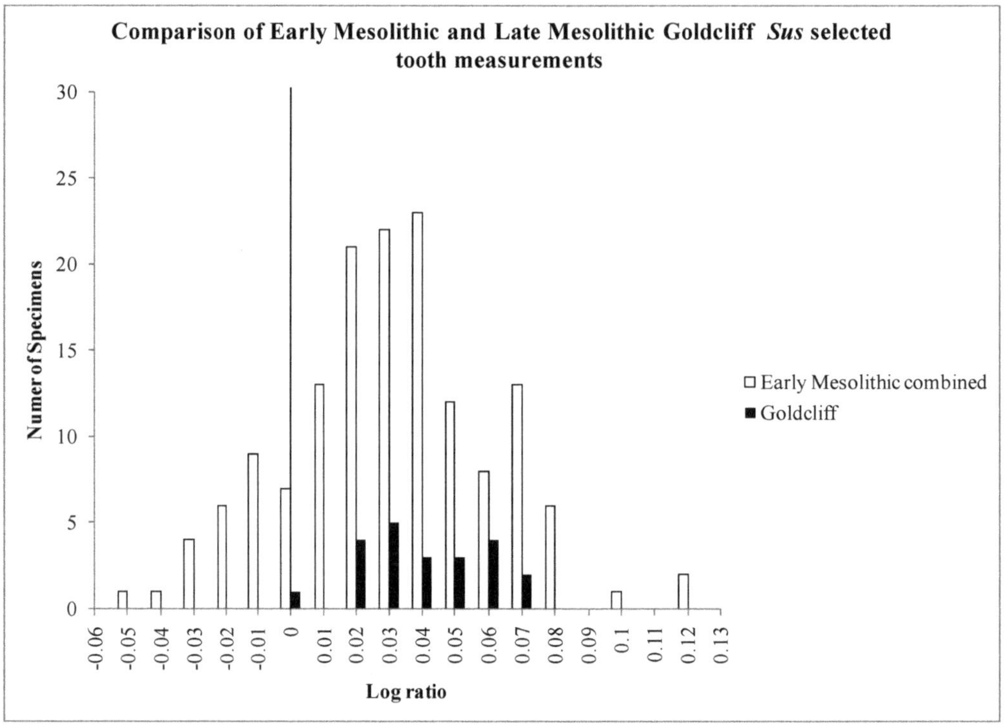

Figure 3.20 Log ratio comparison of *Sus* tooth measurements from the Early Mesolithic and Late Mesolithic (Goldcliff sites).

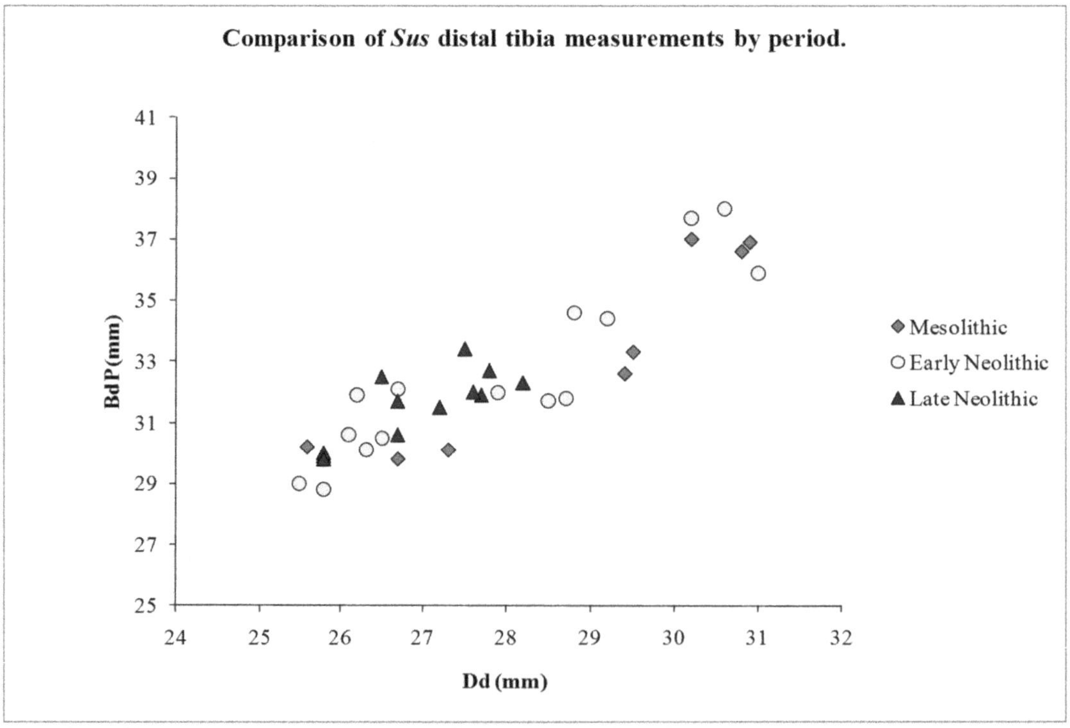

Figure 3.21 *Sus* distal tibia measurements from Mesolithic, Early Neolithic and Late Neolithic sites (Viner 2011, Figure 14.5).

Table 3.1 Results of t-test comparing *Sus* teeth and postcranial measurements. Those marked with a star (*) are considered statistically significant. n = the number of measurements included.

Measurement	Group 1 /n		Group 2 / n		Significance
M_3L	Mesolithic	26	Early Neolithic	24	0.016*
	Mesolithic	26	Late Neolithic	36	0.000**
	Early Neolithic	24	Late Neolithic	36	0.254
M_2L	Mesolithic	30	Early Neolithic	34	0.000**
	Mesolithic	30	Late Neolithic	88	0.000**
	Early Neolithic	34	Late Neolithic	88	0.709
M_1L	Mesolithic	20	Early Neolithic	28	0.012*
	Mesolithic	20	Late Neolithic	66	0.001*
	Early Neolithic	28	Late Neolithic	66	0.765
Tibia BdP	Mesolithic	8	Early Neolithic	17	0.705
	Mesolithic	8	Late Neolithic	13	0.184
	Early Neolithic	17	Late Neolithic	13	0.177
Tibia Dd	Mesolithic	8	Early Neolithic	17	0.474
	Mesolithic	8	Late Neolithic	13	0.118
	Early Neolithic	17	Late Neolithic	13	0.342
Humerus BT	Early Neolithic fused	26	Early Neolithic fusing	17	0.004*
	Late Neolithic fused	13	Late Neolithic fusing	16	0.534
Humerus HTC	Early Neolithic fused	26	Early Neolithic fusing	17	0.047*
	Late Neolithic fused	13	Late Neolithic fusing	17	0.357

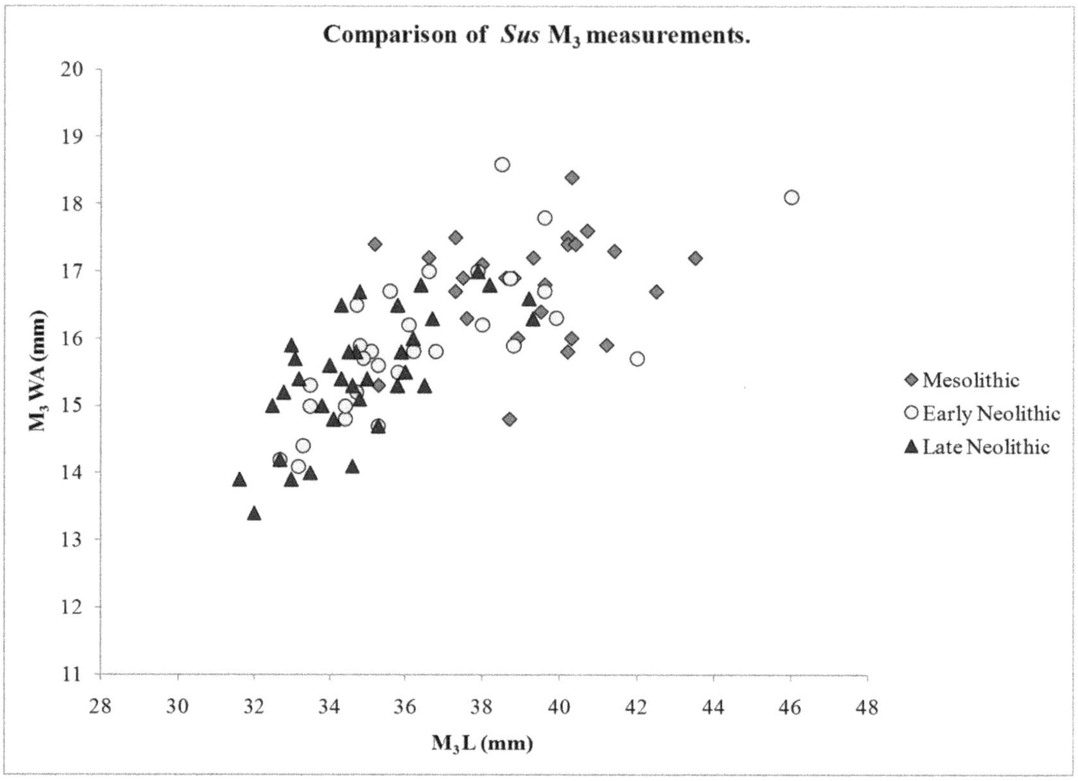

Figure 3.22 Comparison of *Sus* third mandibular molar measurements, greatest length (L) and width of the anterior cusp (WA) (following Payne and Bull 1988). Both loose teeth and those found in jaws have been included.

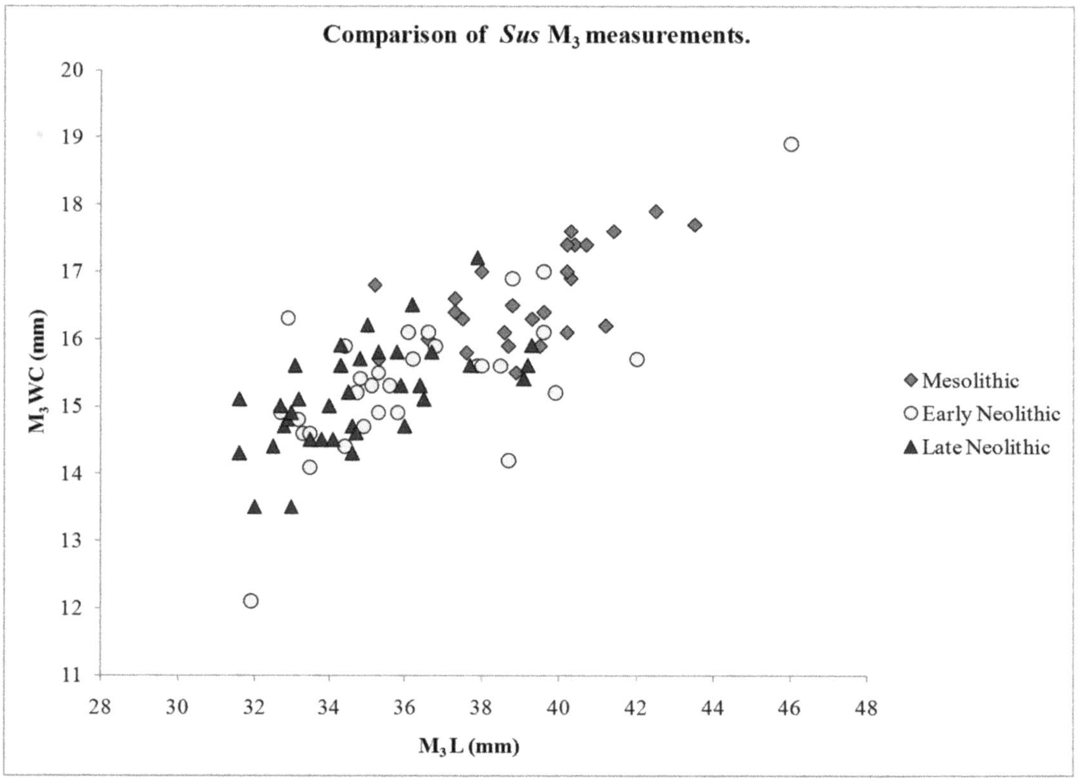

Figure 3.23 Comparison of *Sus* third mandibular molar measurements, greatest length (L) and width of the central cusp (WC) (following Payne and Bull 1988). Both loose teeth and those found in jaws have been included.

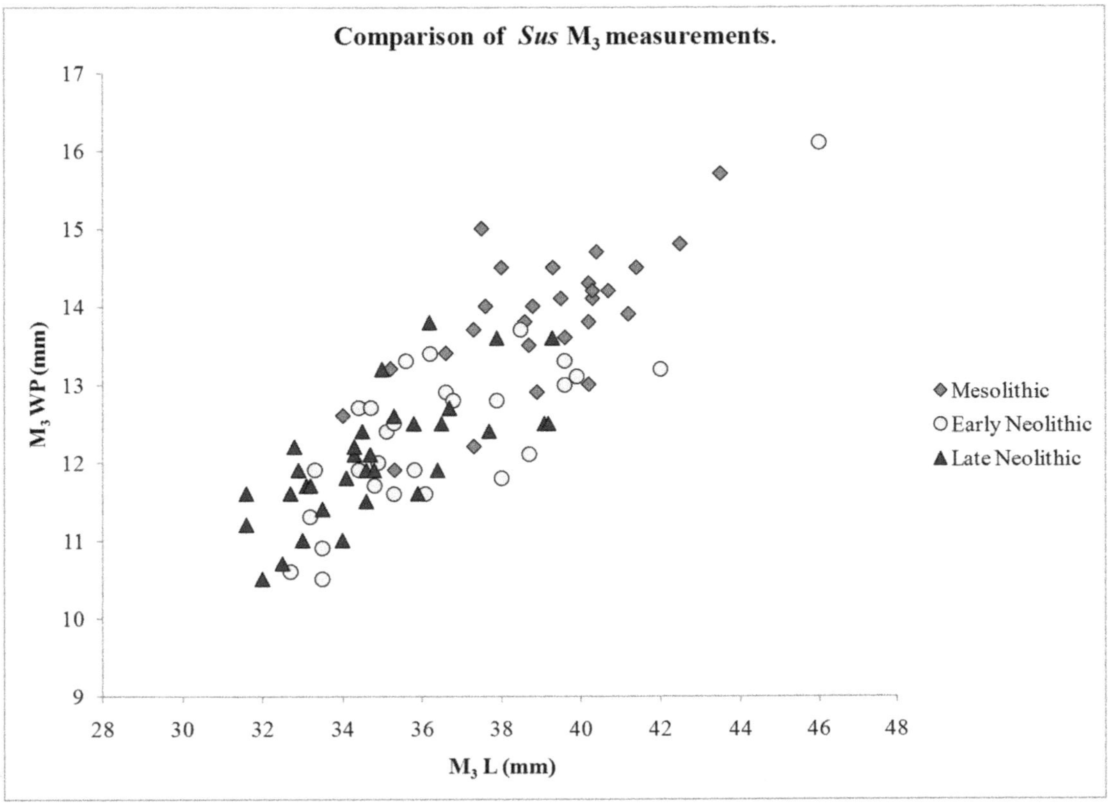

Figure 3.24 Comparison of *Sus* third mandibular molar measurements, greatest length (L) and width of the posterior cusp (WP) (following Payne and Bull 1988). Both loose teeth and those found in jaws have been included.

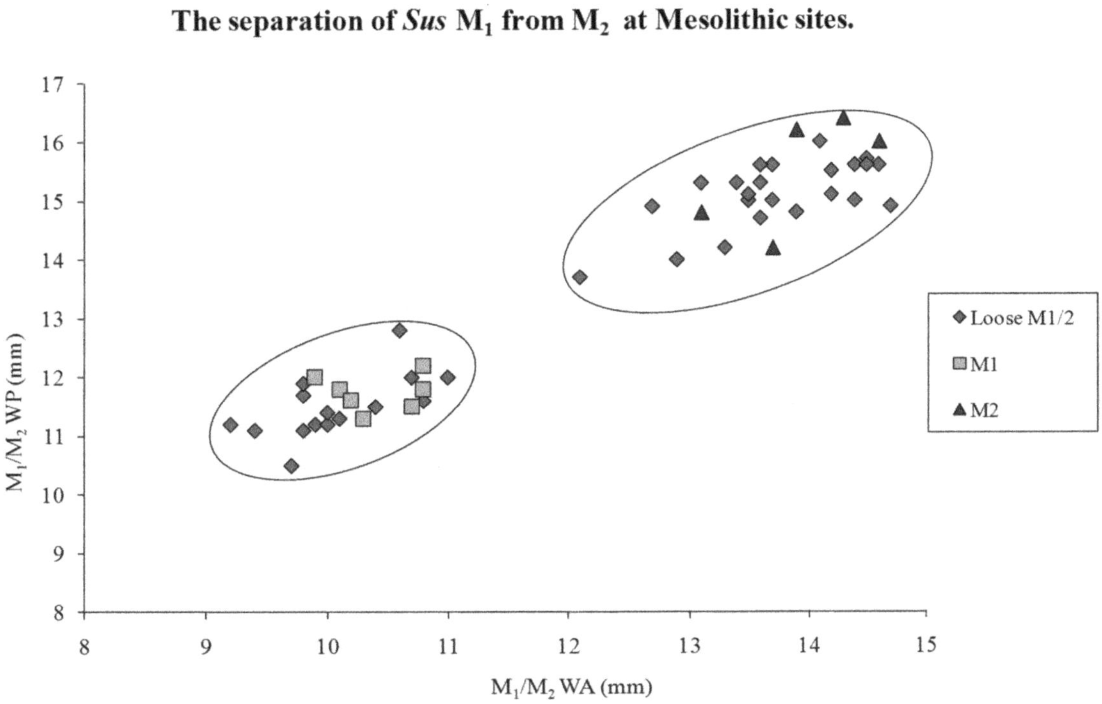

Figure 3.25 The separation of mandibular first and second molars from Mesolithic sites (WA = width of the anterior cusp, WP = width of the posterior cusp, measurements after Payne and Bull 1988).

The separation of *Sus* M_1 from M_2 at Hambledon Hill.

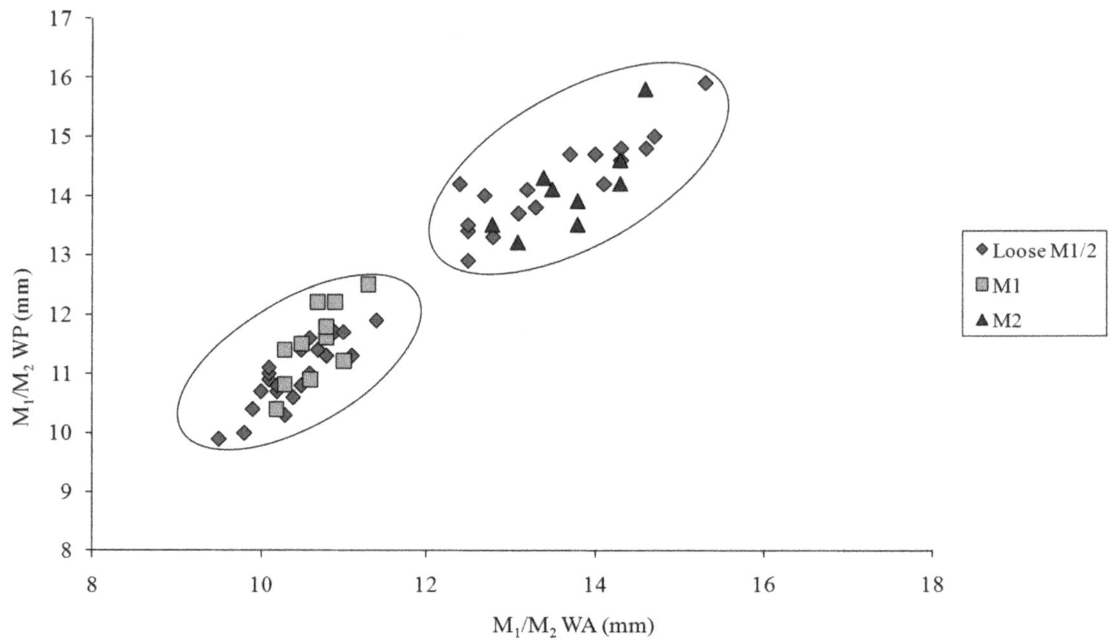

Figure 3.26 The separation of mandibular first and second molars from Early Neolithic Hambledon Hill (WA = width of the anterior cusp, WP = width of the posterior cusp, measurements after Payne and Bull 1988).

The separation of *Sus* M_1 from M_2 at Runnymede Bridge.

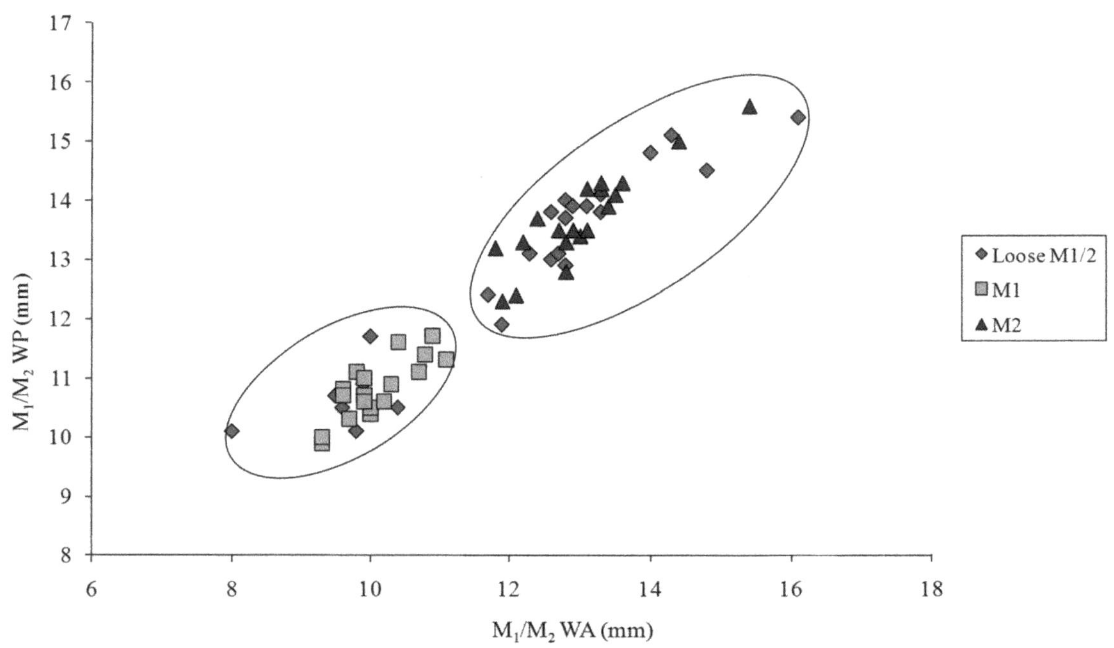

Figure 3.27 The separation of mandibular first and second molars from Early Neolithic Runnymede Bridge (WA = width of the anterior cusp, WP = width of the posterior cusp, measurements after Payne and Bull 1988).

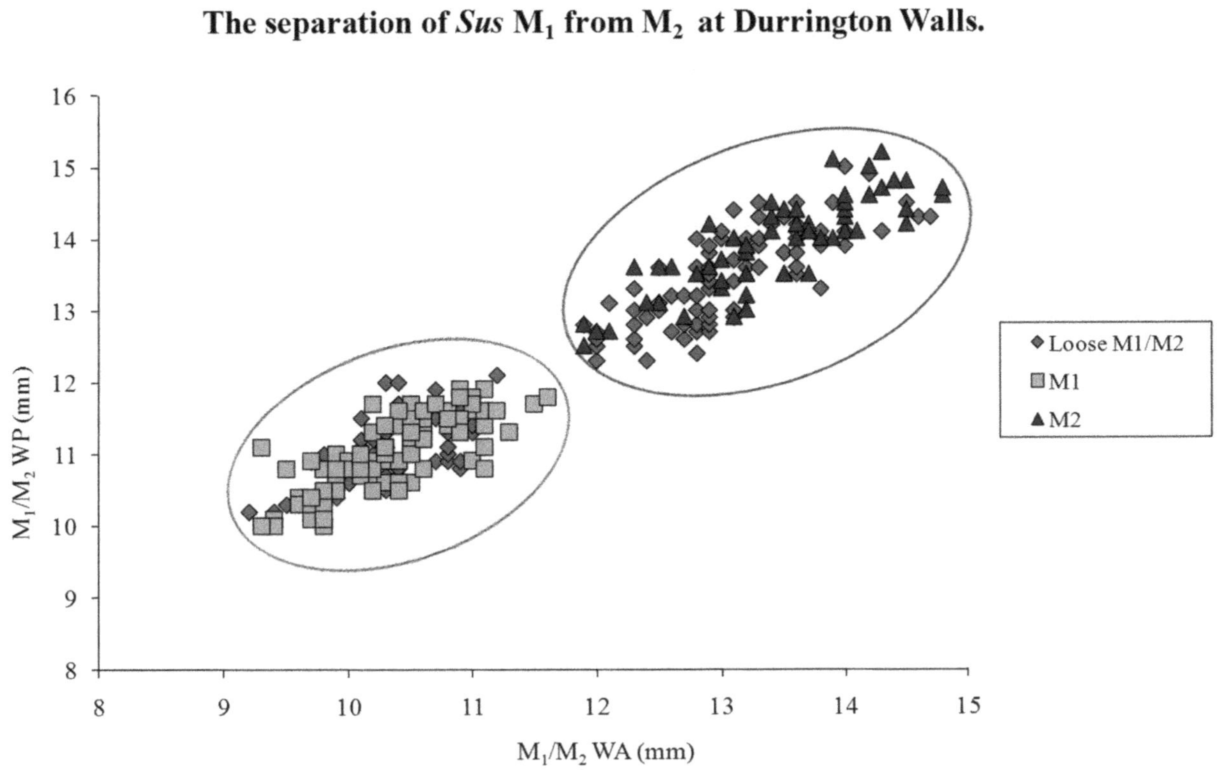

Figure 3.28 The separation of mandibular first and second molars from Late Neolithic Durrington Walls (WA = width of the anterior cusp, WP = width of the posterior cusp, measurements after Payne and Bull 1988).

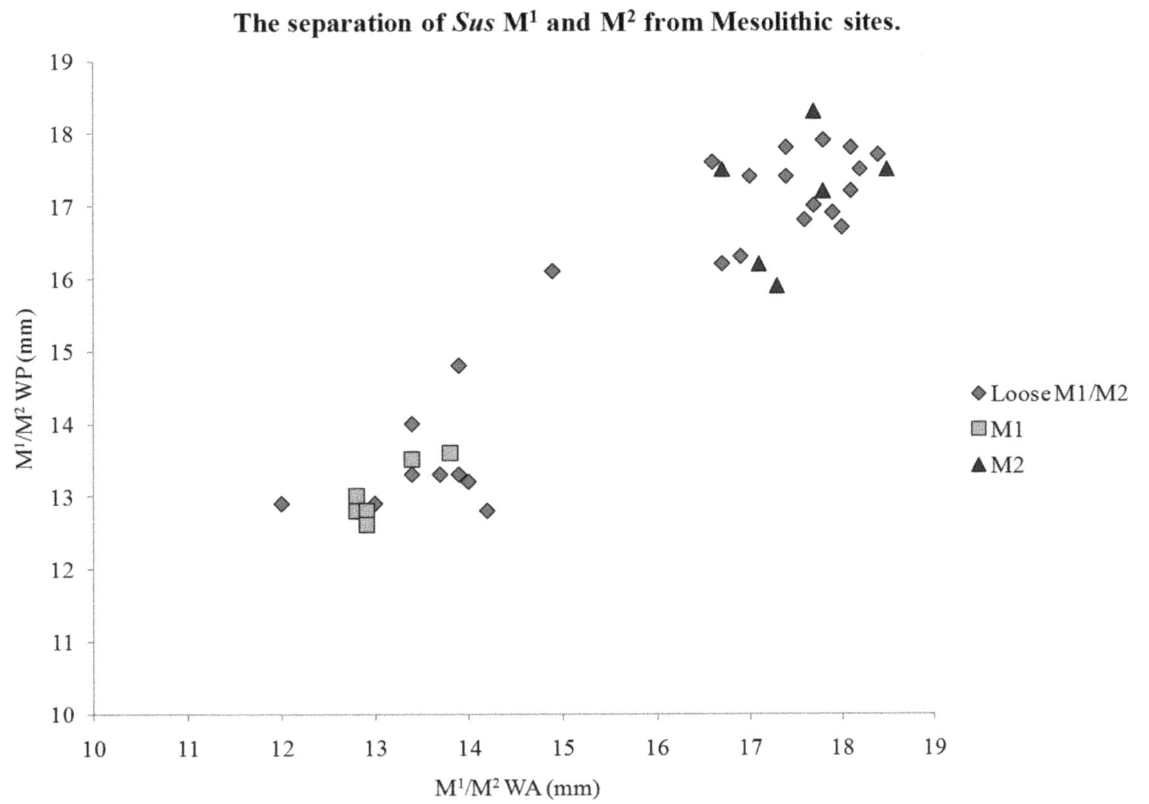

Figure 3.29 The separation of *Sus* maxillary first and second molars from Mesolithic sites (WA = width of the anterior cusp, WP = width of the posterior cusp, measurements after Payne and Bull 1988).

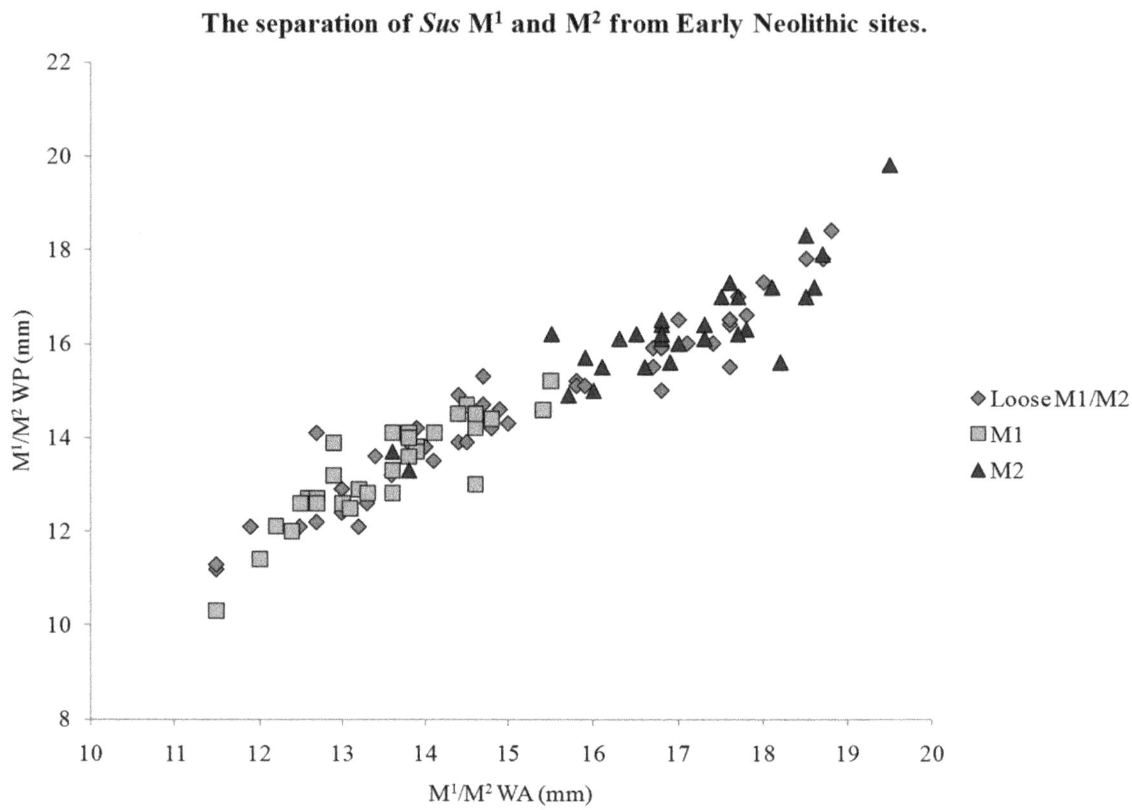

Figure 3.30 The separation of *Sus* maxillary first and second molars from Early Neolithic sites (WA = width of the anterior cusp, WP = width of the posterior cusp, measurements after Payne and Bull 1988).

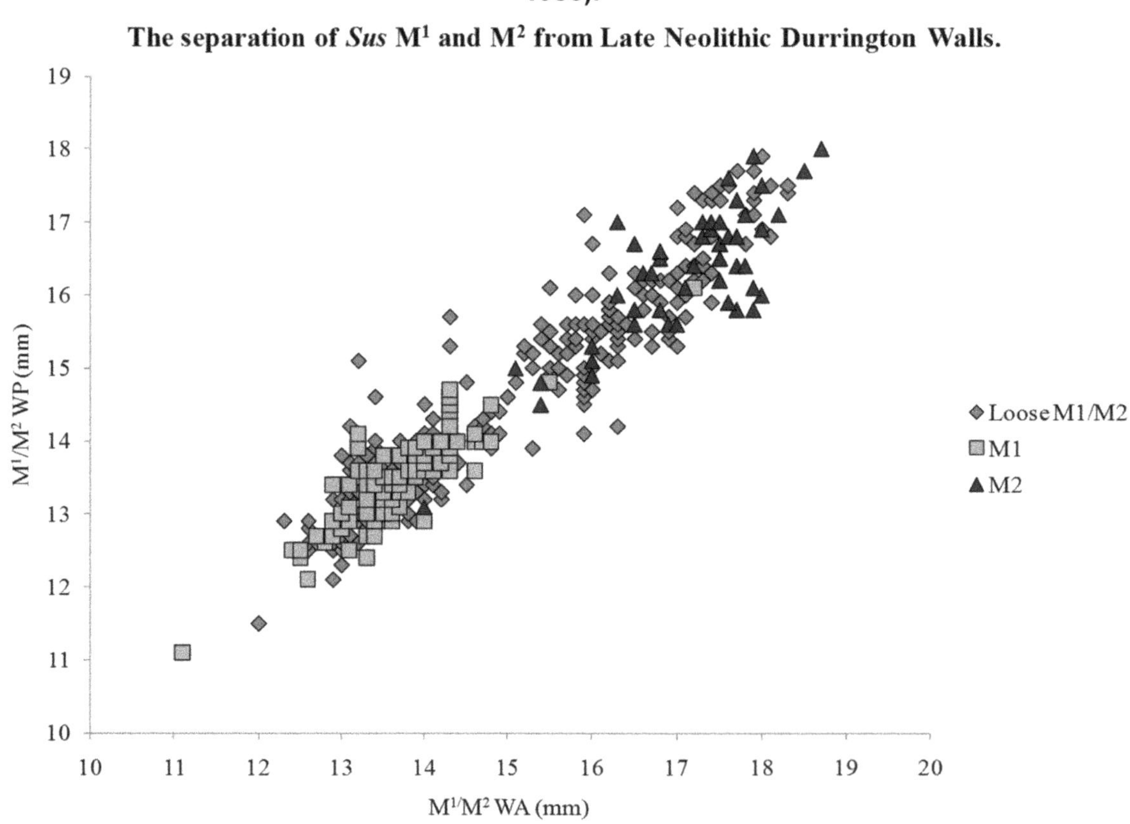

Figure 3.31 The separation of *Sus* maxillary first and second molars from Durrington Walls (WA = width of the anterior cusp, WP = width of the posterior cusp, measurements after Payne and Bull 1988).

It has been mentioned that the third molar is susceptible to environmentally induced change. The second molar is expected to exhibit similar changes with domestication (Mayer *et al.* 1988) as the third molar, but without the same level of susceptibility to environmental variation. Figures 3.32 and 3.33 compare the length and width measurements of *Sus* second molars from all of the periods investigated. The second molar measurements exhibit a reduction in size that is similar to the third molar; there are clear differences between the Mesolithic and Neolithic animals (which is statistically significant), despite some overlap between the tooth measurements from the different periods. Early and Late Neolithic specimens overlap greatly but the Early Neolithic cluster has some large outliers (perhaps wild boar) that are absent in the Late Neolithic period. In contrast with the third molar results that showed a reduction in all of the measurements taken, the smaller size of the second molar is not uniform for all parts of the tooth. In fact, both the length and posterior width of second molars was smaller in the Neolithic than in the Mesolithic (Figure 3.32 and Figure 3.33), but the width of the anterior cusp remained largely unchanged through time. This evidence reinforces the idea that the factors that can cause change in the mammalian skeleton do not affect all parts of a skeletal element uniformly.

Both the second and third molars have provided some indication that the pigs exploited during the Mesolithic were physically dissimilar to those used during the Neolithic. All of the measurements taken from first molar specimens (both anterior and posterior width as well as the greatest length) were, however, by and large chronologically indistinguishable (Figure 3.34 and Figure 3.35), despite t-test results that indicate significant differences between the groups. The Mesolithic and Early Neolithic specimens tend to be larger than those from the Late Neolithic but the overlap is such that, used in isolation, the first molar measurements would suggest very little change in the pigs through time.
The overall impression obtained from the tooth measurements is different from the one obtained from the postcranial bones. Although some minor differences exist between the two Neolithic periods the main change can be observed between the Mesolithic and the Early Neolithic, when a clear reduction in the size of the lower second and third molars occurred.

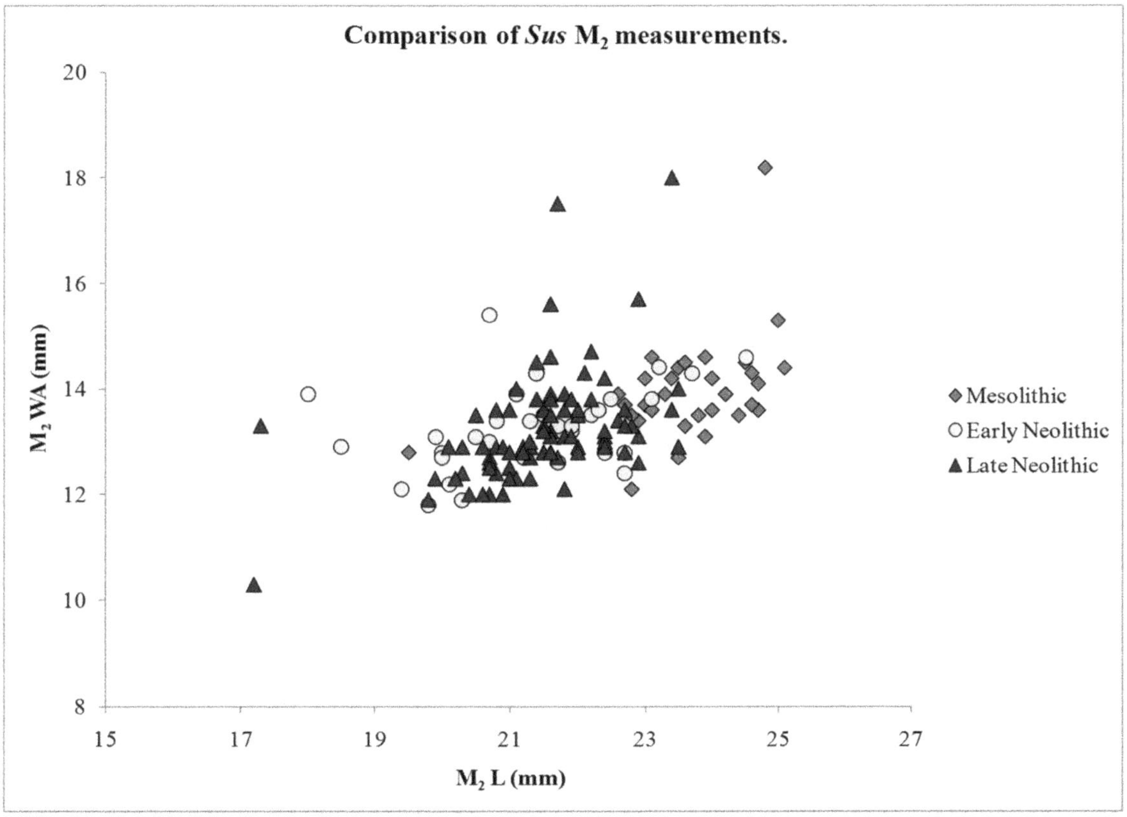

Figure 3.32 Comparison of measurements (after Payne and Bull 1988) from *Sus* second molars (L = length, WA = anterior width). Both loose teeth and jaws have been included.

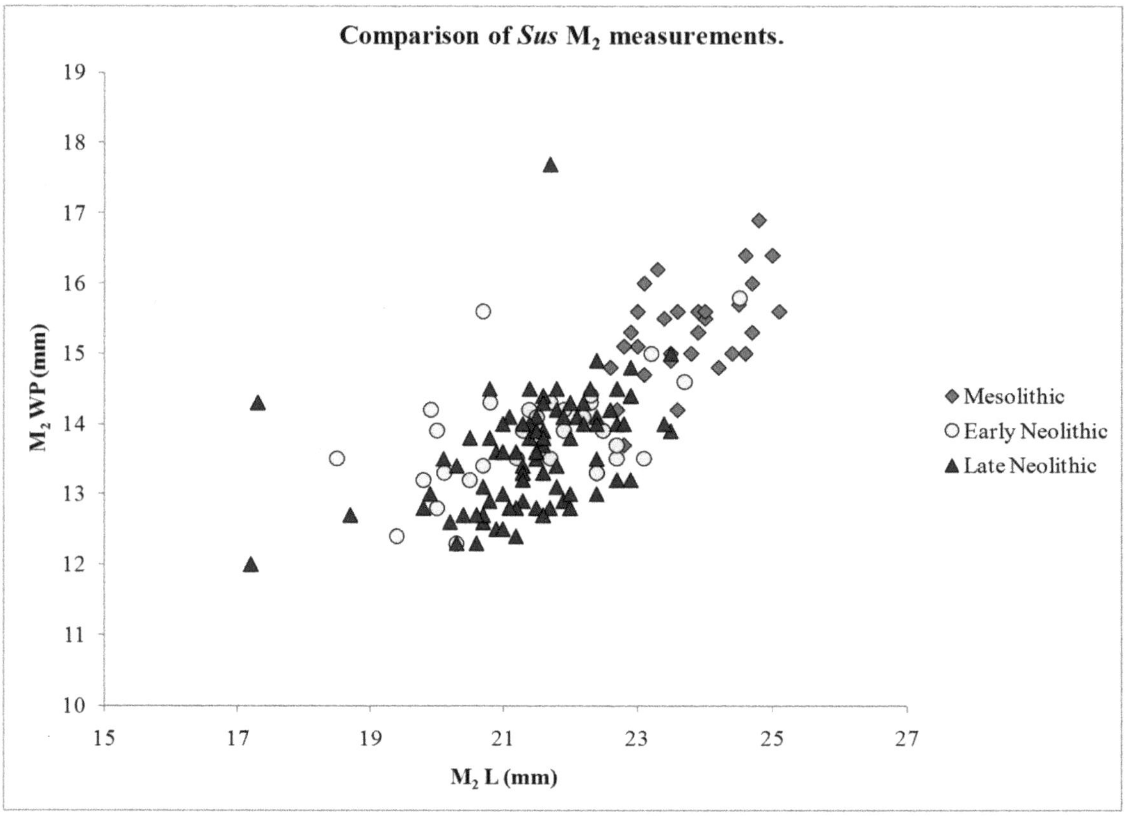

Figure 3.33 Comparison of measurements (after Payne and Bull 1988) from *Sus* second molars (L = length, WP = anterior width). Both loose teeth and jaws are included.

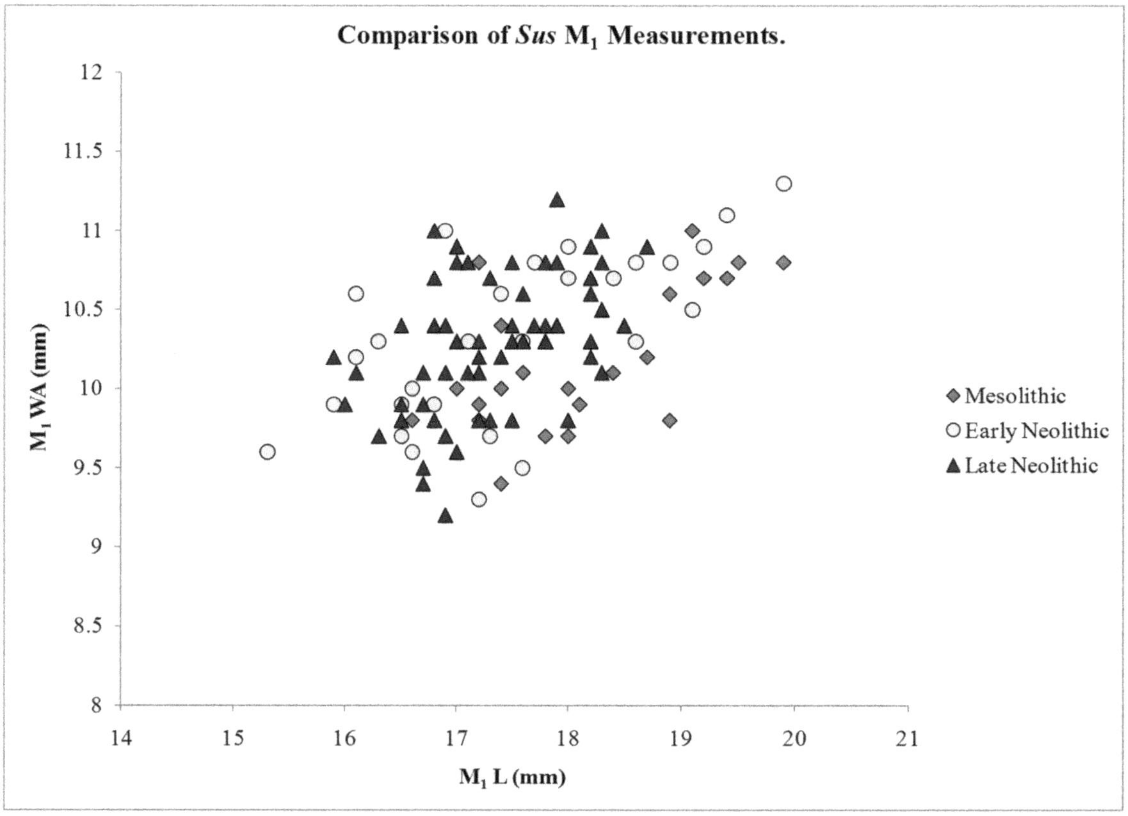

Figure 3.34 Comparison of measurements (after Payne and Bull 1988) from *Sus* first molars (L = length, WA = anterior width). Both loose teeth and jaws included.

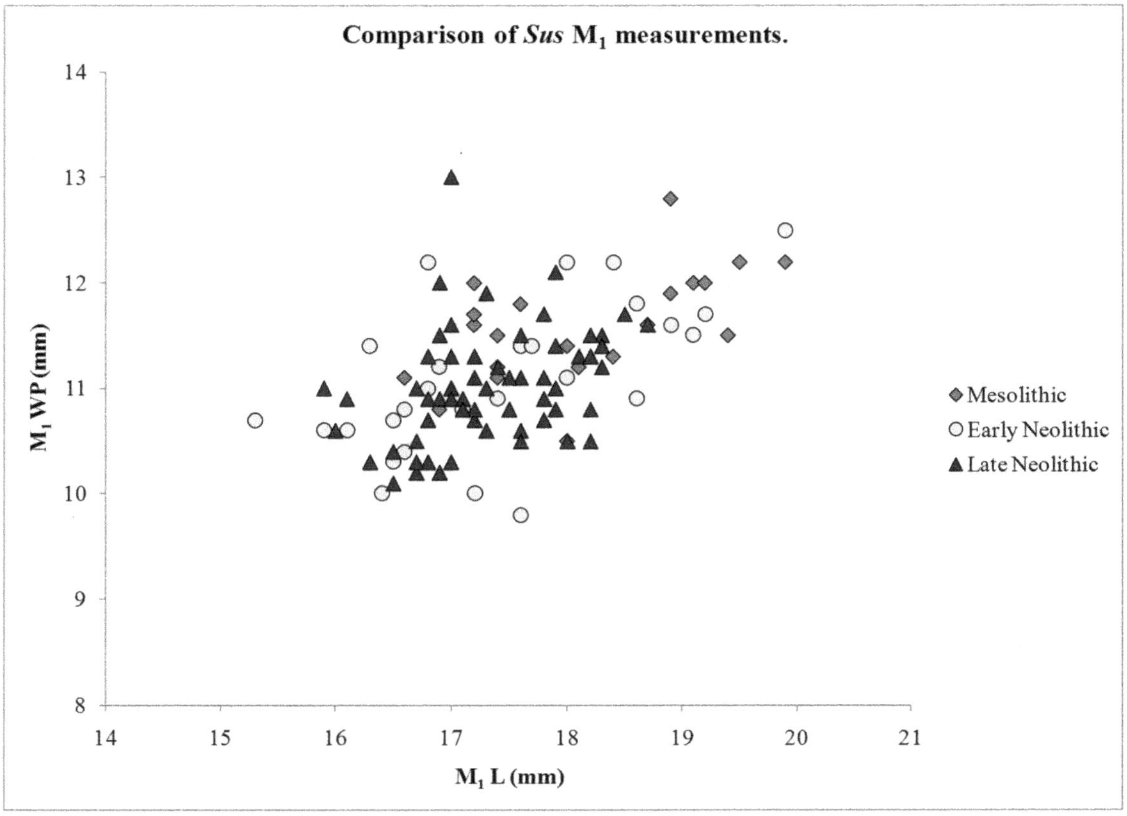

Figure 3.35 Comparison of measurements (after Payne and Bull 1988) from *Sus* first molars (L = length, WP = posterior width). Both loose teeth and jaws included

Maxillary second molar measurements duplicate the patterns observed in the mandibular teeth. Figure 3.36, below, compares maxillary second molar length and anterior width measurements from the Mesolithic, Early Neolithic and Late Neolithic. The Mesolithic sample is small and provides a total of four second molars on which both of the measurements could be taken. Despite the small Mesolithic sample size the overall pattern that appears is similar to that for mandibular second molar measurements. The Mesolithic specimens tend to be of similar length, anterior width (Figure 3.36) and posterior width (Figure 3.37) to the Early and Late Neolithic teeth, but there is a clear tendency for the Mesolithic to plot towards the larger end of the distribution.
Upper first molars provide a different picture. Unlike the upper and lower second and third molars, and the lower first molar, specimens that have been discussed above and in which the general, pervasive trend is towards size reduction with the onset of the Neolithic, in certain dimensions the first upper molar appears to increase from the Mesolithic to the Neolithic.

The shape of the *Sus* mandibular third molar has been presented in Figure 3.40 and Figure 3.41 by plotting the ratio of widths to the tooth length. Figure 3.40 compares the ratio of anterior width/length to central width/length of the third molar. The teeth from presumably wild Mesolithic animals compare well with the later Early and Late Neolithic specimens. The same holds when the posterior width is used (as in Figure 3.41). There is a slight tendency for a greater posterior ratio in the Mesolithic specimens, but this tendency is not significant and in general the shape of the teeth is similar throughout all periods investigated.

Log ratio comparisons of *Sus* remains allowed the problem of small sample size from some sites, periods and skeletal elements to be mitigated. By pooling measurements and comparing them to a standard (established from Late Neolithic Durrington Walls by Albarella and Payne 2005) an idea of the general trends between the Mesolithic, Early Neolithic and Late Neolithic could be approached. When postcranial measurements of pig remains from the study area were compared using the log ratio technique it was found that there was significant overlap between the bones of animals from the Mesolithic and those from the Neolithic periods (Figure 3.42). The range of measurements from the Mesolithic fell entirely within the distribution of the Early Neolithic measurements. Mesolithic animals were generally larger than the standard and most fell to the right of the line, a relatively small proportion plotted close to or overlapped with the standard. This pattern expresses well the overall similarity of and overlap between animals of different periods that was identified as a characteristic of the assemblages in the scatter plots presented above.

Despite the similarity between the Mesolithic and Early Neolithic pig remains, there is some reduction in the size of postcranial elements between the two periods. The Mesolithic bones are towards the larger end of the Early Neolithic distribution (the mean of the Mesolithic sample (0.04) is larger than the Early Neolithic mean (0.02). Overall the difference is rather small, and the high degree of overlap between the pigs from the two periods in terms of their postcranial bones alludes to their physical similarity. The results also indicate that part of the Early Neolithic assemblage was composed of physically larger animals, indicated by the presence of a small number of outliers to the right of the main distribution. The reason for the presence of such large animals during the Early Neolithic will be explored in greater detail in Chapter 7. Potential inter-site variation during the Early Neolithic needs to be assessed before a secure interpretation of the data can be made.

In the Late Neolithic dataset measurements taken from *Sus* postcranial bones are smaller than both the Mesolithic and the Early Neolithic (Figure 3.42). The results indicate that the population was focussed on the standard, a trend that is to be expected as the standard was established from specimens recovered from previous excavations at the same site (Durrington Walls). Animals from the Late Neolithic had comparatively smaller skeletal elements than their counterparts in the Mesolithic and Early Neolithic. Although there is overlap, the reduction in size from the Mesolithic and the Early Neolithic to the Late Neolithic is clear.

Changes in the size of animals with the onset of the Neolithic that continued into the Late Neolithic suggest that there were factors that influenced the pig skeleton during this time period. There is a tendency towards smaller size through time; the Early Neolithic is smaller than the Mesolithic, and the Late Neolithic smaller than the Early Neolithic. The factors that produced this change are unclear, and as yet there is no strong evidence to suggest the introduction of a distinct 'domestic' population during the Early Neolithic. What is clear is that the changes observed to take place with the onset of the Neolithic modified some of the physical characteristics of pigs. In addition, these changes appear gradual and chronologically extensive, occurring over a long period of time from the beginning to the end of the Neolithic.

Teeth and postcranial bones react differently to external influences. Analysis of each has the potential to expatiate understanding of the development of pig exploitation. The comparison of first and second molar measurements using the log ratio technique (Figure 3.43) complements in an interesting way the result obtained from the postcranial bones. When compared to the standard, the focus of Mesolithic specimens is clearly to the right of the line and the high proportion of specimens that were larger than the standard indicates that Mesolithic animals were generally of greater size than in the Neolithic. This said, a number of specimens were very similar to or smaller than the standard, reinforcing the physical comparability of animals from the periods in question. There is a great deal of overlap between the Mesolithic and Early Neolithic, but also a tendency towards smaller toothed animals in the Early Neolithic (an interpretation reinforced by the mean that rests at 0.00 in the Early Neolithic, compared with 0.02 in the Mesolithic). No significant further reduction occurred in the Late Neolithic, the distribution is very similar and the mean is -0.01. The lower mean value of these measurements than the standard indicates that some slightly smaller specimens have been incorporated into the study in comparison to the earlier work from Durrington Walls. The Late Neolithic has a slight tendency to be smaller than the Early Neolithic sample already discussed; however, the divergence is negligible. A number of general conclusions can be drawn from the tooth results. The first is that the divergence between Mesolithic and Neolithic (in terms of the first and second molars) is limited, which may be expected considering that the first molar changed little through time. The likeness of the samples to one another is the most striking characteristic, and whatever size diminution occurs is almost entirely attributable to the second molar. Certainly any attempt to distinguish between different populations based solely on the first and second molars would be unproductive. It would prove difficult even to tell the difference between Mesolithic and Late Neolithic animals from these teeth, despite there being a chronological separation of several thousand years. The second is that (despite the general uniformity), the Early and Late Neolithic have more in common with one another than with the Mesolithic. Although small, attenuation of *Sus* size is a characteristic of the Neolithic. Unlike the results of postcranial log ratio comparison discussed above, the Early Neolithic is closer to the Late Neolithic than the Mesolithic. There is no gradual reduction in the size of teeth. These results suggest that the changes that occurred with the onset of the Neolithic were not the same as those acting during the Early Neolithic/Late Neolithic.

Unlike postcranial bones there are no large outliers in the Early Neolithic tooth plot, but there are some in the Late Neolithic, which, again, may be interpreted as possible wild boar. Unlike bones, however, the large tooth outliers are matched by one similar outlier in the Mesolithic period.

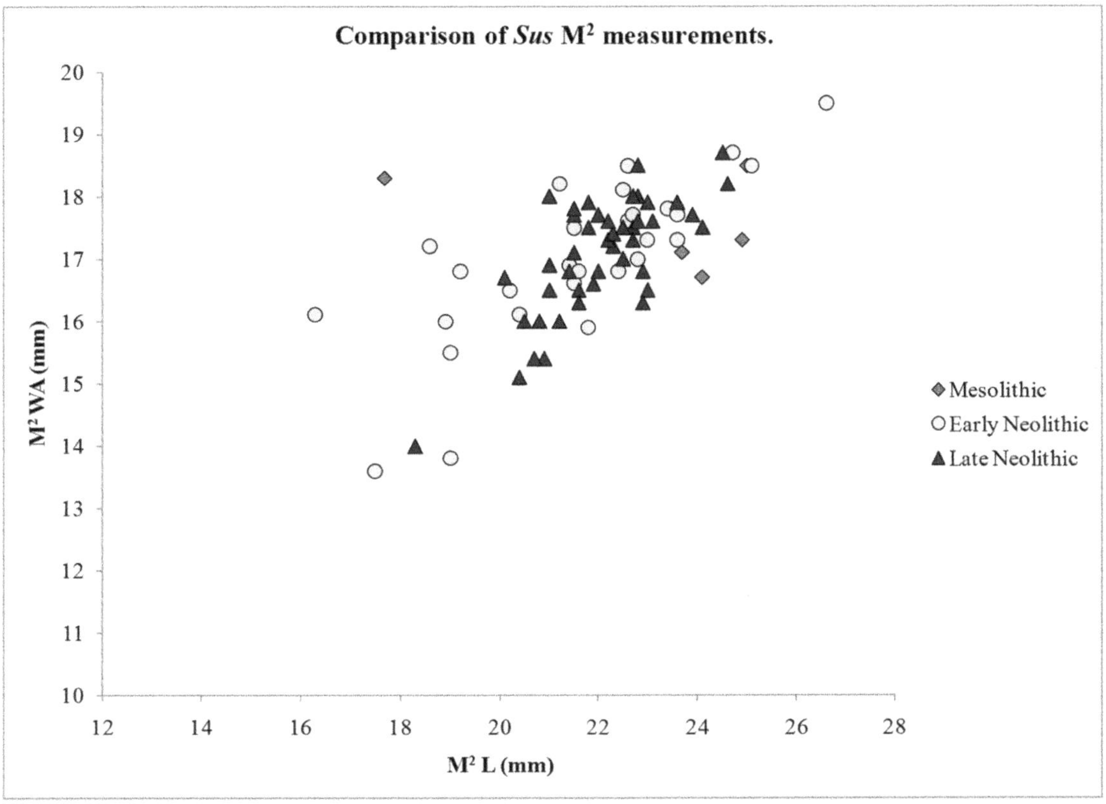

Figure 3.36 Comparison of measurements (after Payne and Bull 1988) from *Sus* upper second molars (L = length, WA = anterior width). Only teeth found in jaws have been included.

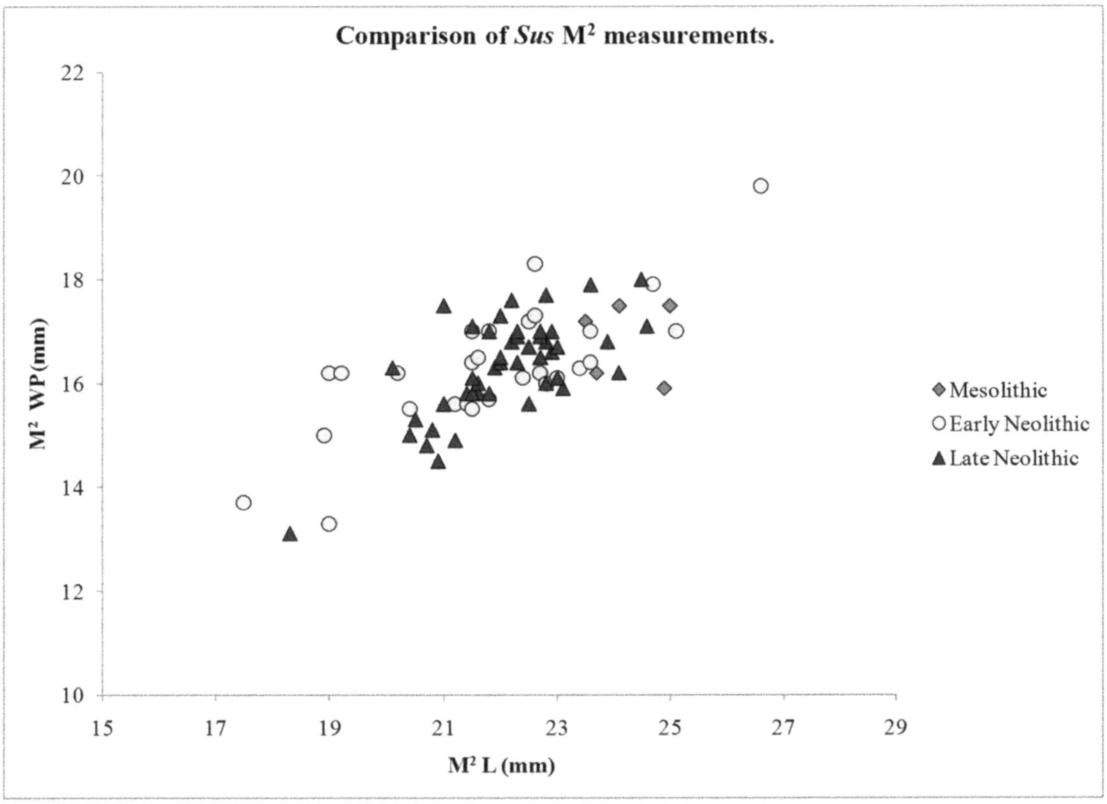

Figure 3.37 Comparison of measurements (after Payne and Bull 1988) from *Sus* upper second molars (L = length, WP = posterior width). Only teeth found in jaws have been included.

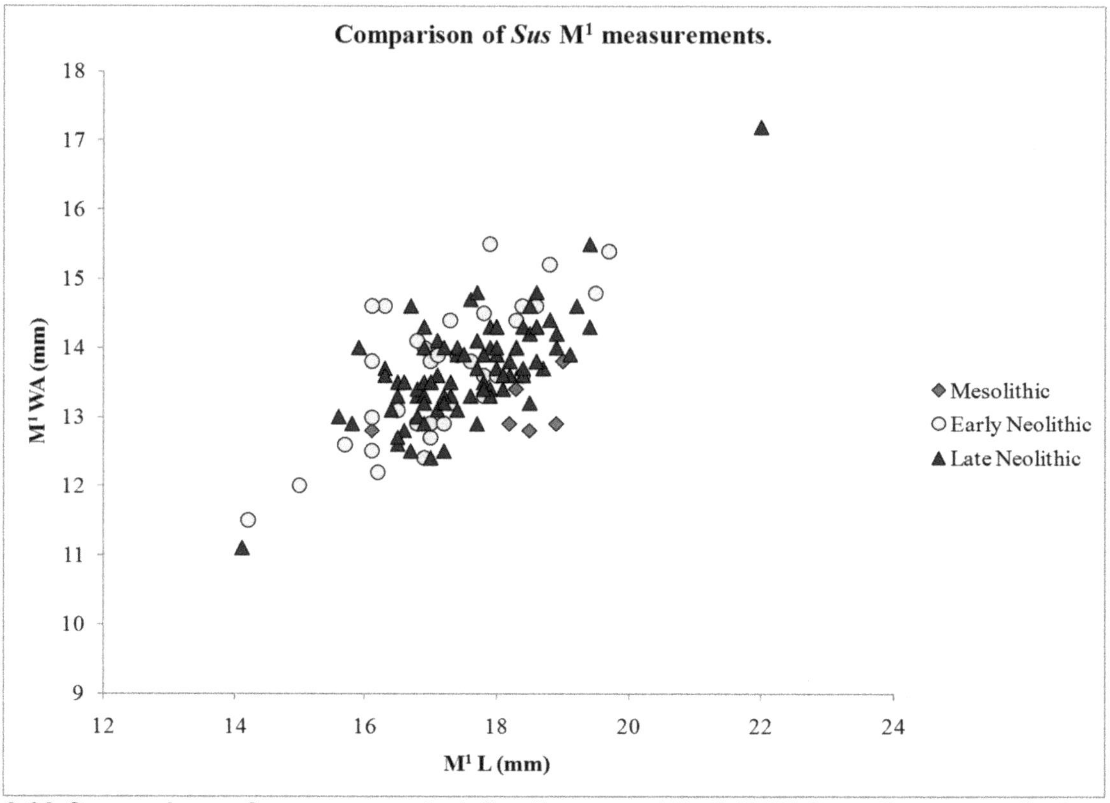

Figure 3.38 Comparison of measurements (after Payne and Bull 1988) from *Sus* upper first molars (L = length, WA = anterior width). Only teeth found in jaws have been included.

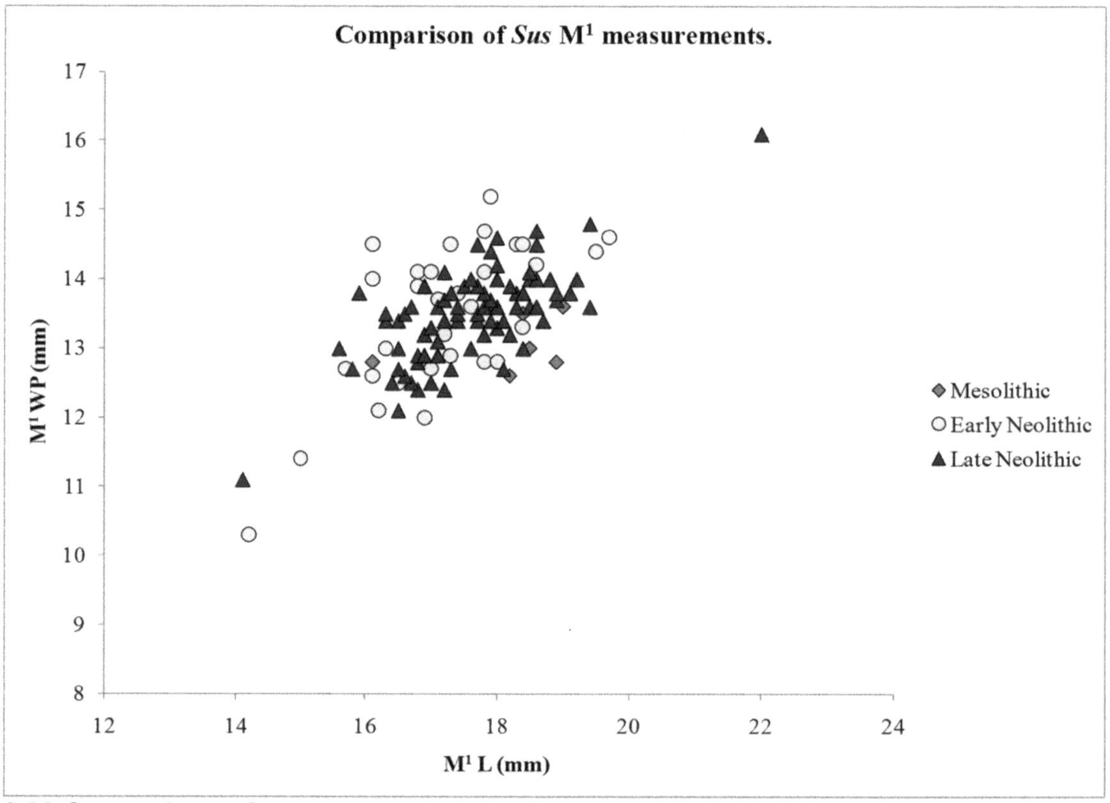

Figure 3.39 Comparison of measurements (after Payne and Bull 1988) from *Sus* upper first molars (L = length, WP = posterior width). Only teeth found in jaws have been included.

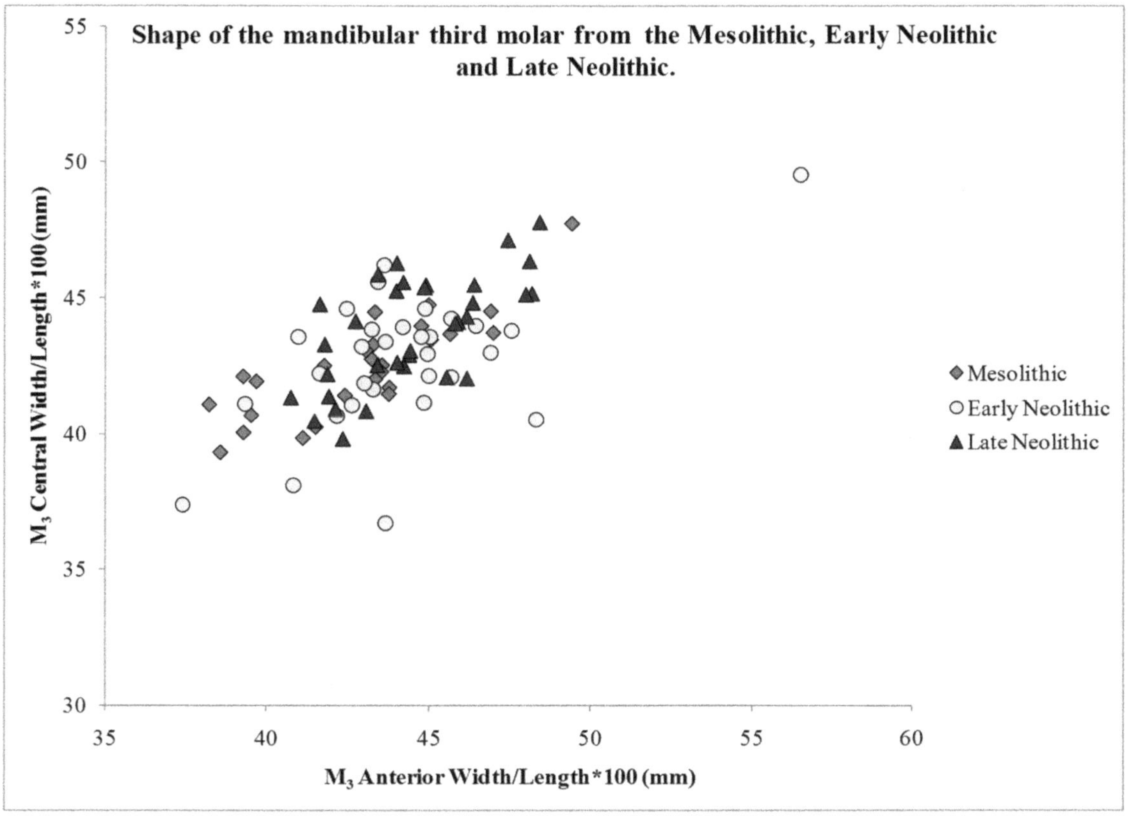

Figure 3.40 The shape of *Sus* mandibular third molars (measurements after Payne and Bull 1988). Both loose teeth and jaws have been included.

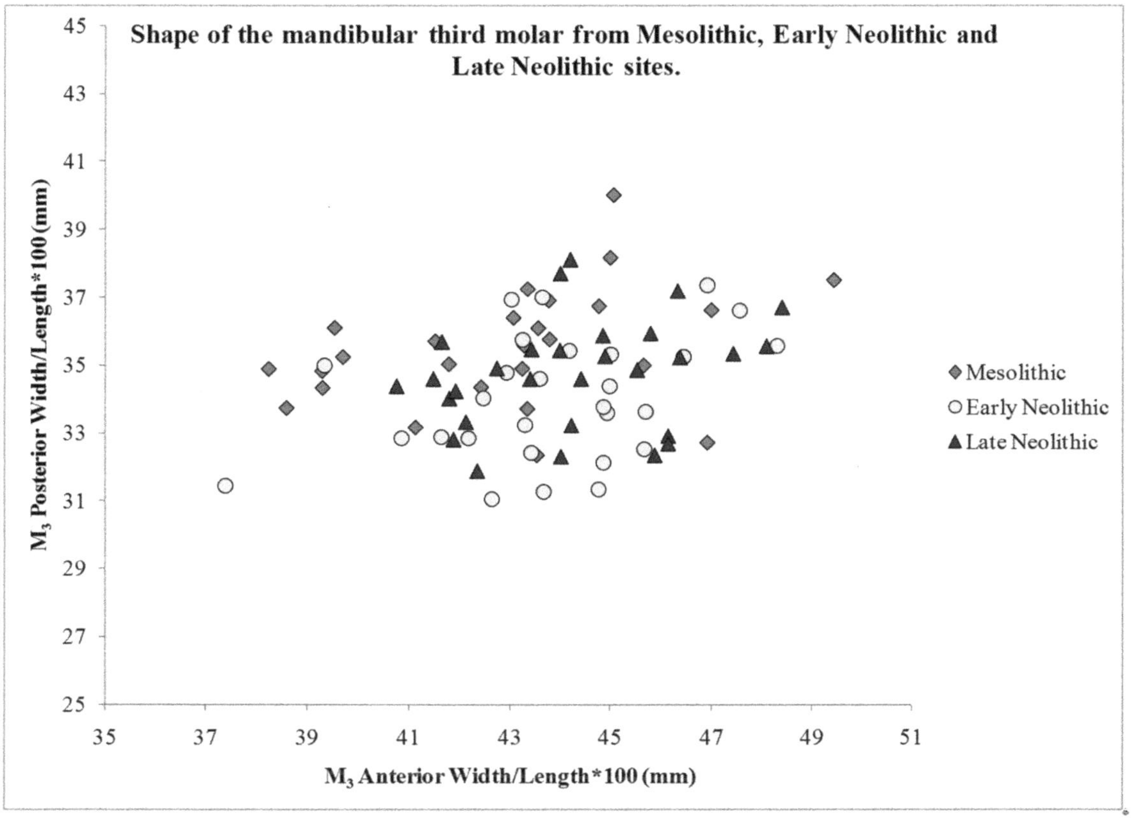

Figure 3.41 The shape of *Sus* mandibular third molars (measurements after Payne and Bull 1988). Both loose teeth and jaws have been included

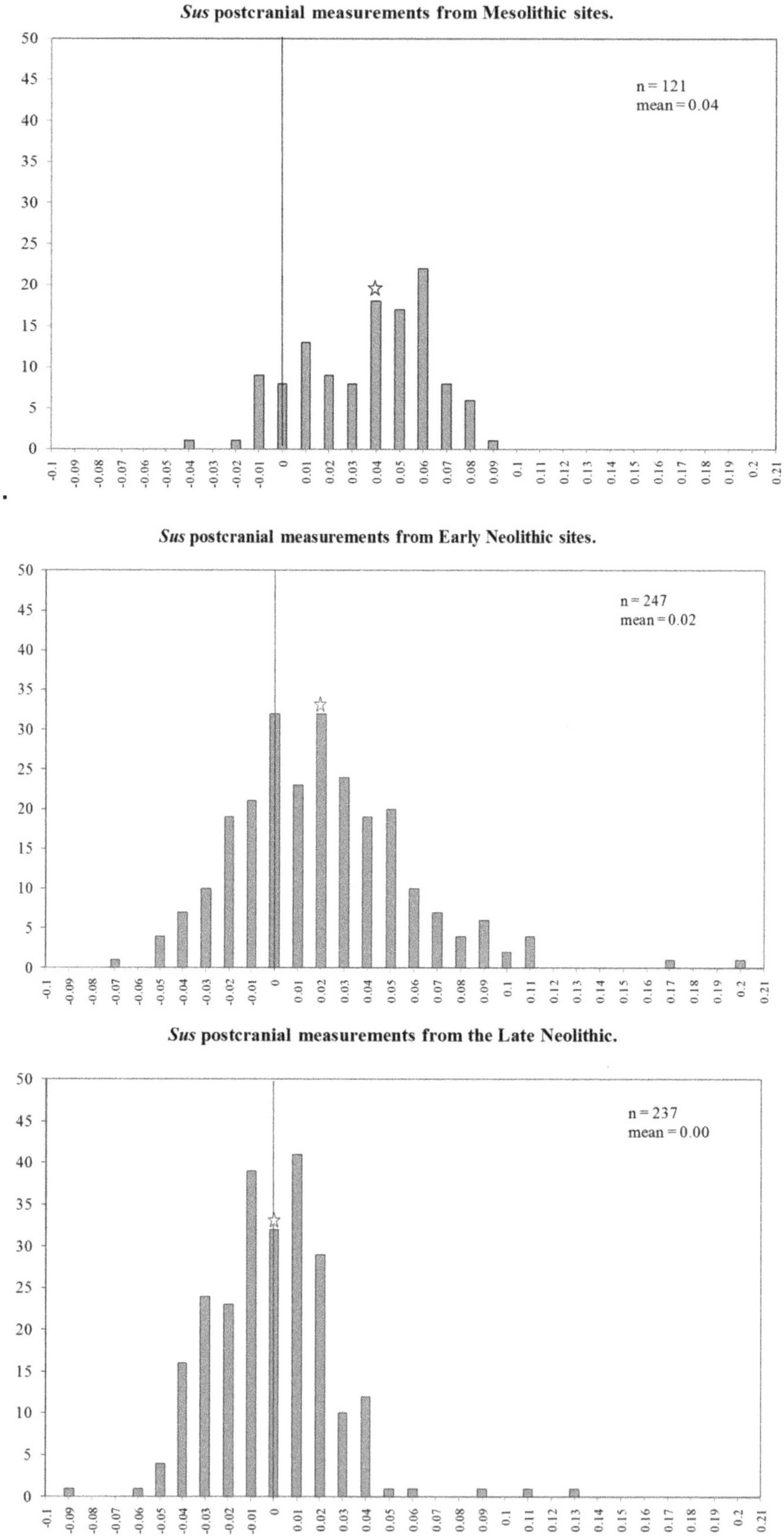

Figure 3.42 Log ratio comparison of *Sus* postcranial measurements from Mesolithic (upper diagram), Early Neolithic (middle diagram) and Late Neolithic (Durrington Walls – lower diagram). Measurements from fused postcranial elements have been included with the exception of the scapula. The mean is marked by a star symbol, and the standard measurement by the vertical black line at 0 (Viner 2011, Figure 14.7).

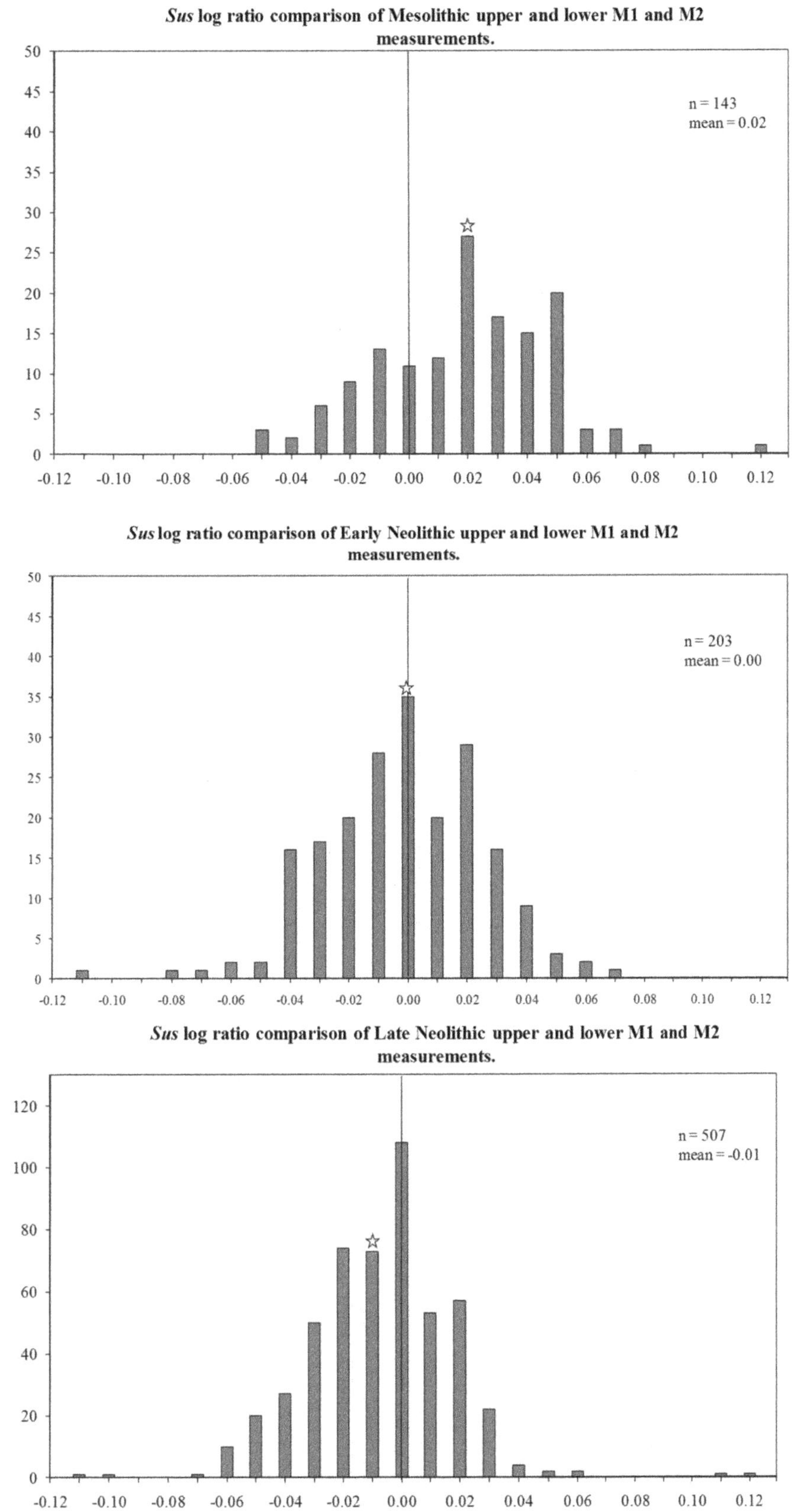

Figure 3.43 Log ratio comparison of *Sus* first and second molar measurements from Mesolithic (upper diagram), Early Neolithic (middle diagram) and Late Neolithic (Durrington Walls – lower diagram). Measurements from upper and lower loose teeth and jaws. The mean is marked by a star symbol, and the standard measurement by the vertical black line at 0 (Viner 2011, Figure 14.6)

When compared with the results of the first and second molars, the third molar data are more suggestive of a significant change in pigs with the onset of the Neolithic (Figure 3.35). In the Mesolithic third molar log ratio comparison measurements generally plot to the right of the standard, but unlike the first and second molar comparison only a small number of measurements are located to the left of the line, indicating that a limited proportion of the animals were smaller than the standard.

In the Early Neolithic the focus had shifted closer to the standard and the mean value was 0.01. Third molars from the Early Neolithic were noticeably smaller than those from the Mesolithic. However, as with other lines of evidence discussed above, there was still a great deal of overlap between the populations. Like the first and second molars there was a reduction in the size of third molars from the Mesolithic to the Early Neolithic.

The Late Neolithic third molar data conform well to that of the Early Neolithic (Figure 3.44), both focus on the standard and are generally smaller in size than the Mesolithic specimens, indicating that there was little or no change in the size of pig teeth in the Late Neolithic. These results concur with those gathered from the scatter plots and log ratio diagrams above, and suggest a high degree of uniformity between the teeth of animals from the Early and Late Neolithic periods. A small reduction in size genuinely occurred, but it was limited.

Differences in the postcranial and dental elements can provide a more detailed picture of change through time. From Figure 3.33, 3.34 and 3.35 it can be seen that there is a difference between the changes that occurred in postcranial elements and teeth (in essence a slower rate of change in bones compared to teeth from the Mesolithic to the Late Neolithic). Postcranial *Sus* remains of the Early Neolithic period were found to be intermediate between those of the preceding and those of the subsequent period (Figure 3.33). Tooth analysis indicated that teeth were reduced in size from the Mesolithic to the Early Neolithic but that the Early Neolithic and Late Neolithic were much more similar, perhaps indicating little change between the two periods. Greater change in the teeth of animals when compared to the postcranial bones is a complex issue, but does provide evidence against the idea of local domestication. The domestication of local wild boar is likely to bring about a prompt reduction in the more plastic bones than in teeth, whereas the evidence we have just discussed points in the opposite direction.

The general change from the Mesolithic to the Neolithic in combination with this evidence against local domestication *in situ* during the Early Neolithic supports the idea that pigs were introduced in a domesticated form from an external source. Some mixing with local wild boar is likely to have occurred, especially given the great overlap between the size of the wild and domestic groups, and will be further explored in the following pages.

3.3 Varying pigs in Early Neolithic Britain – Runnymede Bridge and Hambledon Hill compared.

The differences exhibited in pig remains between the Mesolithic, Early Neolithic and Late Neolithic discussed above hint at changes in the approach to pig exploitation through time. Especially for the Early Neolithic sample (in which two very substantial assemblages provided most of the data) this general approach may mask between site variation in the way that pigs were used. This section will provide a more detailed account of the Early Neolithic evidence and will focus specifically on comparing the two main Early Neolithic assemblages: Runnymede Bridge and Hambledon Hill. Data from other much smaller assemblages will be included when abundant enough.

Age

Age will again be investigated through epiphyseal fusion and the wear of mandibular teeth. Postcranial bones were separated into age categories in the same way as in Section 3.2.1. Mandibular jaws were separated into six categories following O'Connor (1988). No attempt was made to separate the jaws into the ten, more specific, categories used above, as sample sizes were too small.

When considered on the period level, epiphyseal fusion data suggested that pigs of all ages were used during the Early Neolithic period, from very young (*i.e.* those with unfused bones from the Earliest category that must have come from animals that were probably younger than 7-10 months of age), to adult (*i.e.* those with bones from the Late category that were already fused and must have come from animals that were 3-4 years or older). By comparing the epiphyseal fusion data from Runnymede Bridge and Hambledon Hill (Figure 3.45) it is clear that patterns of exploitation at the two sites were, overall, very similar, and that pooling the data from different assemblages has not severely skewed the results. At both of the sites the bones of very young animals were recovered, and the proportion of unfused bones in both the Early and Intermediate categories were comparable (the decrease in the number of unfused bones in the Early category at Runnymede Bridge is anomalous – it is usually expected that the proportion of unfused bones will increase between categories – but is not unprecedented in the zooarchaeological literature (*e.g.* Hongo *et al.* 2007, 116). The proportion of unfused bones from the Late category at Runnymede Bridge is more like the results of the Mesolithic, and Hambledon Hill is more like that of the Late Neolithic (compare Figure 3.1). However, from epiphyseal fusion there is little strong evidence to suggest disparate systems of exploitation at different sites of the Early Neolithic.

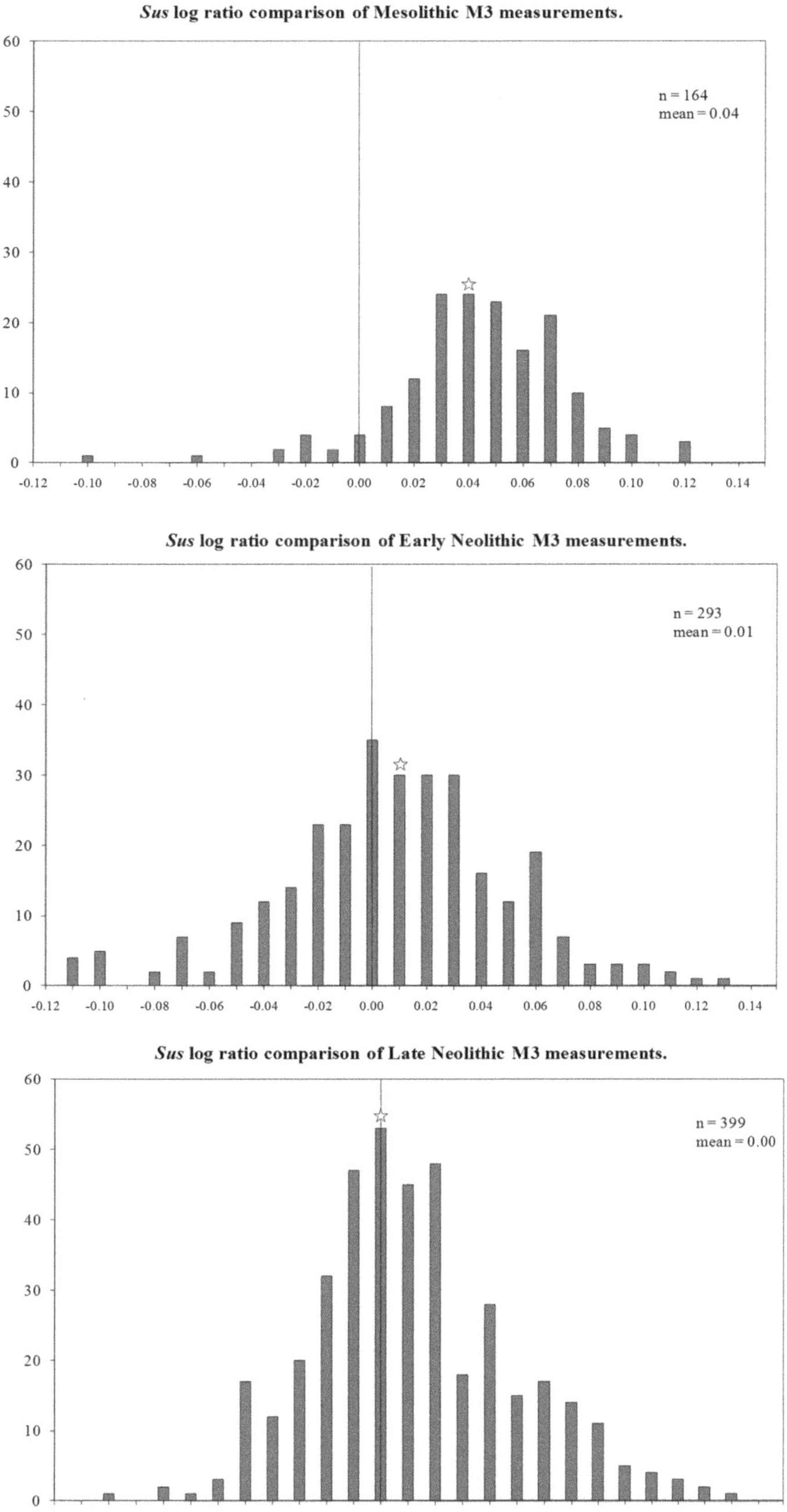

Figure 3.44 Log ratio comparison of *Sus* third molar measurements from Mesolithic (upper diagram), Early Neolithic (middle diagram) and Late Neolithic (Durrington Walls – lower diagram). Measurements from upper and lower loose teeth and jaws. The mean is marked by a star symbol, and the standard measurement by the vertical black line at 0

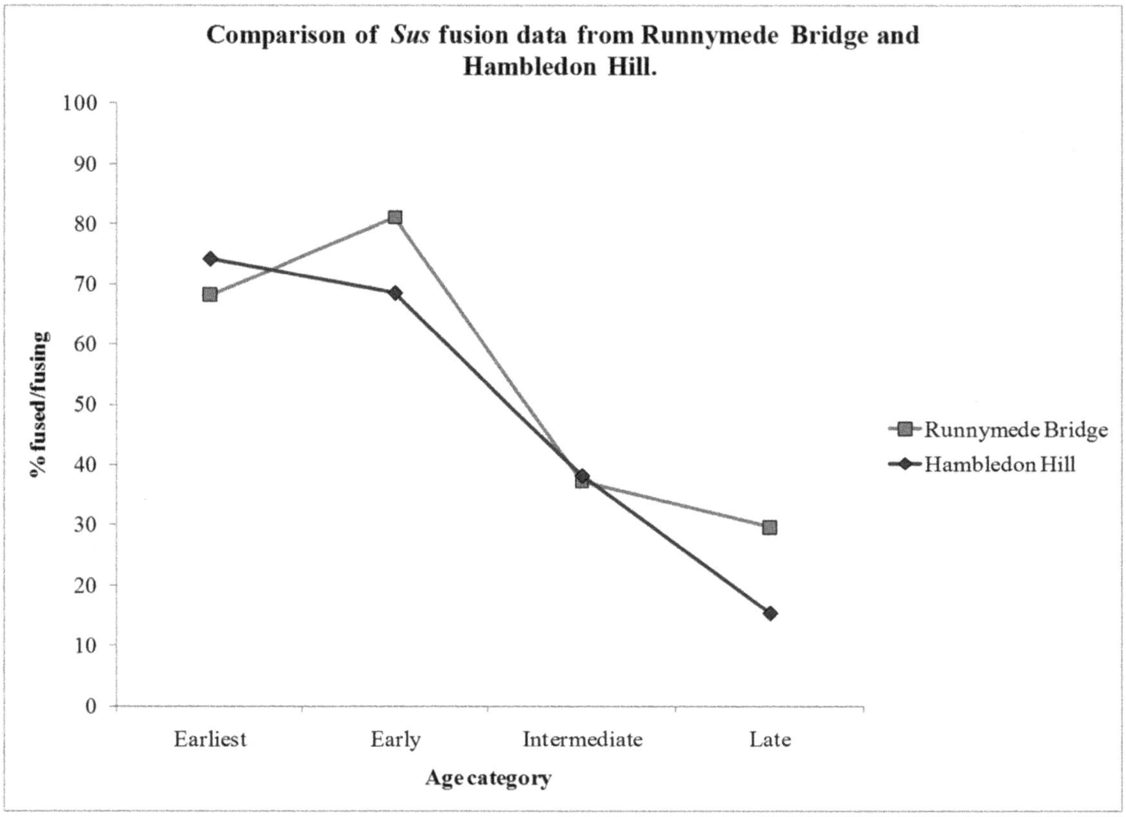

Figure 3.45 Epiphyseal fusion of *Sus* bones from Early Neolithic Runnymede Bridge and Hambledon Hill. Postcranial bones that provided fusion information were included.

Mandibular wear results for Runnymede Bridge are presented in Figure 3.46 and Hambledon Hill in Figure 3.47. At Runnymede Bridge no mandibles came from very young animals, and Immature animals were exploited, but in low frequencies (<10% of the total). Around 40% of the mandibles came from Subadult pigs, and again this coincides with the most efficient period of slaughter if meat was the main emphasis of the economy. Finally, the greatest proportion of mandibles (>50%) came from adult animals. This pattern of gradual increase in exploitation with advancing age is in contrast to the evidence of epiphyseal fusion encountered at the site, which suggested a low level of exploitation of adult animals. It is also distinct from the Hambledon Hill results, in which relatively few adult animals were encountered (see below). The results are at variance to the pattern expected from a purely meat producing economy, and therefore other scenarios need to be explored. Pigs provide no secondary products that are easily identifiable in the archaeological record. They can, however, provide a number of useful resources while still alive. These include manure for fertiliser, blood, consumption of waste (including human excrement), and use for land clearance. The need for any of these products could lead to a pattern in which animals were valued less (or equally) for meat as for other resources. A non-meat use would lead to pigs being kept beyond the point at which they reach their adult body size.

Hambledon Hill provided mandibles from different stages in the pig lifecycle (Figure 3.47). A very small proportion of the mandibles encountered came from Neonatal animals. Immature and Subadult animals were equally well represented in the assemblage, and mandibles attributed to these stages clearly dominate the results. Adults were present, but the proportion was lower than in the earlier categories. This pattern suggests that animals at the site were killed after infancy, but before reaching adulthood. In terms of meat production, this represents the most efficient approach to pig keeping. Once an animal reaches its full body size further investment of food and management no longer provides greater returns (*i.e.* increased meat yield). Small numbers of adult animals, perhaps mostly female and fewer males, would have been kept in order to replenish the population. Evidence of ageing from Hambledon Hill is characteristic of an economy in which pigs were kept primarily for meat. In comparison with epiphyseal fusion data, the evidence from tooth wear provides a different understanding of the Early Neolithic economy. This contrasts with the very general comparison in the first part of this chapter (Section 3.2.1) in which the results of epiphyseal fusion and mandibular tooth wear concurred to a large extent.

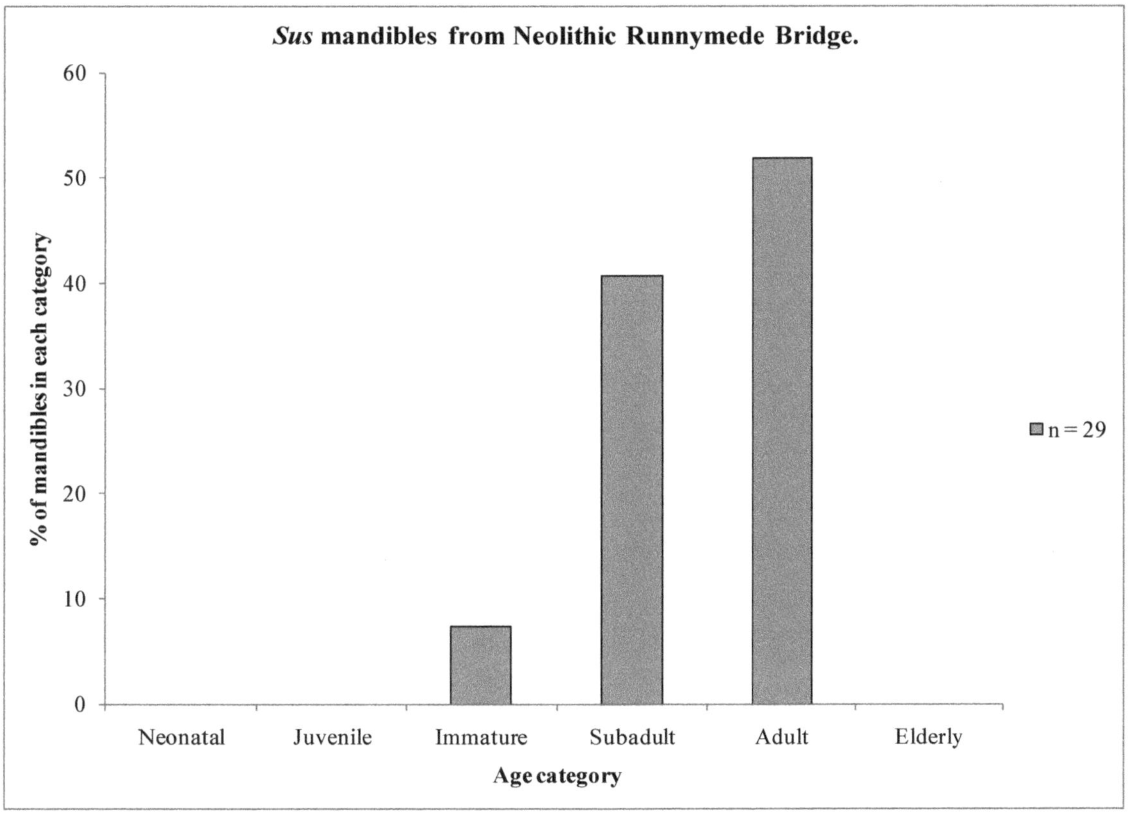

Figure 3.46 Attribution of *Sus* mandibles to age category from Runnymede Bridge. Categories are after O'Connor (1988).

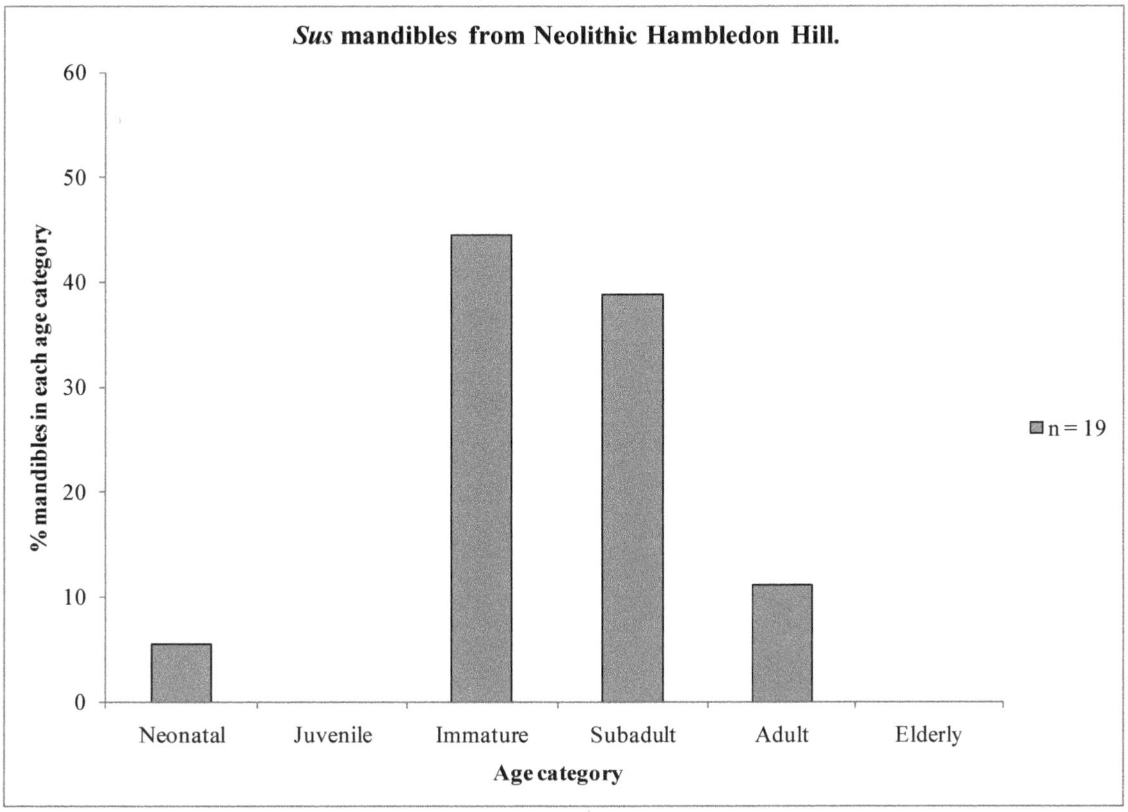

Figure 3.47 Attribution of *Sus* mandibles to age category from Hambledon Hill. Categories are after O'Connor (1988).

Sex

Analysis of canines from Early Neolithic sites considered above suggested that the disparity between female and male representation from the period could be best explained by recovery bias (see Figures 3.9 and 3.12) rather than really differential sex focus. In Figure 3.48 those loose teeth and jaws that could be sexed from Runnymede Bridge and Hambledon Hill have been compared. Female and male specimens from Runnymede Bridge are present in almost equal proportions, including sexed jaws. During the Early Neolithic at Runnymede Bridge the husbandry regime employed did not favour the exploitation of female or male animals, and there is no evidence to suggest that recovery bias played a significant part in the formation of the assemblage.

Pig canines from Hambledon Hill are also included in Figure 3.48. In contrast to the results from Runnymede Bridge, there is a difference in the loose teeth and jaws from Hambledon Hill. Male loose teeth greatly outnumber female ones, but the number of differently sexed jaws is almost equal. Sexed jaws are small in number, and arguably the picture would change if more specimens were added. The data suggest that there was no focus on males at Hambledon Hill, and that the large number of loose male canines compared to loose female canines is a result of recovery bias.

The excavation bias that appears to have affected the recovery of female teeth in the Early Neolithic is therefore likely to be a result of bias at a single site, Hambledon Hill. In addition, there is now no evidence (from canines) to suggest that male animals were used preferentially in the Early Neolithic. Unfortunately the Mesolithic sample of sexed jaws was too small to provide a clear indication of the potential recovery bias acting on the assemblages, therefore it is difficult to know whether the concentration on male pigs was real. The Early Neolithic results suggest that a change in exploitation focus might have occurred in the Late Neolithic. The lack of focus on either sex in the Early Neolithic is not continued into the Late Neolithic (during which there was a focus on the use of male pigs, Figure 3.9).

Biometrical analysis of the distal humerus in the general analysis concurred with tooth data in that faunal assemblages from all periods contained both female and male animals. Figure 3.14 indicated that the Early Neolithic dataset contained two possible size groups, tentatively interpreted as smaller females and larger males. When the dataset was split and the two main Early Neolithic sites were treated separately (as in Figure 3.49) no clear sex grouping can be seen. Distal humerus measurements also testify to the general physical similarity of pigs at different sites during the Early Neolithic as the distribution provides no distinction between Hambledon Hill and Runnymede Bridge. Figure 3.50 and Figures 3.51 provide measurements from unfused and fusing as well as fused specimens. There is no clear distinction between the groups that could be interpreted as sex distinction in either of these plots.

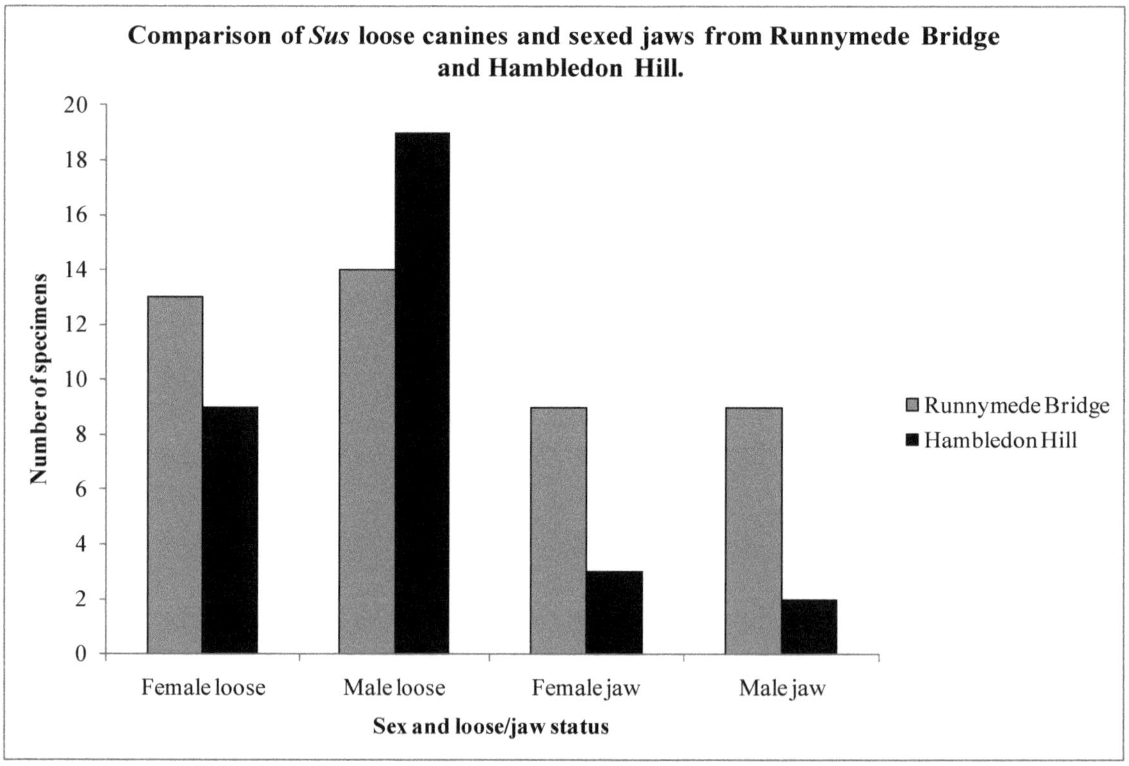

Figure 3.48 The number of *Sus* jaws and loose canines recorded from Runnymede Bridge and Hambledon Hill. Upper and lower specimens have been included

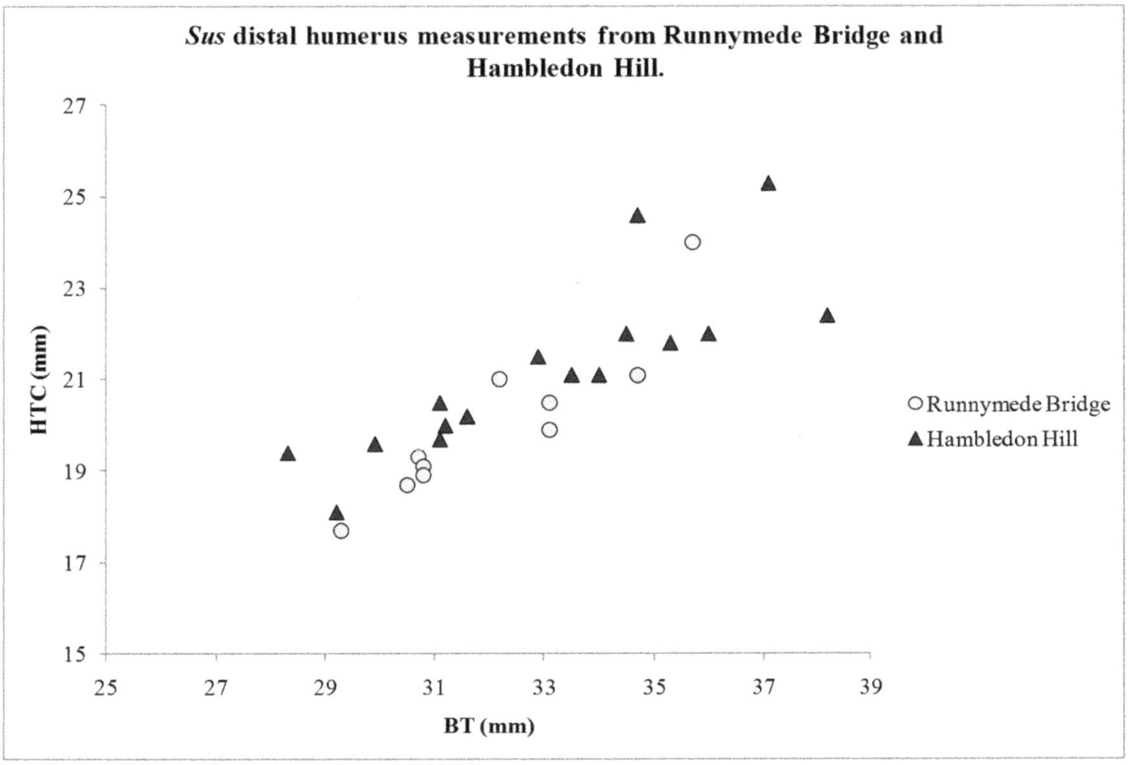

Figure 3.49 Comparison of *Sus* distal humerus measurements (HTC = height of the trochlea, BT = breadth trochlea, von den Driesch 1976) from Runnymede Bridge and Hambledon Hill. Only fused specimens are included.

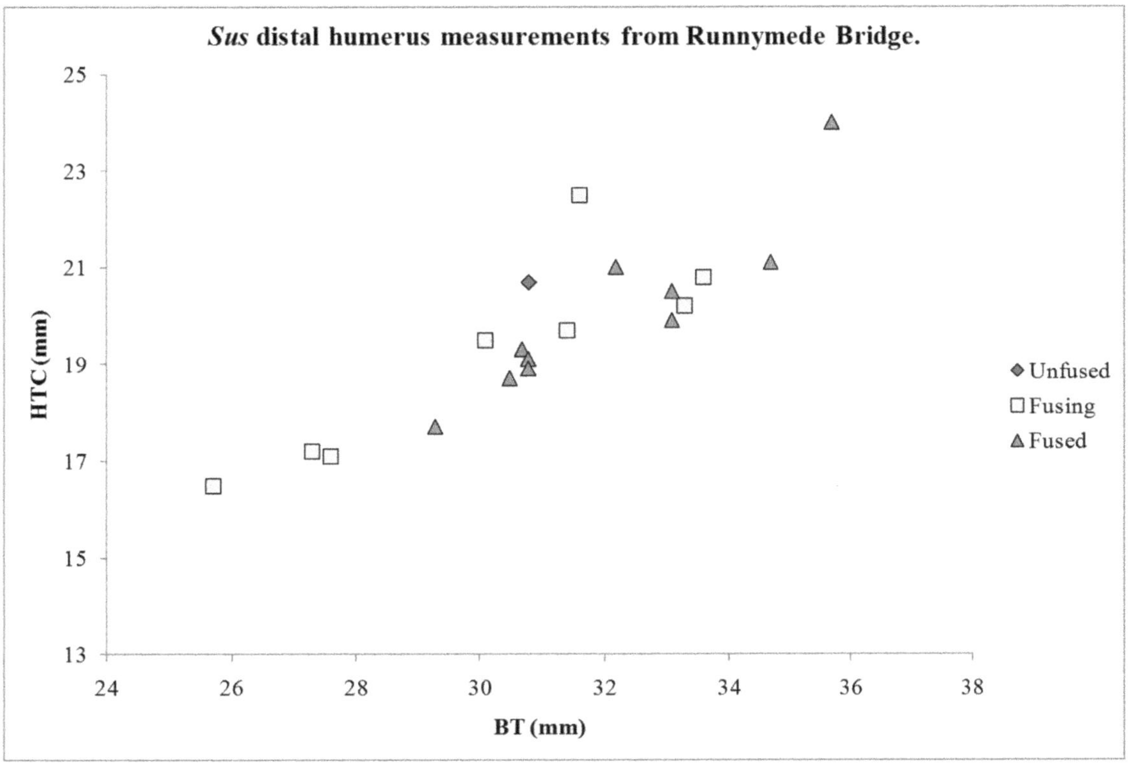

Figure 3.50 Comparison of *Sus* distal humerus measurements (HTC = height of the trochlea, BT = breadth trochlea, von den Driesch 1976) from Runnymede Bridge. Unfused, fusing and fused specimens have been included.

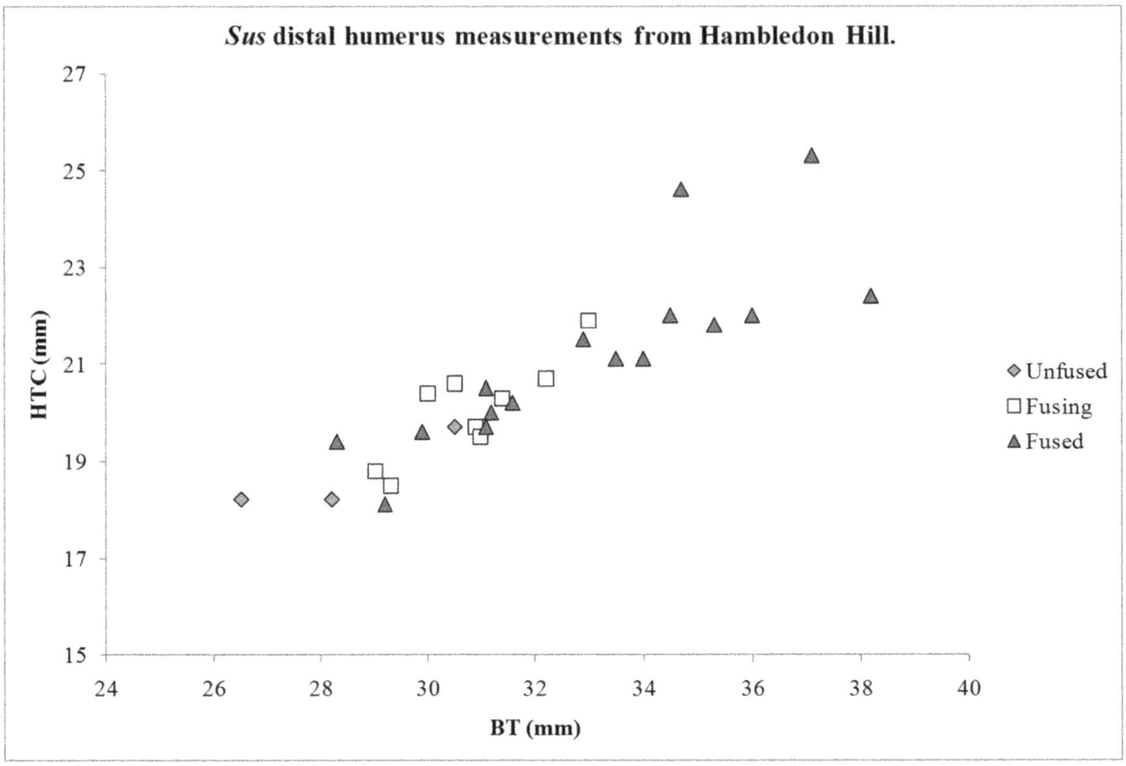

Figure 3.51 Comparison of *Sus* distal humerus measurements (HTC = height of the trochlea, BT = breadth trochlea, von den Driesch 1976) from Hambledon Hill. Unfused, fusing and fused specimens have been included.

Biometrical Analysis

The biometrical analysis conducted above found changes from the Mesolithic to the Early Neolithic, and from the Early Neolithic to the Late Neolithic in terms of the physical size of some parts of the pig skeleton. This section aims to determine whether there were also differences within the Early Neolithic by comparing results from the two main Early Neolithic sites.

Any large scale difference in astragalus measurements is likely to be the result of different populations and/or conditions. When the astragalus measurements from Runnymede Bridge and Hambledon Hill were plotted (Figure 3.52) no clear difference between astragali from the two sites could be seen (a fact supported by statistical analysis, see Table 3.2).

The third molar, in common with other teeth and postcranial elements, exhibited change over time in British prehistory. From the Mesolithic to the Early Neolithic there was a clear reduction in third molar length, width of the anterior cusp, central cusp and posterior cusp, with the latter providing the most significant reduction. Figures 3.53, 3.54 and 3.55 compare the third molar measurements from Runnymede Bridge and Hambledon Hill, and provide evidence for difference between the two sites. The specimens overlap to a large extent, but the Runnymede Bridge specimens tend to cluster towards the smallest end of the Hambledon Hill distribution. The length, and to a lesser extent the anterior width, seem to be different between the two sites, whereas the central and posterior cusps are more comparable. Small sample size might be an issue, but if the difference is regarded to be genuine, the implication is that animals at Runnymede Bridge had smaller teeth than those at Hambledon Hill. Log ratio investigation of the assemblages will be employed to explore this trend in greater detail.

Table 3.2 Results of t-test comparing *Sus* teeth and postcranial measurements. Those marked with a star (*) are considered statistically significant. n = the number of measurements included.

Meas.	Group 1 /n		Group 2 / n		t-test
Ast. GLl	Runnymede Bridge	20	Hambledon Hill	18	0.680
Ast. GLm	Runnymede Bridge	20	Hambledon Hill	18	0.499
M3L	Runnymede Bridge	15	Hambledon Hill	18	0.010*
M3WA	Runnymede Bridge	15	Hambledon Hill	18	0.053*
M3WC	Runnymede Bridge	15	Hambledon Hill	18	0.132
M3WP	Runnymede Bridge	16	Hambledon Hill	12	0.074

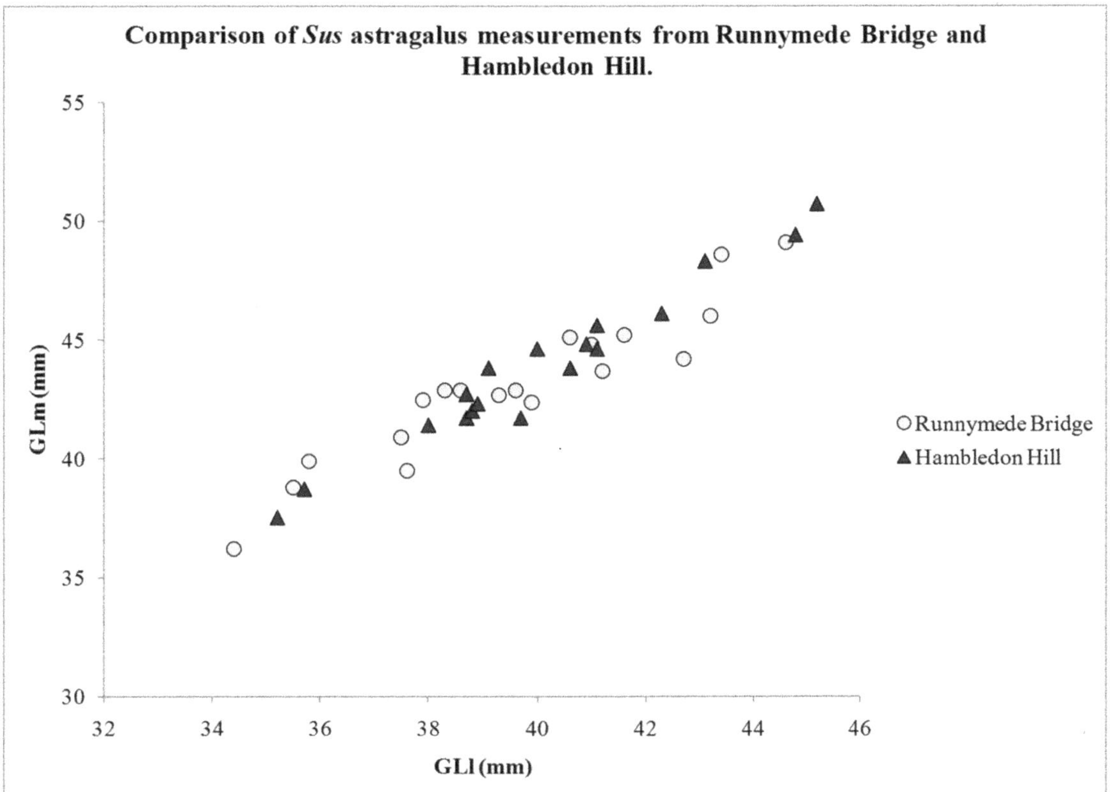

Figure 3.52 Comparison of *Sus* astragalus measurements (GLl = greatest lateral length, GLm =greatest medial length, von den Driesch 1976) from Runnymede Bridge and Hambledon Hill. Only measurements from specimens recorded as 'normal' have been included.

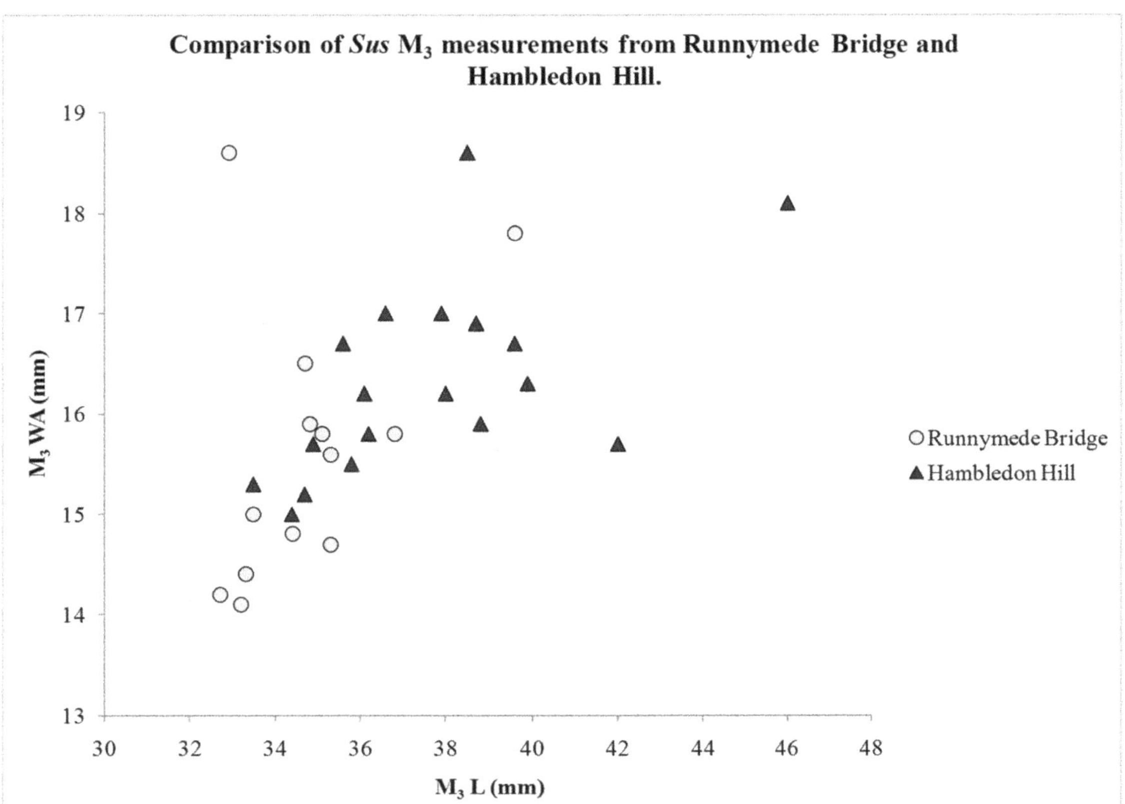

Figure 3.53 Comparison of *Sus* third molar measurements (M_3L = length, M_3WA = width of the anterior cusp, Payne and Bull 1988) from Runnymede Bridge and Hambledon Hill. Heavily worn teeth have been excluded.

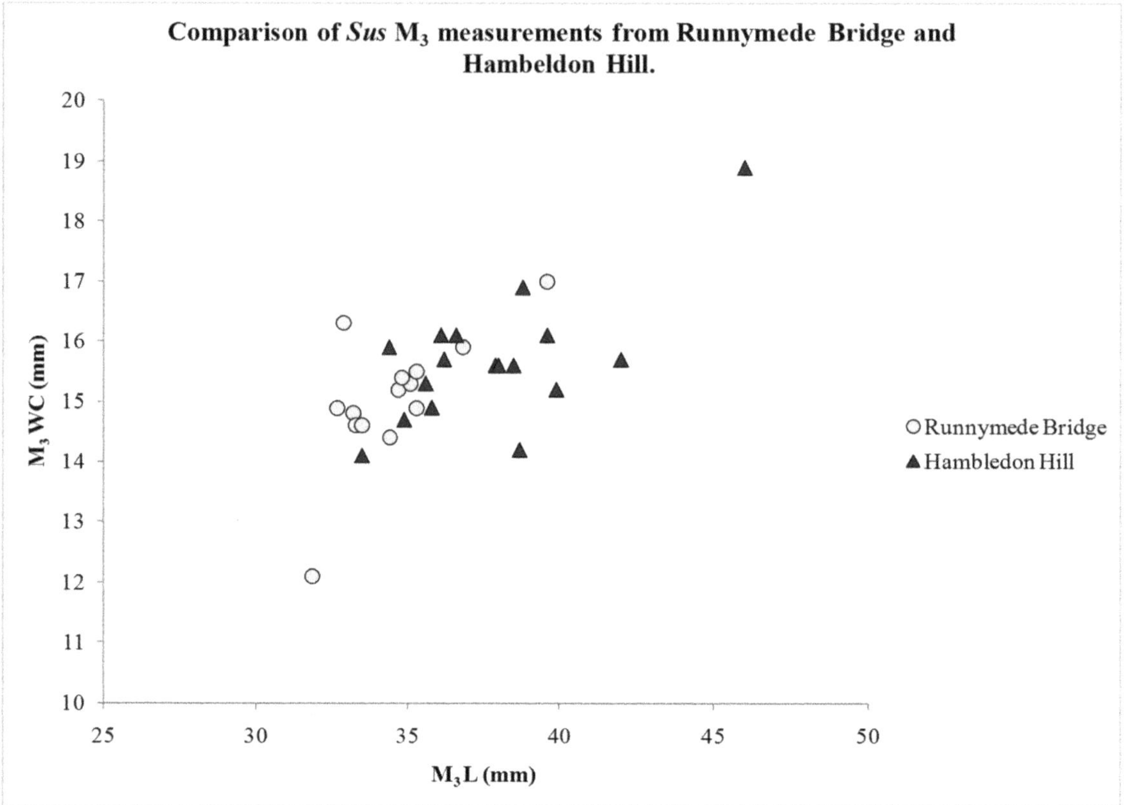

Figure 3.54 Comparison of *Sus* third molar measurements (M_3L = length, M_3WC = width of the central cusp, Payne and Bull 1988) from Runnymede Bridge and Hambledon Hill. Heavily worn teeth have been excluded.

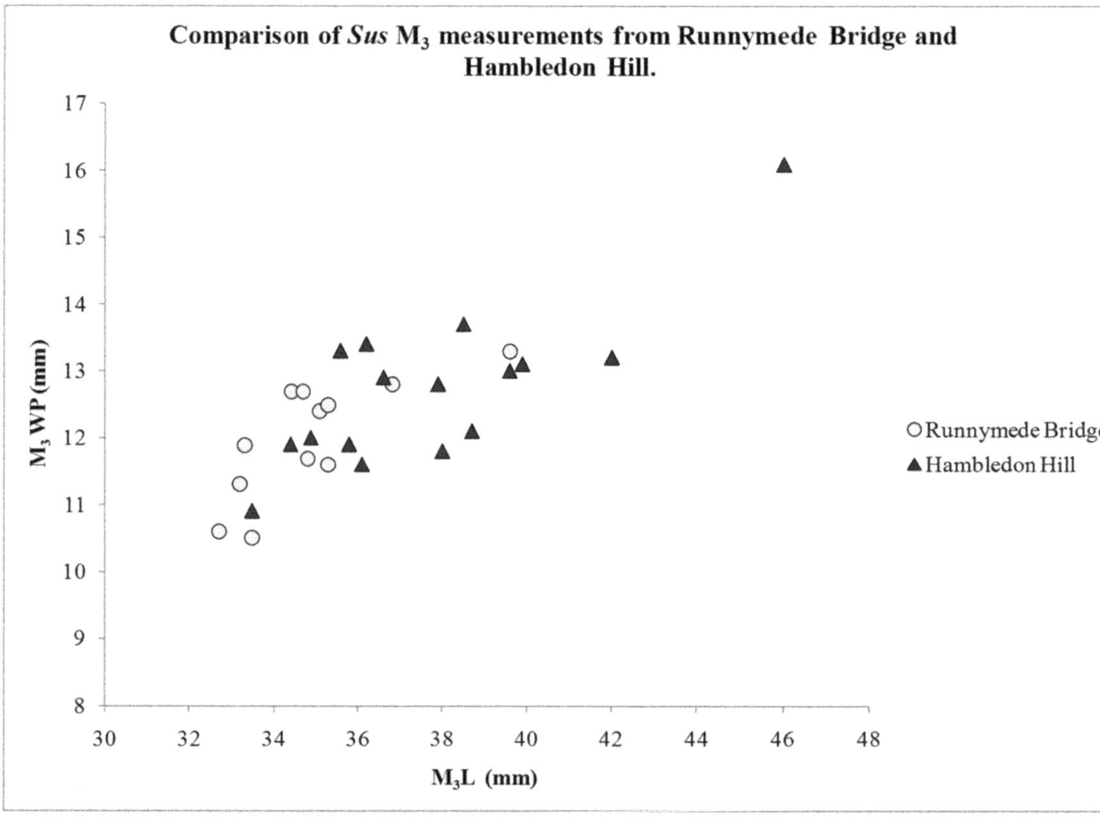

Figure 3.55 Comparison of *Sus* third molar measurements (M_3L = length, M_3WP = width of the posterior cusp, Payne and Bull 1988) from Runnymede Bridge and Hambledon Hill. Heavily worn teeth have been excluded.

Astragalus measurements found no difference between specimens from Hambledon Hill and Runnymede Bridge, while the third molar measurements hinted at differences in the size of pigs from the two sites. In this section measurements from the Early Neolithic will be compared using the log ratio technique in an attempt to determine whether there were differences between pigs at Runnymede Bridge and Hambledon Hill.

In Figure 3.56 pig postcranial measurements from different sites have been compared using the log ratio technique. Postcranial remains from the two sites are extremely similar. Measurements from Runnymede Bridge and Hambledon Hill are distributed in the same way and have an identical mean. Both plots continue the trend identified above in which the postcranial remains are smaller than those from the Mesolithic and larger than those from the Late Neolithic. In addition, both of the sites provided a small number of larger outliers. These outliers are of greater size than the postcranial bones from the Mesolithic (see Figure 3.32). It appears then that occasionally, very large animals, presumably wild boar, were exploited at both of the Early Neolithic sites investigated.

In Figure 3.57, which shows the log ratio comparison of tooth measurements from Runnymede Bridge and Hambledon Hill, the Runnymede Bridge specimens are of clearly smaller size than those from Hambledon Hill. Hambledon Hill measurements indicate a population which was generally larger than the standard, whereas the mean of the Runnymede Bridge measurements was lower than the standard indicating generally smaller tooth measurements.

The log ratio comparisons presented in Figures 3.56 and 3.57, reinforce the conclusions drawn from astragalus and third molar measurements. That is, there was no difference between the size of pigs at Runnymede Bridge and Hambledon Hill in terms of their postcranial remains, but that the teeth of pigs from Runnymede Bridge tended to be smaller than those from Hambledon Hill. Since smaller tooth size relative to bones has been suggested to distinguish Mesolithic wild boar from (presumably) mainly domestic Early Neolithic pigs, the difference between the two sites is of some interest. The Hambledon Hill animals have characteristics that are closer (but not identical) to the Mesolithic and this may indicate a greater wild boar component in the assemblage. Considering, however, that distinct wild boar outliers have been identified, a more likely scenario is a difference in husbandry practices between the two sites. At Hambledon Hill domestic pigs may have been allowed to mix with wild boar more freely than at Runnymede. We also cannot exclude the possibility that Hambledon Hill includes a component of locally domesticated wild boar, which could be smaller or absent at Runnymede.

3.4 Conclusion

The aim of this chapter was to identify changes through time and between sites in the husbandry practices and physical characteristics of *Sus*. As presented here, the results have provided important information about the complex transition from the Mesolithic to the Early Neolithic. The type and extent of variation that occurred as a result of domestication was not uniform, and skeletal elements were affected at different rates. Direct comparisons of epiphyseal fusion data and mandibular tooth wear can be problematic, but it has been possible to draw a number of conclusions from the data. During the Mesolithic exploitation of wild boar lacked focus, and use was made of animals from the earliest stages of life to adulthood. Elderly animals were conspicuous in their absence, but this perhaps reflects a scarcity of extremely old animals in the wild populations. With the onset of the Neolithic, exploitation changed resulting in variation in the animals themselves. Animals of all ages (with the exception of very Elderly pigs) continued to be used but according to epiphyseal fusion there was a greater tendency towards the use of younger animals. Detailed investigation of the mandibles indicated that pigs were killed very soon after adulthood.

The Late Neolithic saw further changes in the exploitation of pigs, with a greater tendency towards the use of younger animals at Durrington Walls. The Late Neolithic was also the only period during which there is evidence for the presence of Elderly animals.

Biometrically, there is no evidence, from the admittedly limited dataset, to suggest that wild boar varied from the Early to the Late Mesolithic. Goldcliff (Late Mesolithic) animals all fell within the range of Early Mesolithic animals from the other sites studied. That said, comparison with other Mesolithic *Sus* remains in the following chapters will allow a more detailed understanding of the relative size of pigs from the period. Wild boar appear to have been used into the Early Neolithic, but some of these animals were apparently larger in size than the Mesolithic wild boar recorded. The results could suggest an increase in the size of British wild boar through time.

The physical similarity of animals from the Mesolithic and Early Neolithic periods is attested by the results of biometrical analysis, but there are some clear points of dissimilarity that indicate change through time. There were also differences in the way that bones varied through time compared to the way that teeth varied. Length and width measurements from the second and third molars (and to a limited extent the first molar) showed clearly a physical reduction in pigs from the Mesolithic of the kind that would be expected if those from the Neolithic were introduced domesticates. Postcranial bones also diminished in size, but to lesser degree. The evidence points to an introduction of domesticated types from an external source rather than local domestication of pigs in Britain.

Early Neolithic sites were investigated in isolation to determine whether the perceived changes were common to all sites of the Early Neolithic, or whether variation between sites was important. Particularly informative is the difference between the reduction in tooth size and postcranial size at early Neolithic sites (postcranial remains at both Runnymede Bridge and Hambledon Hill are comparable, but Runnymede Bridge tooth measurements are smaller than those at Hambledon Hill). The results suggest a higher degree of mixing between native wild boar and potentially introduced domestic pigs in the Hambledon Hill assemblage.

All of the issues raised here need to be understood in the context of other British and European evidence. Following the presentation of cattle results in Chapter 4, Chapter 5 will compare the results from this study with evidence from other published sources. A detailed discussion of all of the evidence will be undertaken in Chapter 7.

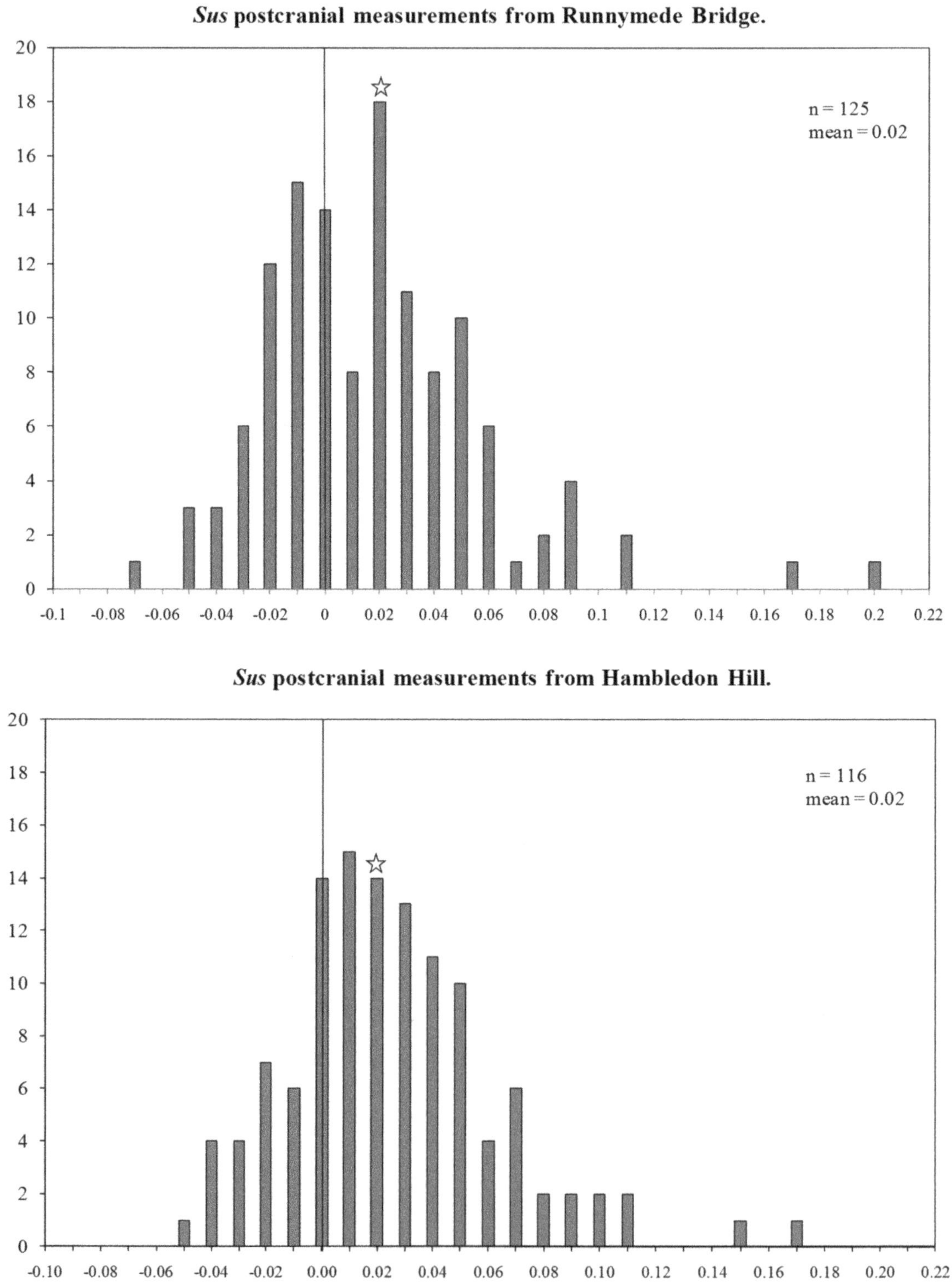

Figure 3.56 Log ratio comparison of *Sus* postcranial measurements from Runnymede Bridge (upper diagram) and Hambledon Hill (lower diagram). Measurements from fully fused specimens only. The mean is marked by a star symbol, and the standard measurement by the vertical black line at 0 (Viner 2011, Figure 14.8).

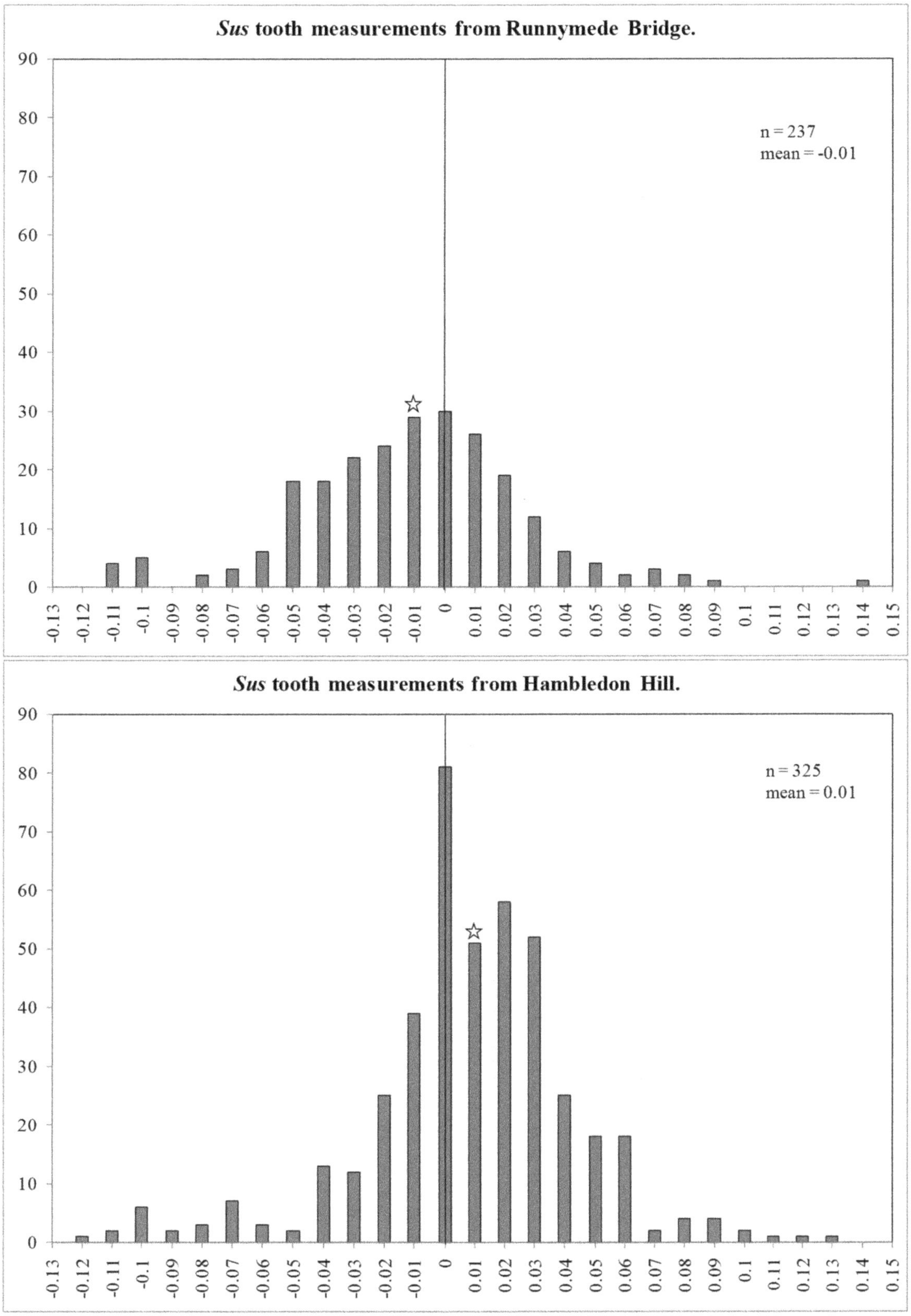

Figure 3.57 Log ratio comparison of *Sus* tooth measurements from Runnymede Bridge (upper diagram) and Hambledon Hill (lower diagram). Measurements from upper and lower teeth have been included (excluding heavily worn specimens). The mean is marked by a star symbol, and the standard measurement by the vertical black line at 0 (Viner 2011, Figure 14.9).

Chapter 4
Age, sex and biometrical analysis of cattle remains

4.1 Introduction

This chapter will begin by comparing cattle remains from the Mesolithic, Early Neolithic and Late Neolithic periods. This first general section aims to highlight changes in *Bos* exploitation and physical characteristics over time, and will also attempt to provide a baseline of Mesolithic aurochs measurements with which domestic types can be compared. In the second part of the chapter results from different Early Neolithic sites will be compared in order to identify any variations in the approach to cattle exploitation that may have existed at contemporary sites. Information about the age of animals during each period of the research will be presented first, and followed by the results of the biometrical aspect of the investigation. Sex is an important aspect of husbandry practices and will be investigated here through biometrical information collected from both postcranial and dental elements. Ageing and biometrical analysis (including the use of biometry to elucidate sex distribution) are essential to the understanding of the nature of cattle exploitation and will form the core of both the general and specific results that follow.

4.2 Variation in cattle exploitation through time

Age

Investigation of the age at death of cattle included the use of two methods: epiphyseal fusion and mandibular tooth eruption and wear. These approaches aimed to highlight any differences in the way cattle were exploited both through time, and between sites. The results of these two methods are not always in accordance and can be difficult to integrate, but an attempt will be made to achieve a holistic impression of the age structure through time.
Cattle bones were grouped into one of four age categories (Earliest, Early, Intermediate and Late) depending on their position in the fusion sequence of the cattle skeleton. The epiphyses in each category can be seen in Table 2.3. Once grouped, the proportion of fused/fusing bones was plotted as in Figure 4.1.

The fusion of cattle postcranial bones provided evidence of a significant difference between the Mesolithic and the subsequent Early Neolithic and Late Neolithic periods (Figure 4.1). In the Mesolithic, unfused bones in the Earliest and Early fusing categories were almost entirely absent. The Intermediate category had a slightly higher number of unfused elements, but still, more than 90% were fused or fusing. The frequency of unfused bones increased somewhat in the Late category, which included more than 80% fusing or fused epiphyses. This means that most of the bones in all age categories were fused/fusing, and indicates that the bones of very young and young aurochsen were absent from most of the assemblages studied. This trend is strong enough to suggest that the system employed by Mesolithic humans in their exploitation of aurochs focussed on the use of adult animals, and that there was an avoidance of young animals. Such avoidance may have been deliberate, or could have been dictated by the circumstances of hunting. The possibility of an active avoidance of young aurochs in the Mesolithic is intriguing and will be explored further in the following sections. More analysis is also needed to eliminate taphonomic biases as the main contributing factor to the age pattern in the Mesolithic, and a comparison with the tooth data (below) will help to tackle the issue.

Results from Early and Late Neolithic sites provided a different impression from those of the Mesolithic. In the Early Neolithic a small number of bones from the Earliest and Early categories were unfused (10-15%), an indication that some very young animals were represented at the sites studied. The proportion of unfused bones in Early Neolithic assemblages changed between the Early and Intermediate categories, with evidence to suggest that 15-20% more animals were killed before the bones in the Intermediate category fused. About another 20% of animals appear to have been killed between the Early and Intermediate stages (*i.e.* reaching 2-3 years of age). Finally, in the Early Neolithic plot, between 60% and 65% of bones from the Late category were fused. Therefore, more than a third of the animals died by 2-3 years of age (compared with <5% of animals in the Mesolithic).

A similar pattern persisted into the Late Neolithic, with some variation. Fewer bones from very young animals were present than in the Early Neolithic (evidenced by the less frequent occurrence of unfused Earliest/Early category bones). More than in the Early Neolithic (*c.*40%) animals were slaughtered before the bones in the Intermediate category became fused (*i.e.* before reaching 2-3 years of age). A slightly smaller proportion of the Late Neolithic population survived into full adulthood compared with the Early Neolithic.

A distinct pattern that emerges from this analysis is the virtual absence of very young animals in the Mesolithic. Secondly, there is no evidence to suggest a significantly different approach to cattle exploitation between the Early and Late parts of the Neolithic. In both periods most cattle reached adulthood, though not to the same extent as in the Mesolithic. In the Neolithic, some slaughter occurred in the early stages of life, and with a greater frequency before 2-3 years of age, particularly in the Late Neolithic. It will be necessary to compare the epiphyseal fusion results from different Early Neolithic sites to establish whether this pattern applies to the entire period.

Further ageing evidence was provided by the state of mandibular tooth eruption and wear. Samples of appropriate mandibles were small in all cases (27 specimens in the Early Neolithic, and 17 specimens in the Late Neolithic), and must be considered in combination with other lines of evidence. Chapter 2 provided a discussion of the method used to attribute cattle mandibles to a category, and details of the specimens included can be seen in Appendix VI. The Mesolithic sample was too small to be presented graphically, but the few mandibles that could be placed into an age category (three in total) came from either Adult or Elderly aurochsen, which supports the results of the epiphyseal fusion.

Mandibles of all age categories were present in the Early Neolithic and Late Neolithic samples. In the Early Neolithic period mandibles from the Juvenile, Immature and Subadult categories were relatively few in number and each category provided less than 10% of the total specimens studied (Figure 4.2). The greatest contribution was made by Adult animals (55%), with a smaller proportion being designated as Elderly (less than 30%) in the Early Neolithic. Again these results support those from epiphyseal fusion data discussed above. Mandibular tooth wear suggests that most cattle were slaughtered as adults, but that a small number of animals died while young.

In the Late Neolithic period, too, use was made of animals of all ages, with a clear emphasis on Adults (Figure 4.3). Juveniles made up a small proportion of the sample (less than 10%) a result that is similar to that from the Early Neolithic. The absence of very young animals in the epiphyseal fusion evidence is therefore likely to be an artefact of taphonomy, though the small sample size should caution against interpreting them as important in the Late Neolithic. A greater proportion of Immature mandibles were found, and this probably reflects the higher number of specimens with unfused bones at the Intermediate stage discussed above. The high number of Adult (and the smaller sample of Elderly) cattle compared to the younger age categories might suggest that in addition to meat production, cattle were important for a product that they continued to provide for some time after reaching their full adult size.

The age categories established by O'Connor, and used here, allow the more detailed investigation of cattle age through mandibular wear. In Figure 4.4 those *Bos* mandibles from the Early Neolithic and Late Neolithic sites that were attributed to the Adult category have been separated further into Adult 1, Adult 2 and Adult 3 categories (see Table 2.4 which follows O'Connor 2003). The evidence from *Bos* mandibles in Figure 4.4 suggests that there was no period of intense slaughter at a specific point during the Adult period, but instead cattle were killed throughout adulthood, and into the Elderly period of their lives on both Early and Late Neolithic sites.

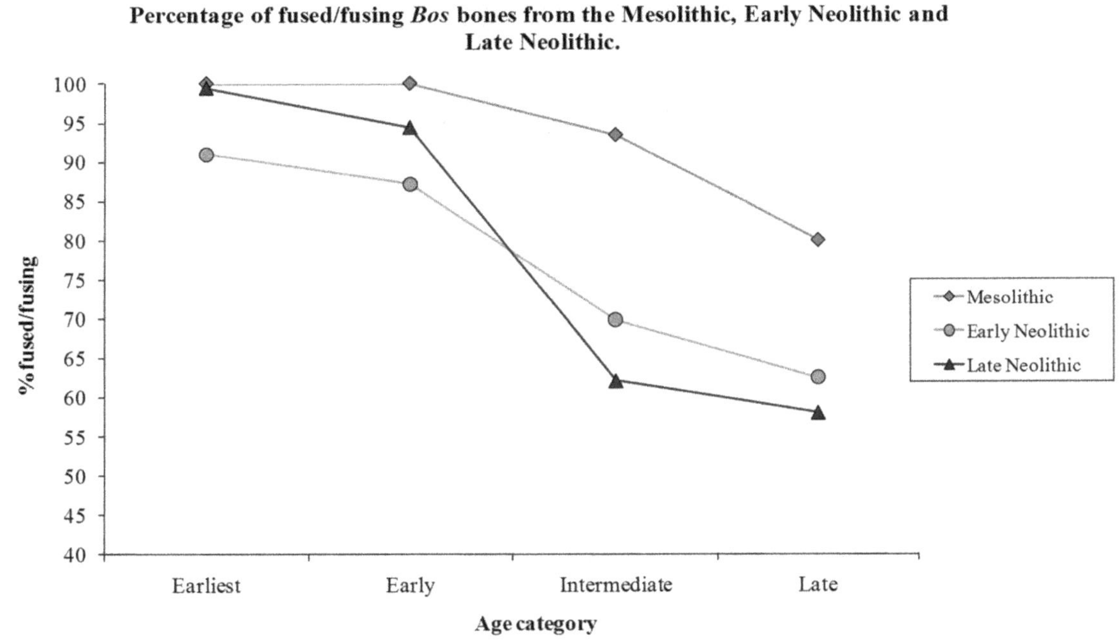

Figure 4.1 Fusion of *Bos* bones from the Mesolithic, Early Neolithic and Late Neolithic. Postcranial bones that provided fusion information have been included (Viner 2011, Figure 14.10).

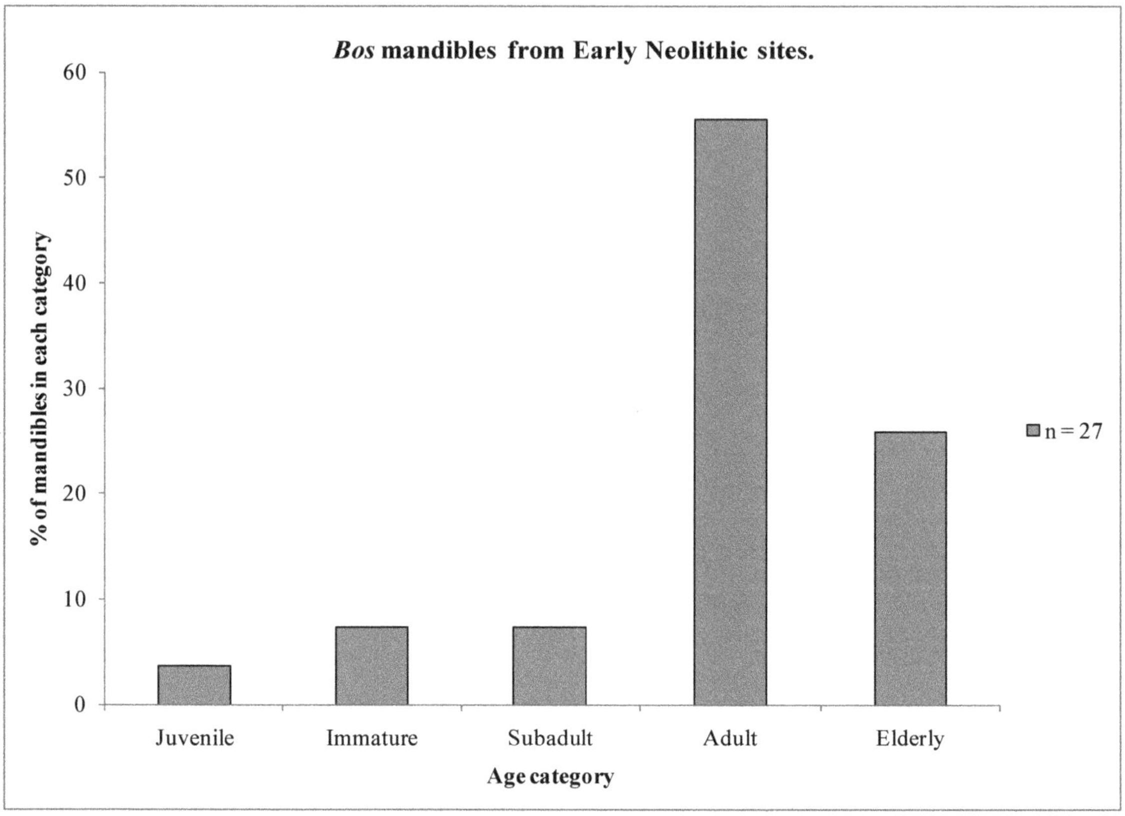

Figure 4.2 Attribution of *Bos* mandibles to age category from Early Neolithic sites. Categories are after O'Connor (1988).

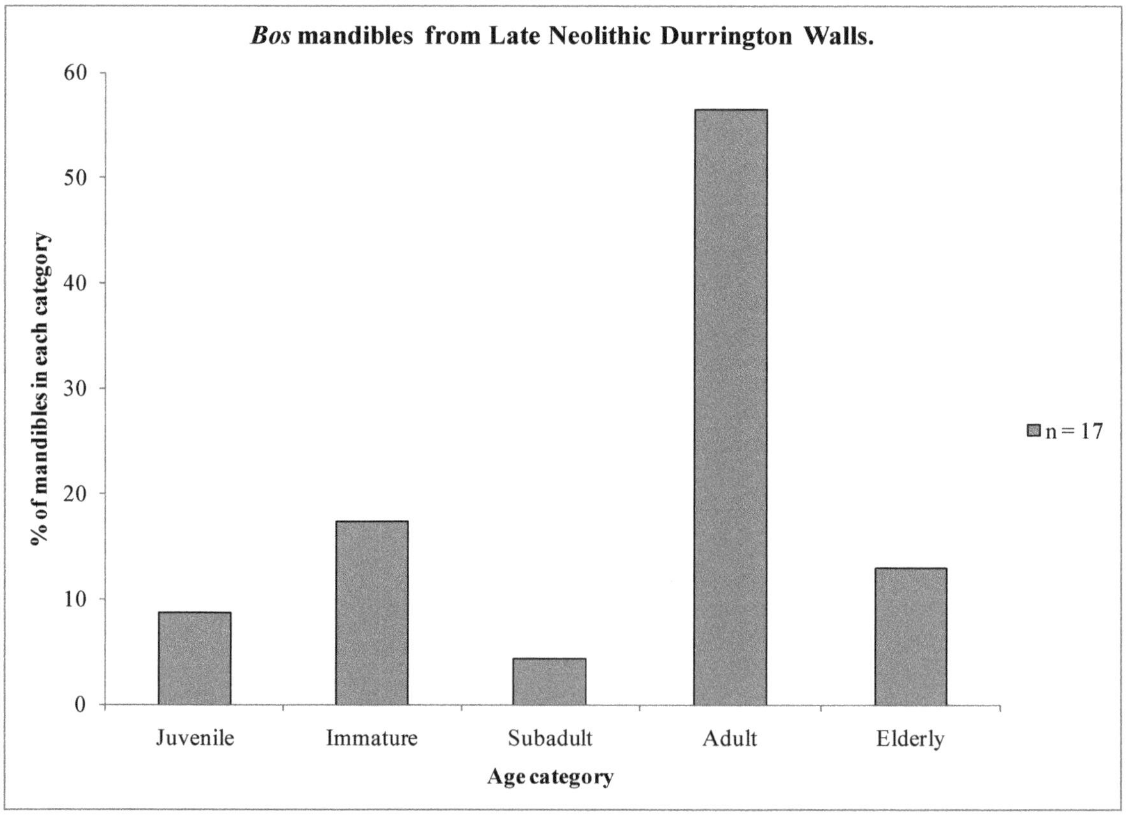

Figure 4.3 Attribution of *Bos* mandibles to age category from Late Neolithic Durrington Walls. Categories are after O'Connor (1988).

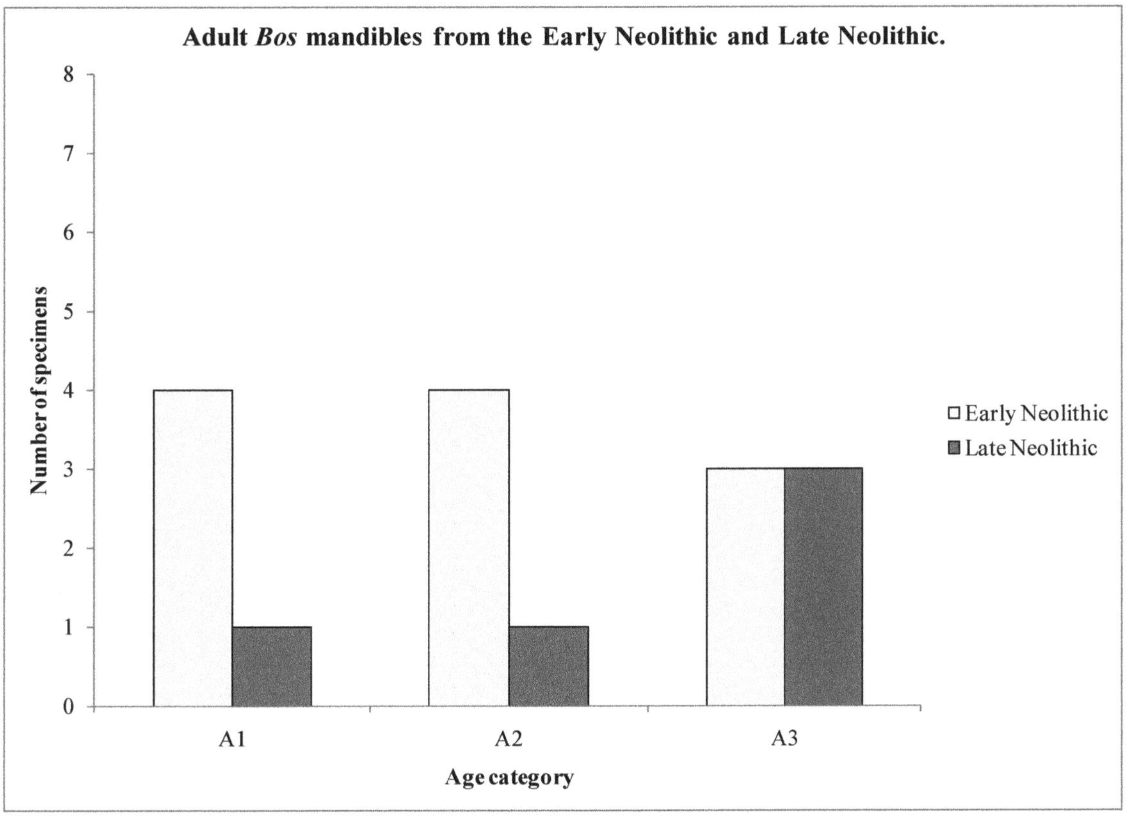

Figure 4.4 Adult *Bos* mandibles from the Early Neolithic and Late Neolithic. Age categories follow O'Connor 2003.

The main differences in mortality patterns between the Mesolithic and the Neolithic seen in mandibular tooth wear concur with those of epiphyseal fusion. The absence of very young animals in the Mesolithic is evident in the results of epiphyseal fusion, and supported by mandibular tooth wear. The death of around 20% of cattle in the Early Neolithic, and 30% of cattle in the Late Neolithic before age 2-3 years (according to epiphyseal fusion), is consistent with mandibular tooth wear data in which the majority of mandibles fell into the Adult category.

Sex

Information regarding sex distribution is an essential aspect of understanding exploitation of wild animal resources and husbandry practices. With the move from wild to domestic exploitation changes in the proportion of female and male animals, and perhaps the appearance of castrates, are to be expected. Changes in sex distribution can be a reflection of the changing motivation for using an animal species. *Bos* sex has been investigated here by using biometry and under the assumption that the significant sexual dimorphism of aurochs and domestic cattle may be visible zooarchaeologically (Degerbøl and Fredskild 1970; van Vuure 2005).

Biometrical analysis has provided detailed information regarding patterns of female and male *Bos* exploitation during the Mesolithic, Early Neolithic and Late Neolithic. In Figure 4.5 measurements of the astragalus (the greatest lateral length and the distal breadth) of animals from British Mesolithic and Neolithic sites have been plotted. In addition, published data provided by Degerbøl and Fredskild (1970) from aurochs remains of known sex has been included in the plot. Specimens from the British Mesolithic form two distinct groups of differently sized animals. When compared to the animals from Denmark, the two distinct British groups correlate well with the female and male groupings of the Danish material. Therefore, the sample of British aurochsen presented here is best interpreted as consisting of both female and male specimens. This distinction was identified by Rowley-Conwy and Legge (1988) in their comparison of Early Mesolithic Star Carr aurochs remains and the data presented by Degerbøl and Fredskild (1970), and is reinforced by the other Mesolithic specimens added here. There is almost no overlap between the distributions of the two sexes. Male specimens from both British and Danish sites form a tight cluster in the top right of the plot. The female specimens are spread over a larger area of the plot and compared to the males have a wider distribution; the majority of the specimens form a cluster, but there are also a number of much smaller outliers that challenge the integrity of the grouping. The possibility that these specimens are from intrusive domestic types cannot be ruled out, but the presence of much smaller female aurochsen in the Danish material does suggest that they could also be genuine wild animals. It is, however, unclear whether the measurements published by Degerbøl and Fredskild were from adult specimens, and these smaller Danish specimens could come from juvenile animals.

When compared to the Mesolithic distribution the Neolithic astragalus measurements are obviously distinct (Figure 4.5). Apart from the small Mesolithic outliers mentioned previously there is little overlap between the main body of Mesolithic specimens and the main body of Neolithic specimens. The implication of these results is that in the Mesolithic exploitation of aurochsen did not focus on either sex, both female and male animals contributed to the diet of Mesolithic people and this is reflected in the presence of both on archaeological sites of the period. Therefore the smaller animals found on Neolithic sites are unlikely to be either female or male aurochs. The presence of larger animals on Mesolithic sites and smaller animals on Neolithic sites cannot be explained as a shift in sex emphasis with the onset of the Neolithic. A real change in practice between the two periods is the most likely explanation of the different size of animals between the periods in question. Thus a Mesolithic system that used male aurochs compared with a Neolithic one that exploited only the smaller females can be ruled out. It will be argued here that this change in practice is from Mesolithic wild to Neolithic domestic exploitation, and supports the conclusions of many previous researchers.

Sex differences in the astragalus that were apparent in the Mesolithic plot are not continued into the Neolithic. Figures 4.6 and 4.7 show the astragalus measurements separated into Mesolithic, Early Neolithic and Late Neolithic specimens. The Early Neolithic specimens all plot within a single group showing no clear separation. Two groups can be distinguished in the Late Neolithic plots, but the distinction is not strong and the sample is perhaps too small to be meaningful. There could be a number of explanations for the lack of sex groupings on Neolithic sites. The first is that only one sex is represented on sites of the period, but this seems unlikely. The second is that sexual dimorphism has been reduced to such an extent that female and male animals are no longer clearly biometrically distinguishable. Decreasing sexual dimorphism has been suggested as a characteristic of *Bos* domestication (Guintard 1999) and identified at sites along the middle Euphrates (Helmer *et al.* 2005), and is also a possibility in the Early Neolithic of Britain. Finally the use of castrates could confuse the results. This final suggestion is difficult to rule out.

Figure 4.8 compares the greatest lateral and medial lengths of the astragalus. These measurements tend to provide similar information as they are both length measurements, but plotting them allows the inclusion of specimens for which other measurements are not available. The plot reiterates what has been suggested for other astragalus measurements above, namely that the Mesolithic sample can be separated into female and male specimens, that astragali from the Neolithic tended to be much smaller than those of the Mesolithic, that Early and Late Neolithic astragali are very similar to one another, and that sexual dimorphism is not as easily identified in Neolithic cattle as it is in Mesolithic aurochs. An important result of plotting the two length measurements is that a very large Late Neolithic specimen can be identified in the top right of the distribution. This specimen falls well within the Mesolithic plot, although it is unclear whether it comes from a female or a male animal. As a result of the astragalus measurements two undoubted aurochs specimens can be identified in the Late Neolithic. The largest Early Neolithic specimens could potentially belong to female aurochs or large domesticated cattle.

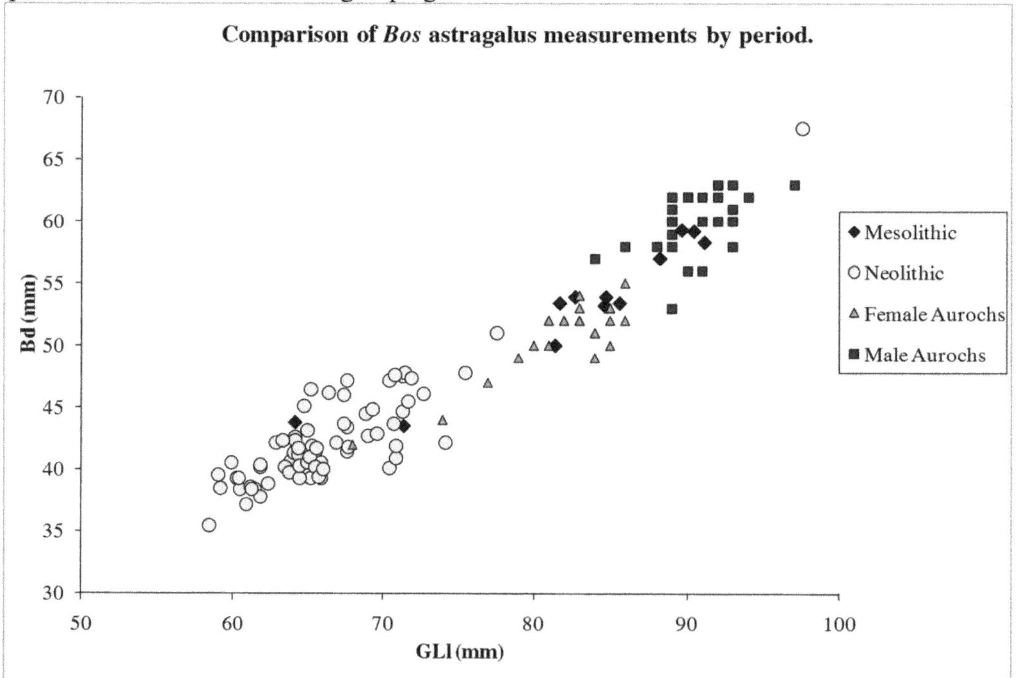

Figure 4.5 ***Bos*** **astragalus measurements from Mesolithic and Neolithic sites, and aurochs measurements (Degerbøl and Fredskild 1970) from Danish known sex remains (Bd = breadth of the distal end, GLl = greatest length, after von den Driesch (1976) (Viner 2011, Figure 14.11).**

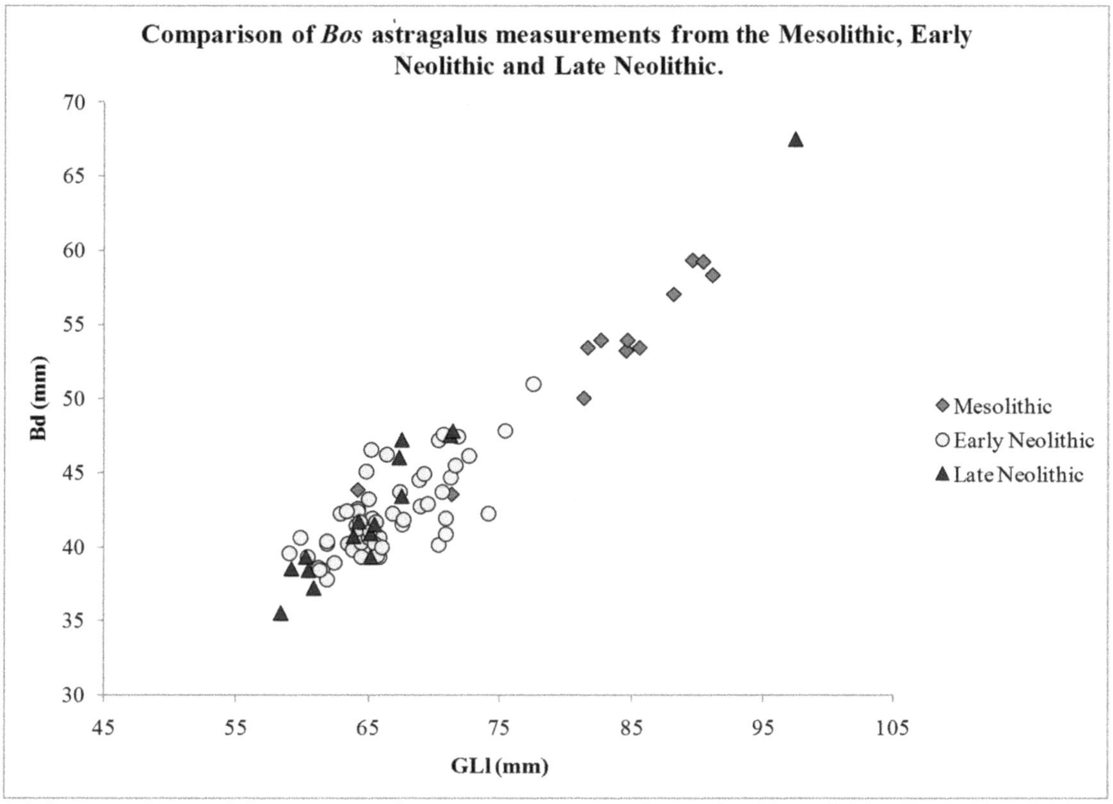

Figure 4.6 Comparison of *Bos* astragalus measurements from the Mesolithic, Early Neolithic and Late Neolithic (Bd = breadth of the distal end, GLl = greatest lateral length, after von den Driesch (1976).

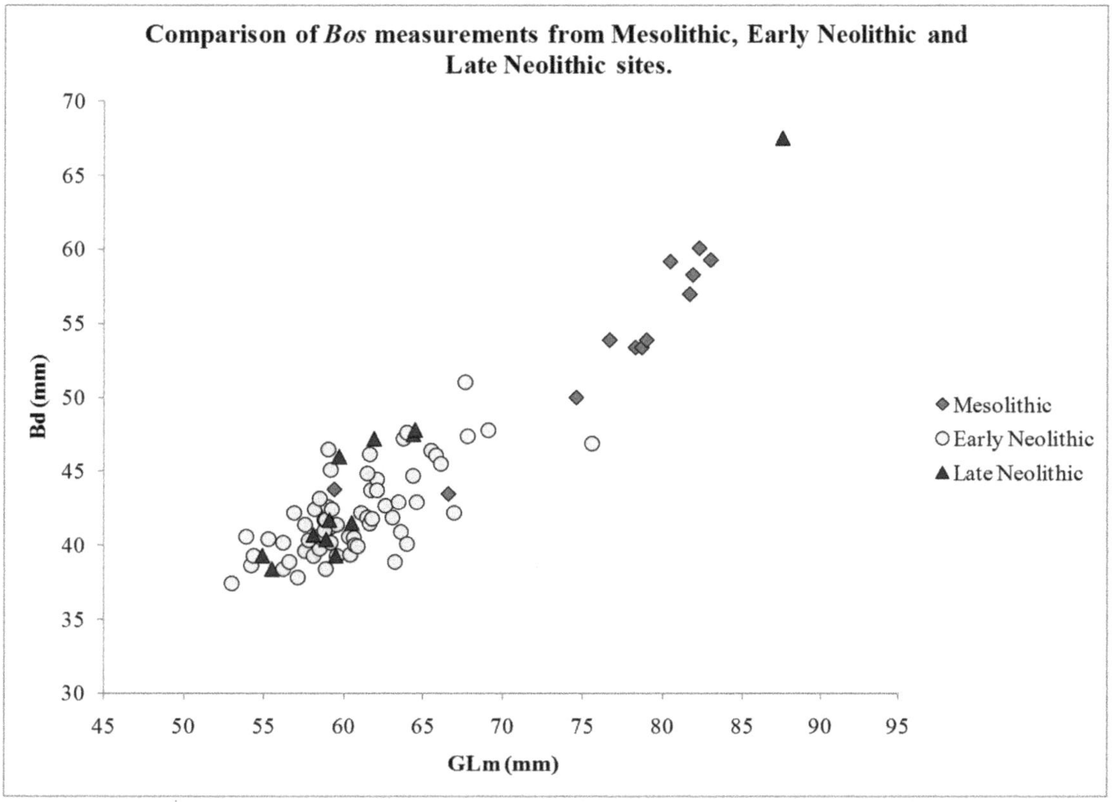

Figure 4.7 Comparison of *Bos* astragalus measurements from the Mesolithic, Early Neolithic and Late Neolithic (Bd = breadth of the distal end, GLm = greatest medial length, after von den Driesch (1976).

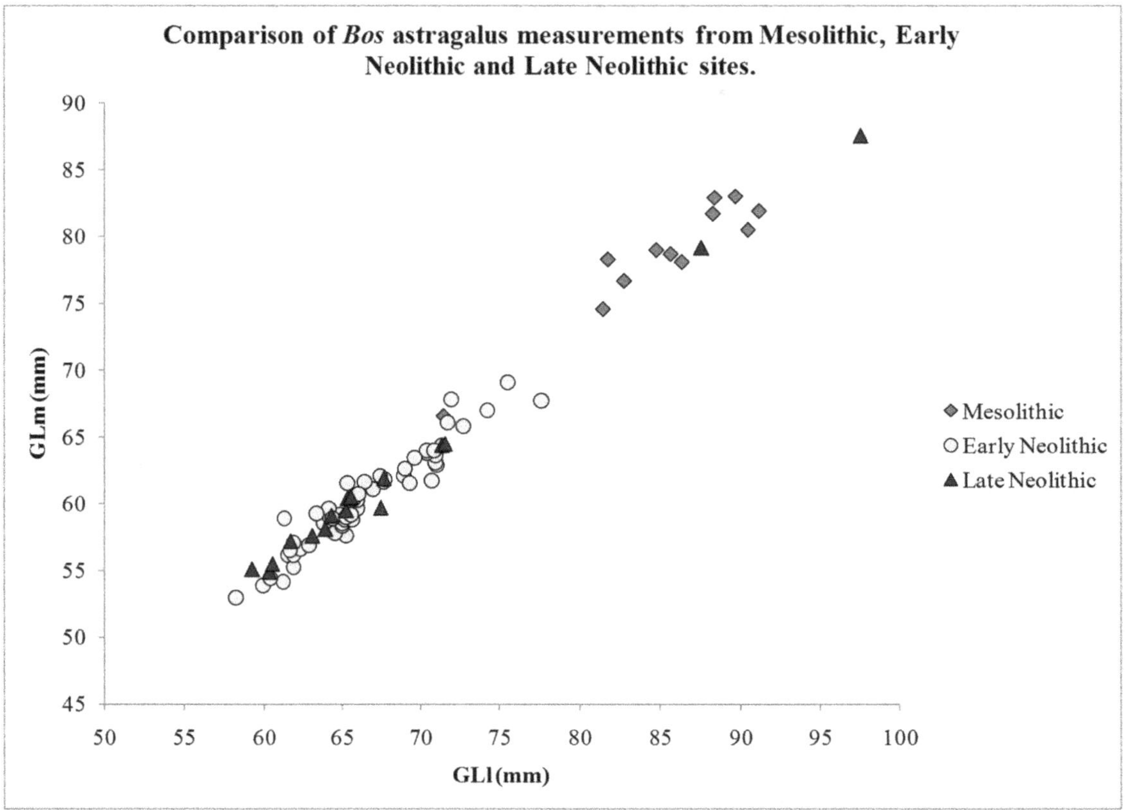

Figure 4.8 Comparison of *Bos* astragalus measurements from the Mesolithic, Early Neolithic and Late Neolithic (GLl = greatest lateral length, GLm = greatest medial length, after von den Driesch (1976).

Other skeletal elements have the potential to provide evidence of sex differences in the Neolithic. Measurements of the distal humerus have been plotted in Figure 4.9 (fused specimens only). There is no obvious difference between specimens from the Early Neolithic and those from the Late Neolithic in Figure 4.9, the distribution of the two overlap and they appear to have a similar range. Horizontally, two groups (that have been separated by a line on the figure) appear that do not respect the temporal distinction; individuals from the Early and Late Neolithic are found in both groups. The group furthest to the right of the plot consists of a small number of large specimens. The other group, which has more members, is made up of smaller specimens. These groups could be interpreted as a small group of large adult males, and a bigger group of small adult females. In light of the astragalus results above, and the single Mesolithic specimen included in the plot, there is little chance that these represent larger wild animals and smaller domesticated ones. In addition, the breadth of the humerus is known to be more sexually dimorphic than the height of the trochlea in other mammals, for instance pigs (Payne and Bull 1988). Lending more support (albeit tentatively) to the suggestion that these are different sex groups.

Distal humerus measurements would therefore suggest that female cattle were more abundant on Early Neolithic and Late Neolithic sites than male ones (as the group of smaller specimens has more members in both of the Neolithic periods investigated). If this interpretation is correct it can also be concluded that female cattle are of similar size in the Early Neolithic and Late Neolithic, as evidenced by their overlapping distribution in Figure 4.9. The specimens in the male group are slightly larger in the Late Neolithic than in the Early Neolithic, but the sample size is very small and more data are therefore necessary to make any firm interpretations.

Measurements of the distal humerus have been used to suggest that both female and male cattle were present on Early and Late Neolithic sites, and that the number of females was greater than the number of males. One explanation for this trend might be related to the ageing information presented above, in which between 20% and 30% of cattle did not reach adulthood. Rather than a real disparity between the number of females and males at the Neolithic sites, the pattern could be explained by the death of males at a younger age, a pattern that would be missed in the analysis of the distal humerus as unfused and fusing specimens were excluded. In pigs measurements of the distal humerus only increase to a limited extent after fusion (Albarella and Payne 2005; Payne and Bull 1988). It is conceivable that the same applies to cattle, and therefore a comparison of the measurements from unfused, fusing and fused specimens could provide more evidence of the female/male balance at these archaeological sites. Figure 4.10 presents the distal humeri measurements from the Early Neolithic separated by fusion stage, though no unfused specimens are available. The number of animals in both the female and male groups has increased, but the pattern expressed above, showing that the smaller animals greatly outnumber the larger ones still persists. The proportional distribution of fusing specimens between the two clusters is also similar to that of the fully fused bones. In the Late Neolithic, too, the addition of fusing specimens does not change the distribution of animals between the two groups, but there are only two fusing specimens to be added (one of which plots with the 'female' group and the other with the 'male' group, Figure 4.11). The addition of fusing specimens therefore adds little to the sex ratio that was suggested on the basis of fully fused specimens, as the 'missing' males cannot be found in the age group with fusing distal humeri. Very young males might have been represented by unfused distal humeri that were destroyed by taphonomic processes due to their fragile nature. An alternative explanation is that few males were brought to the Neolithic sites that have been used for the study. This possibility could be linked to the function of the sites – most of them being of ritual nature rather domestic activity.

Bones from castrated animals tend to vary somewhat in shape from complete animals, and it would be expected that their presence within a sample would be apparent by their distinction in size and shape, but no third group is present in the astragalus or humerus samples from Neolithic sites that would indicate the occurrence of large numbers of castrated animals. Nevertheless these may be difficult to detect and their size and morphology will also be very much dependent on the age of castration, as the earlier an animal is castrated the more female like its bones will appear (Davis 2000). It is therefore possible that the larger number of females compared to males is due to the addition of castrated animals within the size range of the female cattle.

Evidence from the Neolithic period when compared to the Mesolithic highlights the nature of the changes that were taking place at the beginning of the period. In the Mesolithic fully grown male and female aurochsen are represented in approximately equal proportions. With the onset of the Neolithic, there is evidence that the change was from the exploitation of aurochs to the utilisation of domestic cattle. In the Neolithic different sex groups can still be identified, though not as distinctively as in the Mesolithic. In general, female cattle outnumbered male animals in both periods, which has some important implications for husbandry strategies in this period. The occurrence of castrates cannot be ruled out, but in the absence of a large enough sample of metapodials measurements (the elements typically used for this kind of sex distinction), we have no positive evidence for their presence.

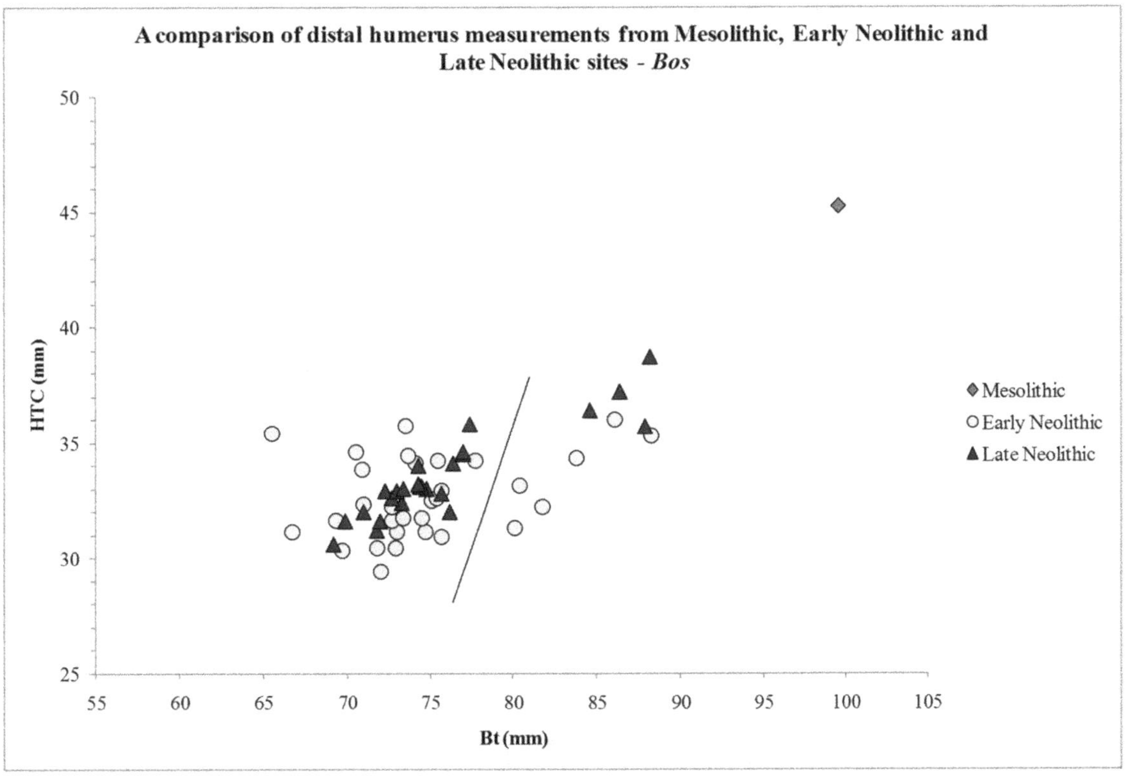

Figure 4.9 *Bos* distal humerus measurements (HTC = height of the trochlea, BT = breadth of the trochlea, von den Driesch 1976) from Mesolithic, Early Neolithic and Late Neolithic sites. Only fully fused specimens have been included (Viner 2011, Figure 14.12).

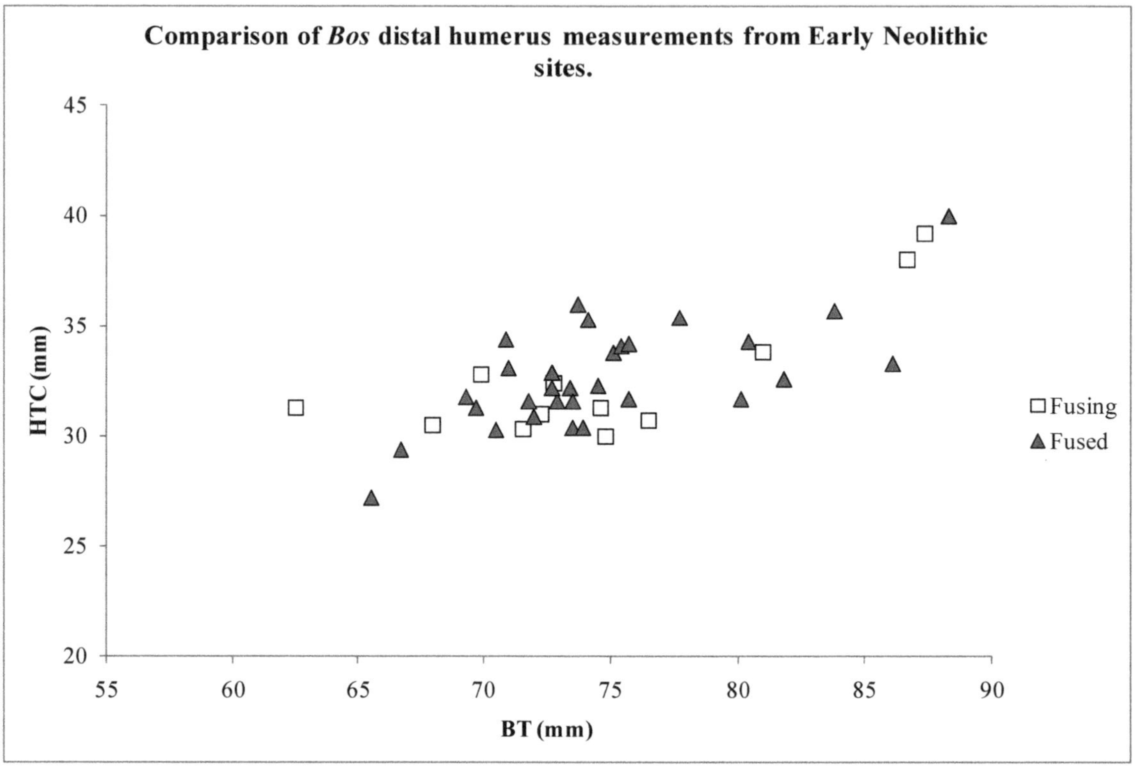

Figure 4.10 ***Bos*** **distal humerus measurements (HTC = height of the trochlea, BT = breadth of the trochlea, von den Driesch 1976) from Early Neolithic sites. Fusing and fused specimens have been included.**

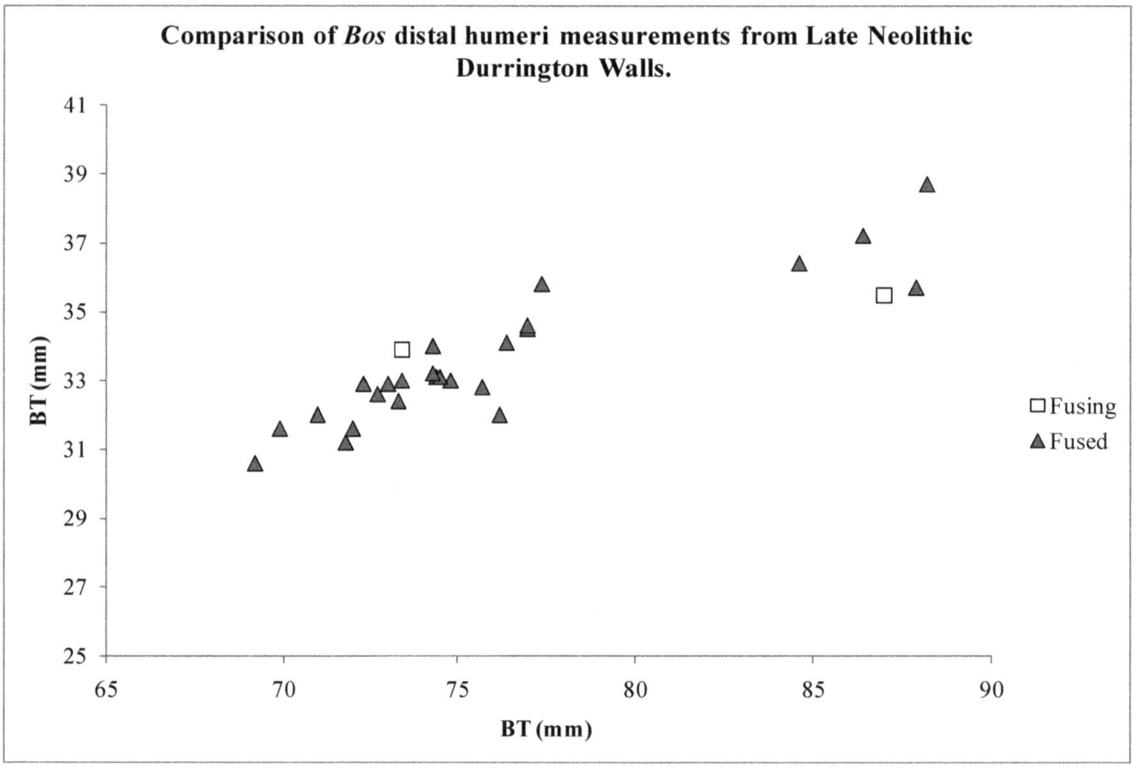

Figure 4.11 ***Bos*** **distal humerus measurements (HTC = height of the trochlea, BT = breadth of the trochlea, von den Driesch 1976) from Late Neolithic Durrington Walls. Fusing and fused specimens have been included.**

Size and shape

In addition to dealing with sex ratios in cattle samples, the biometrical aspect of this research aimed to identify variation in the size and shape of animals that could be attributed to changing human influences or other factors. Measurements chosen for each part of the analysis aimed to minimise, wherever possible, the influence of fluctuating environmental conditions, age and sexual dimorphism. In cattle of the Neolithic period the distal tibia varies little with age as evidenced by Figure 4.12 which compares measurements from unfused, fusing and fused specimens that all share the same distribution. There is no statistically significant difference between the fused, unfused and fusing specimens (Table 4.1).

In pigs, the distal tibia varies little between females and males (Payne and Bull 1988), and if it can be assumed that the same is true for cattle measurements of the distal tibia can be informative of differences between populations. In the Mesolithic period (Figure 4.13) there is evidence of two distinct groups within the (admittedly small) distal tibia collection. Sexual dimorphism is probably still an important enough factor in distal tibia variation, at least for aurochsen, and these groups are likely to represent female and male aurochsen. Stark differences between Mesolithic and Neolithic *Bos* are apparent; there is almost no overlap between the specimens from the Mesolithic and those from the Neolithic, confirming the impression that populations of aurochs were significantly larger. However, as with the astragalus, a small number of Neolithic specimens potentially belong to aurochs. The largest Neolithic outlier is certainly from an aurochs recovered from the Severn estuary. Based on these two elements (the astragalus and the distal tibia) alone there is little evidence to suggest a gradual change from the Mesolithic to the Early Neolithic, or the presence of animals that are clearly intermediate between the two groups. However, the significant time gap between the Late Mesolithic and Early Neolithic (as discussed in Chapter 2) needs to be considered. From these results it is also unlikely that any considerable mixing took place between wild and domestic cattle in the Neolithic.

Minor differences emerged in the pattern from the Early and Late Neolithic, but in general the pattern from the two periods was similar. The size of the Late Neolithic sample is small and should therefore be treated with caution. Elements from the Early Neolithic and Late Neolithic are similar in terms of the measurements obtained, and the distribution of the two periods overlap. Late Neolithic specimens are much more tightly clustered than in the preceding period all fall towards the middle and larger end of the Early Neolithic distribution. There are no specimens comparable with the smallest of the Early Neolithic. However, no great difference in the coefficient of variation exists between the Early and Late Neolithic specimens (Table 4.2).

Four Early Neolithic and one Late Neolithic specimen are separated from the main distribution of distal tibia measurements in Figure 4.13. One possibility is that these specimens come from male animals while the smaller ones come from female cattle and, when considered in conjunction with the distal humerus evidence that provided evidence of exploitation of both sexes, this scenario seems likely. However, it is clear that the specimens in question are also very similar in size to female aurochs and the possibility that they originated from wild aurochs cannot be excluded at present.

Teeth and postcranial bones vary differently under the influence of external factors and therefore both need to be considered when trying to determine the relationship between animals of different periods. Unlike pigs in which most undamaged teeth could be measured, the morphology of cattle teeth meant that many specimens had to be excluded from biometrical analysis (only specimens of known position were measured, see Chapter 2 for a discussion). As a result of small sample size, the analysis of individual teeth is limited to the third molar. Lower third molar measurements have been plotted in Figure 4.14, and upper third molars in Figure 4.15 from the Mesolithic, Early Neolithic and Late Neolithic. Most of the Early and Late Neolithic mandibular third molar specimens plot in the same part of the graph, a further indication that husbandry practices were relatively constant between the two periods. A number of the largest Late Neolithic specimens were intermediate between the main Late Neolithic distribution and the single Mesolithic specimen (Figure 4.14). These teeth could potentially come from aurochs and at this point cannot be identified with certainty. The maxillary third molar plot (Figure 4.14) shows once again that the Early and Late Neolithic are indistinguishable from one another, and the small number of larger specimens are likely to be domestic cattle but could come from aurochs.

So far biometrical analysis has focused on single elements, and although informative, individual bones and teeth can only partly elucidate overall trends in cattle skeleton variation. Log ratio comparison of *Bos* remains has the advantage of producing larger samples. It will also help to identify whether changes in physical characteristics were specific to certain skeletal elements or part of a more general overall change. The diagrams presented in this section have all been facilitated by the use of a standard, which in this case was established from measurements of aurochs bones from Ilford (discussed in Chapter 2). The aim of biometrical analysis in this context is to identify changes in the physical proportions of animals through time in the study area. It is expected that the animals from Ilford will be substantially larger than any of the animals from the periods under study and the bones from the Mesolithic, Early Neolithic and Late Neolithic will mainly appear to the left of the standard.

When a log ratio comparison of Mesolithic postcranial bones was employed the results revealed, as expected, a population of animals that were substantially smaller than the standard established from Pleistocene aurochs (Figure 4.16), although a small number of specimens do overlap with the Ilford standard. The Mesolithic results concur with the findings of other research in which a general reduction in the size of aurochs from the Pleistocene to the Holocene has been identified in Europe (*e.g.* Guintard 1999; van Vuure 2005). Mesolithic astragali (Figure 4.4 above), in addition to the information presented by Degerbøl and Fredskild (1970) indicated that female aurochs specimens can overlap substantially with the domestic cattle of the Neolithic period. The same pattern is substantiated by log ratio comparison of cattle postcranial remains. Those from the Early Neolithic plot some distance to the left of the standard and are therefore significantly smaller than aurochs from the Pleistocene. The Early Neolithic sample peaks some way to the left of the Mesolithic sample. Early Neolithic animals were noticeably smaller than those found in the preceding period an indication of a substantial change from the Mesolithic to the Early Neolithic, and there is no indication of continuity between the periods.

The results from the Late Neolithic postcranial sample are remarkably similar to those from the preceding Early Neolithic sites. The mean of the Early Neolithic distribution is -0.19, compared to -0.18 in the Late Neolithic. Distribution of measurements within the sample from the two periods was also very similar, with the exception of some larger outliers in the Late Neolithic plot. These very large outliers are likely to be aurochs specimens, and their inclusion in the sample helps to explain the slightly larger mean of the Late Neolithic sample compared to the Early Neolithic. Log ratio diagram of the Early Neolithic compared with the Late Neolithic suggests a high level of continuity through time with regard to the physical size of the cattle. By extension this might also indicate a similarity in the methods of exploitation and continuation of the economic emphasis that maintained them.

The log ratio comparison of tooth measurements is shown in Figure 4.17. The Mesolithic sample is small (14 specimens), the measurements were generally smaller than the standard, and larger than the Early Neolithic with a mean of -0.04. Mesolithic tooth measurements do not contradict the results of postcranial analysis, as they are larger or overlap with those from the Neolithic. The smallest of the Mesolithic specimens most likely derive from small female aurochs. Like the postcranial bones discussed above, teeth from the Early Neolithic are much smaller than the standard. In contrast with the postcranial bones, there are two Early Neolithic specimens that are undoubtedly from wild animals. Continuing to the Late Neolithic, there is little evidence to suggest a change in the size of specimens from the Early Neolithic. The plots are very similar, and have the same mean at -0.10. The Late Neolithic plot lacks the very large specimens that were found in the Early Neolithic and in the postcranial plot (Figure 4.16). The largest specimens in the plot (specifically those at -0.02) could belong to either large domestic males or small aurochsen. The presence and characteristics of aurochsen in the Neolithic of Britain will be discussed in the following section (4.2.4).

It is clear from the log ratio diagrams that changes in the *Bos* skeleton were more pronounced in bones than they were in teeth. In all cases the teeth of aurochs and cattle plot closer to the Ilford standard than the bones from the same period do. In addition the teeth of Mesolithic and Neolithic animals are more similar than their postcranial remains.

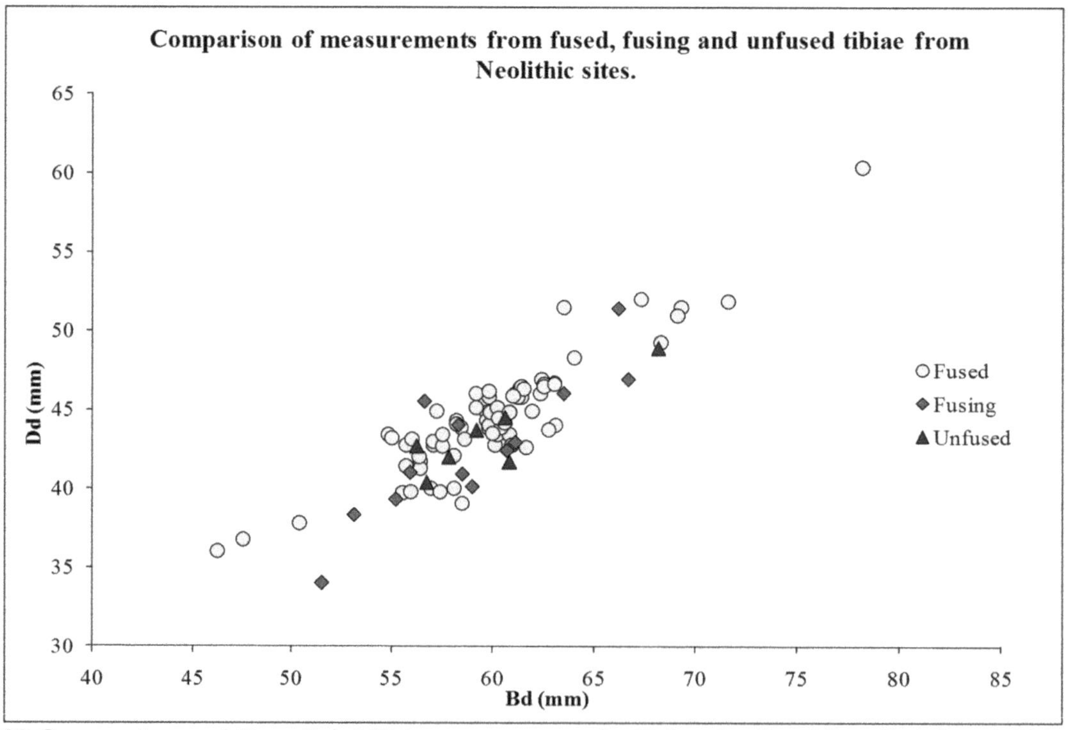

Figure 4.12 Comparison of *Bos* distal tibia measurements (Dd = depth of the distal end, Bd = breadth of the distal end, von den Driesch 1976) from fused, fusing and unfused specimens from the Early Neolithic.

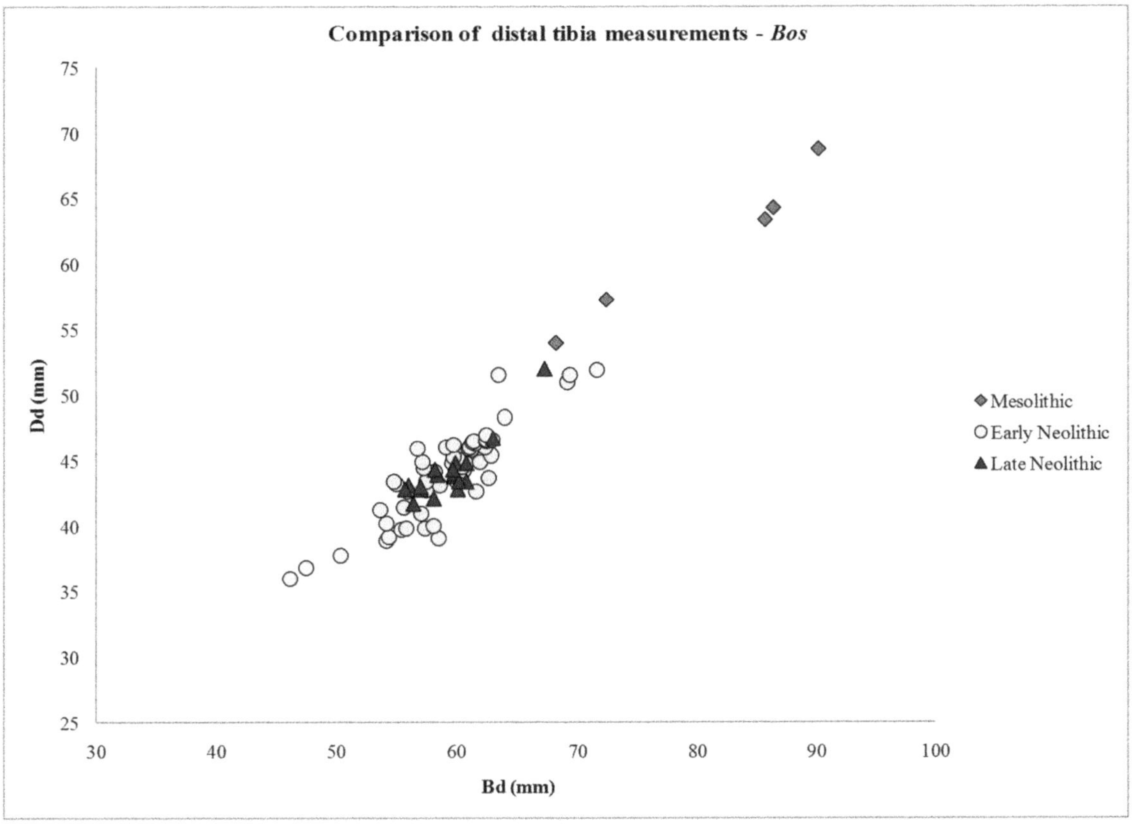

Figure 4.13 Comparison of *Bos* distal tibia measurements (Dd = depth of the distal end, Bd = breadth of the distal end, von den Driesch 1976) from the Mesolithic, Early Neolithic and Late Neolithic. Only fully fused specimens have been included.

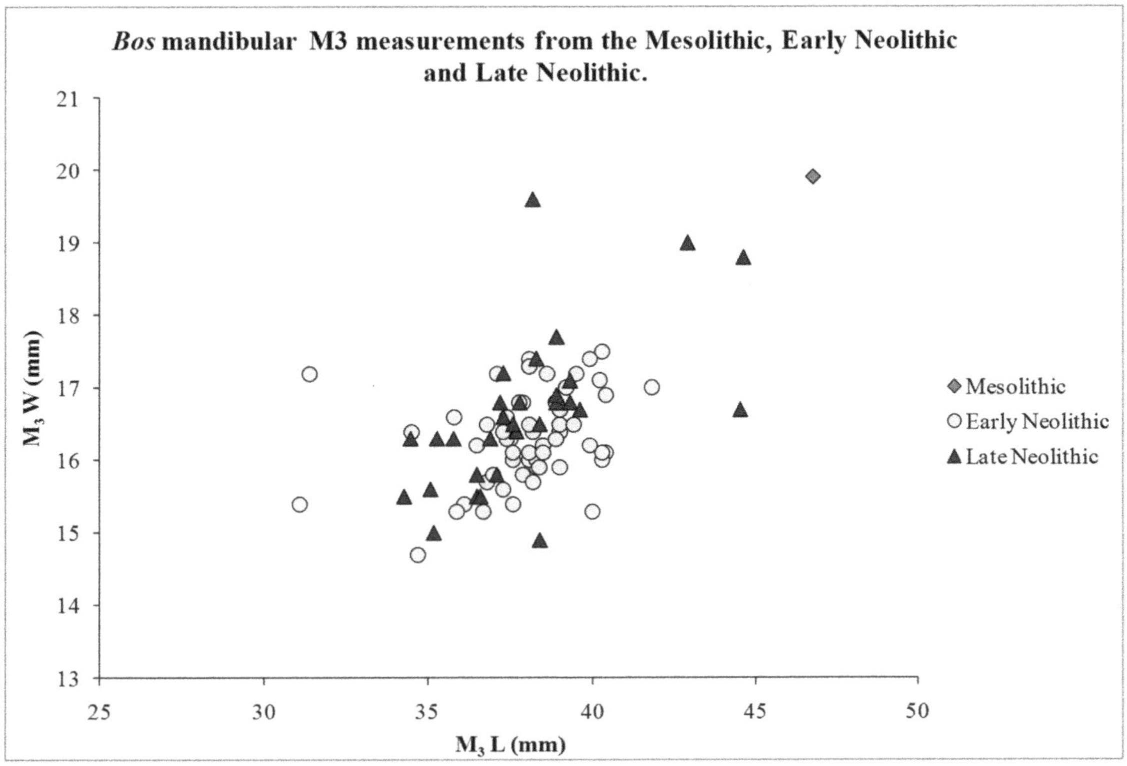

Figure 4.14 *Bos* mandibular third molar measurements (M_3L = length, M_3W = width) from the Mesolithic, Early Neolithic and Late Neolithic. Both jaws and loose teeth have been included.

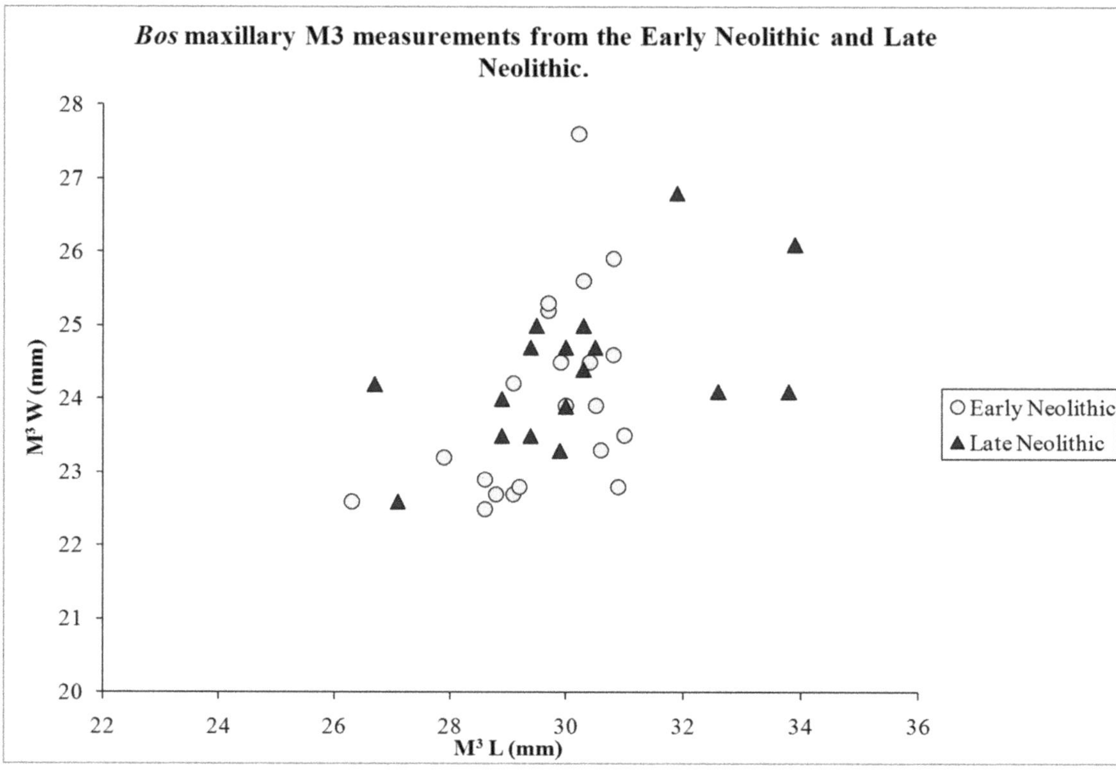

Figure 4.15 *Bos* maxillary third molar measurements (M^3L = length, M^3W = width) from the Early Neolithic and Late Neolithic. Both jaws and loose teeth have been included.

The presence of aurochsen in the Neolithic of Britain

Reduction in the size of aurochs from the Pleistocene to the Holocene has been identified by a number of researchers (for instance Degerbøl and Fredskild 1970; Guintard 1999; van Vuure 2005). The generally smaller size of the Mesolithic aurochsen investigated here suggests that such a size reduction was present in the populations that lived in Britain. This known reduction in size of wild aurochsen, along with the possibility that change has been masked by the long time gap between the available Mesolithic and Neolithic zooarchaeological assemblages, makes it essential to determine whether the small size of cattle during the Neolithic could be a result of size reduction in wild animals rather than the introduction of domesticates.

In Chapter 3, the remains of pigs from Late Mesolithic Goldcliff were compared with the bones of animals from the Early Mesolithic and helped to provide evidence that there was no significant reduction in the size of wild boar from the Early to the Late part of the period. Unfortunately, no aurochs remains were available from Goldcliff, and it is difficult to establish whether size reduction in aurochs occurred from the Early Mesolithic onwards. Relatively little is known about the physical characteristics of the aurochs in the British Neolithic, due in large part to the infrequency with which they are found on archaeological sites after the Mesolithic period. Looking beyond the archaeological record it is possible to gather some information regarding the physical proportions of the aurochs through the use of palaeontological specimens. One such specimen is 'Alice' (a feminine name although there is no evidence that the animal was female), a partially complete, extremely well preserved aurochs skeleton retrieved from the Severn estuary that has been dated to the Early Neolithic, and is now stored at Newport museum. The animal was adult when it died as all specimens were recovered in a fully fused state. When compared to the standard measurements (Figure 4.18) postcranial elements from Alice tend to plot to the left of zero, and are overall smaller than the Pleistocene standard. All of the measurements fall within the range of the Mesolithic postcranial specimens in Figure 4.16, suggesting that there was no great difference between aurochsen in the Mesolithic and Neolithic. Teeth from 'Alice' were larger than any of the Mesolithic tooth measurements (Figure 4.18) but available measurements were too few to allow any firm conclusions to be drawn. It would appear, then, that Neolithic aurochsen were similar in size to their Mesolithic predecessors and there is no evidence to suggest that there was a significant reduction in the size of wild cattle from the Mesolithic to the Neolithic. A larger body of Neolithic material is required to appropriately assess the physical size of aurochs in the Neolithic, and an attempt will be made to compare these data with other available published data in Chapter 5.

It has been shown above that there is a strong distinction between Mesolithic aurochs remains and those from Neolithic cattle, and, if the remains of Alice are representative of Neolithic aurochsen, that wild cattle from the Neolithic were comparable to those from the Mesolithic. Therefore, it would appear that Neolithic cattle were also distinct from Neolithic aurochs, and the large outlying specimens in the Early and Late Neolithic do indeed confirm this. The changes in age structure between the Mesolithic and Neolithic also support the view that we are dealing with populations interacting with humans in different ways. Aurochs bones seem to be present in small numbers (both teeth and postcranial specimens) on sites of the Early and Late Neolithic. The most likely explanation is that the hunting of aurochs continued sporadically into the Neolithic as has been suggested by other researchers of the period (Cotton *et al.* 2006). An alternative, but perhaps less likely explanation, given the biometrical separation of aurochs from domestic cattle, is that these animals were captured by Neolithic herders for breeding purposes (*cf.* Guintard 1999, and others).

Evidence from the Early Neolithic and the Late Neolithic has suggested a small number of specimens (both teeth and postcranial elements) that could be either large domestic cattle or small aurochsen. These few ambiguous specimens cannot be securely identified and for the time being we are forced to accept the existence of this grey area between the main domestic cluster and the clearly wild outliers.

4.3 Evidence for variation in the Early Neolithic

The following section will explore the evidence for homogeneity or difference between the uses of cattle at archaeological sites during the Early Neolithic period in Britain. The two main sites, Hambledon Hill and Runnymede Bridge, which provided substantial faunal assemblages will form the focus of this section. The other sites included here (Early Neolithic Hazleton North and Early Neolithic Coneybury Anomaly) were less abundant in terms of cattle remains but have been included here when sample sizes were large enough to provide meaningful results. Therefore, although a number of sites have contributed to the analysis, some of the results will be interpreted with the caveat that sample sizes were extremely small. Coneybury Anomaly, in particular, will be included wherever possible due to its importance as the earliest Neolithic site included in the study. Durrington Walls data have also been included in some cases. The age, sex and biometrical evidence from different Early Neolithic sites will be compared. This comparison is an important aspect of the analysis as combining the evidence from all Early Neolithic sites might mask differences in exploitation at different sites.

Age

The major trends observed in the generalised epiphyseal fusion plot (Figure 4.1) were repeated to some extent across all Neolithic sites included in the study when each site was plotted individually (Figure 4.20). Figure 4.20 also includes Late Neolithic Durrington Walls for comparative purposes. Despite the general similarity between different sites, some differences can be identified. For instance at both Hambledon Hill and Hazleton North there is hardly any evidence for calves, a pattern that contrasts sharply with their occurrence Runnymede Bridge, albeit in relatively small numbers (although the extremely small sample size from Hazleton North should be considered). This lack of very young animals in certain sites in the Early Neolithic could indicate the existence of consumer and separate producer sites at which breeding took place. Runnymede Bridge, the only site considered to be exclusively habitation in character, provided a larger number of bones from young cattle. Other sites (including Late Neolithic Durrington Walls that has been described as a consumer site (Parker Pearson *et al.* 2006), have evidence of ritual/funerary activity and it could be argued that cattle were imported to the site later in their lives. The possibility of cattle mobility in the Late Neolithic will be explored in Chapter 6.

All sites show a similar shift from the Early to the Intermediate category, though the intensity of slaughter between these two categories varies somewhat, between 20% and 50% of bones were unfused in the Intermediate category. Hambledon Hill had more cattle that survived until the Intermediate bones were fused than any of the other sites analysed. Hazleton North had the lowest survival in the Intermediate category, but because of the very small sample size any interpretation must be cautious. In all of the assemblages the difference between the Intermediate and Late category fusion was limited. According to the proportion of fused/fusing bones in the Late category, much use was made of adult animals, with varying intensity from site to site. For instance there was greater emphasis on Adults at Hambledon Hill (>70% fused/fusing) than Runnymede Bridge and Hazleton North (*ca.*50% fused/fusing). The same pattern has been identified by previous investigators of the faunal material from Hambledon Hill, in particular Legge (2009) interpreted the age profile as indicative of a dairying economy. Bearing in mind these differences, uniformity of mortality curves is the overarching pattern and this suggests that the aims and motivation for cattle husbandry were somewhat similar at different sites in the Early Neolithic.

Relatively few mandibles were complete enough to be included in the inter site comparison of mandibular wear. Mandibles from Runnymede Bridge (Figure 4.21), Hambledon Hill (Figure 4.22) and Coneybury Anomaly (Figure 4.23) were separated into Juvenile, Immature, Subadult, Adult and Elderly categories. A number of mandibles were recorded for which only a minimum age category could be suggested, these have not been included in the plot but can be seen in Appendix VI.

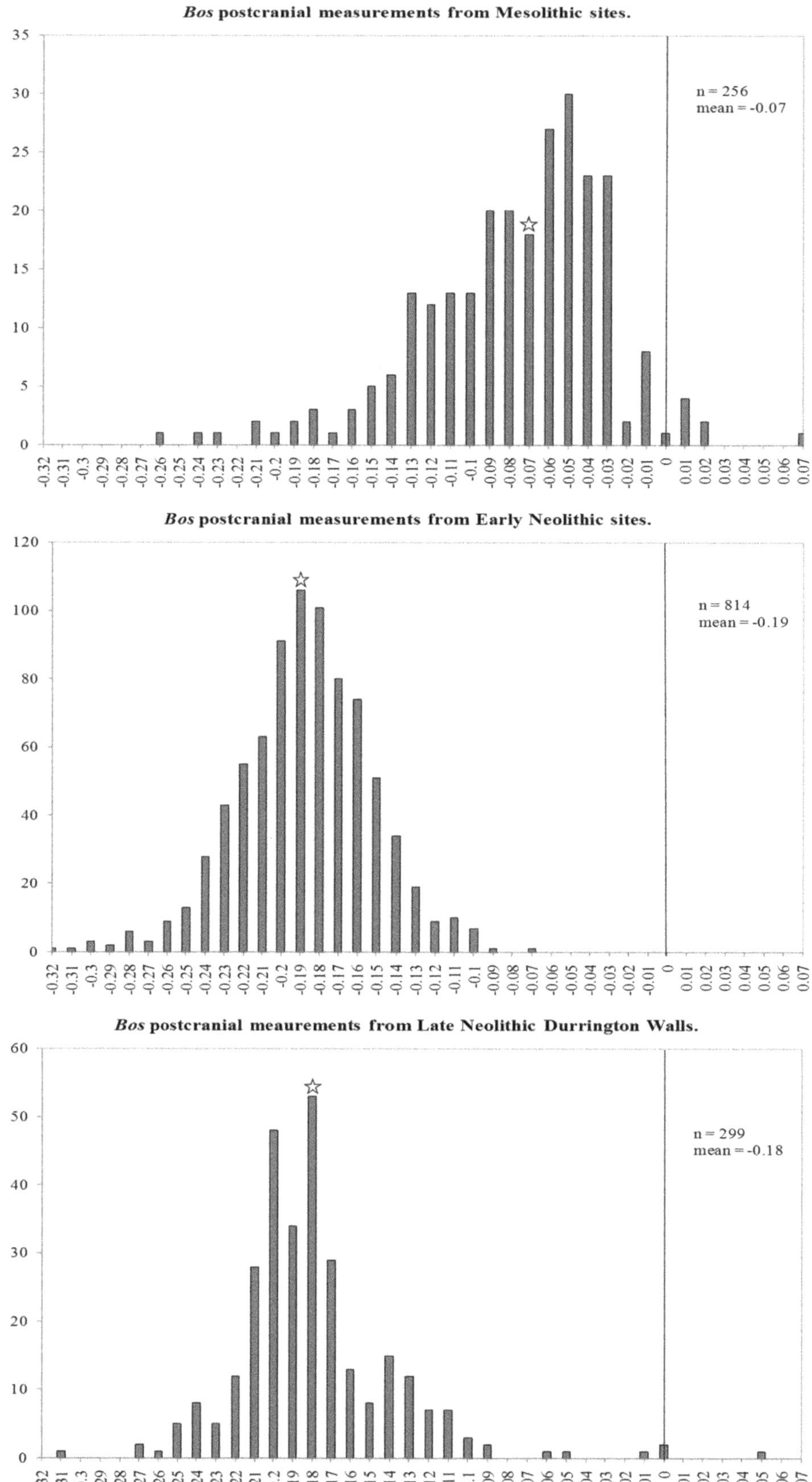

Figure 4.16 Log ratio comparison of *Bos* postcranial measurements from the Mesolithic (upper diagram), Early Neolithic (middle diagram) and Late Neolithic (Durrington Walls – lower diagram). Only measurements from fused specimens (excluding the scapula and horncores) have been included. The mean is marked by a star symbol, and the standard measurement by the vertical black line at 0 (Viner 2011, Figure 14.14).

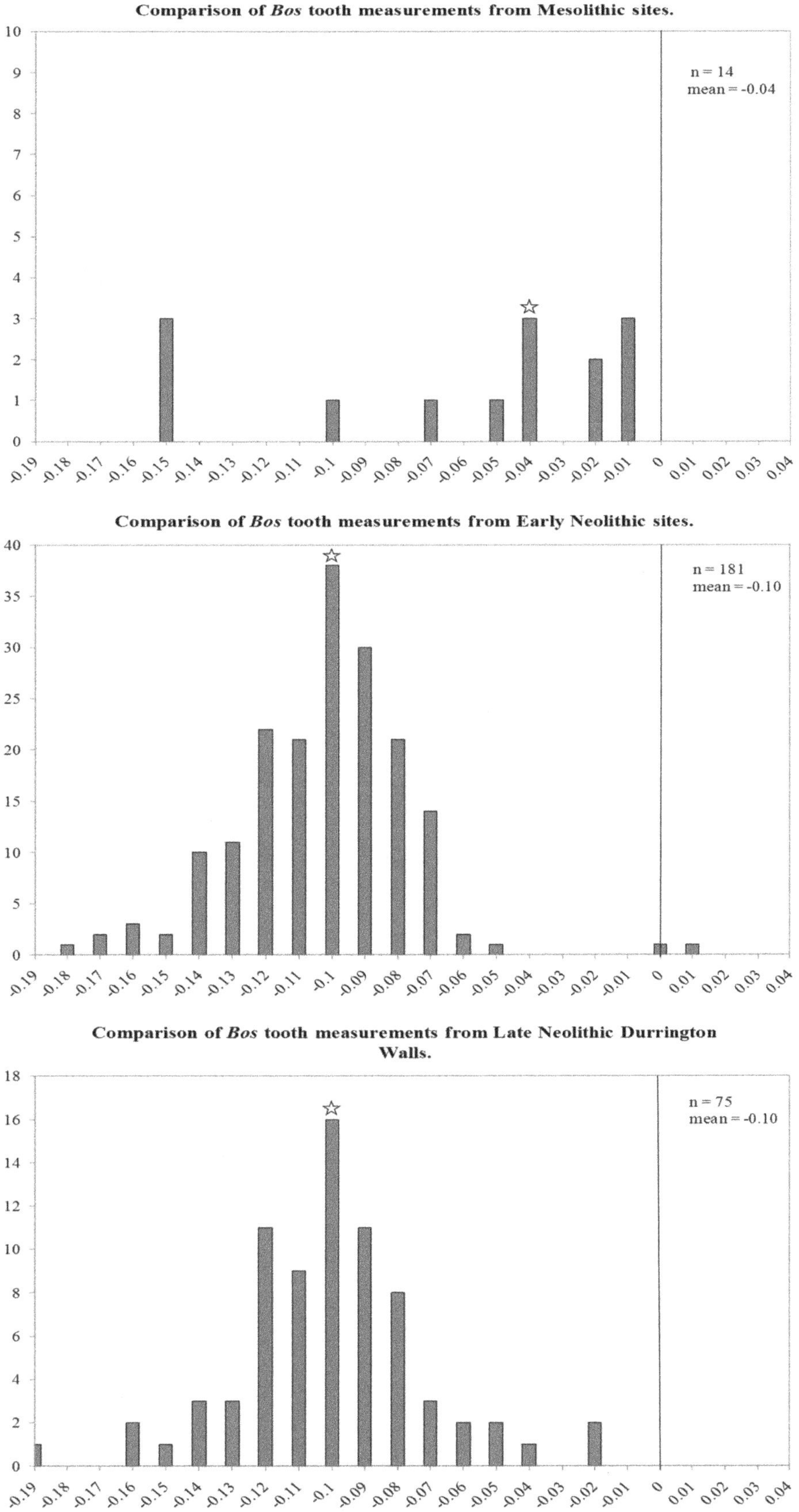

Figure 4.17 Log ratio comparison of *Bos* tooth measurements from the Mesolithic (upper diagram), Early Neolithic (middle diagram) and Late Neolithic (Durrington Walls – lower diagram). Measurements from heavily worn specimens have been excluded. The mean is marked by a star symbol, and the standard measurement by the vertical black line at 0 (Viner 2011, Figure 14.13).

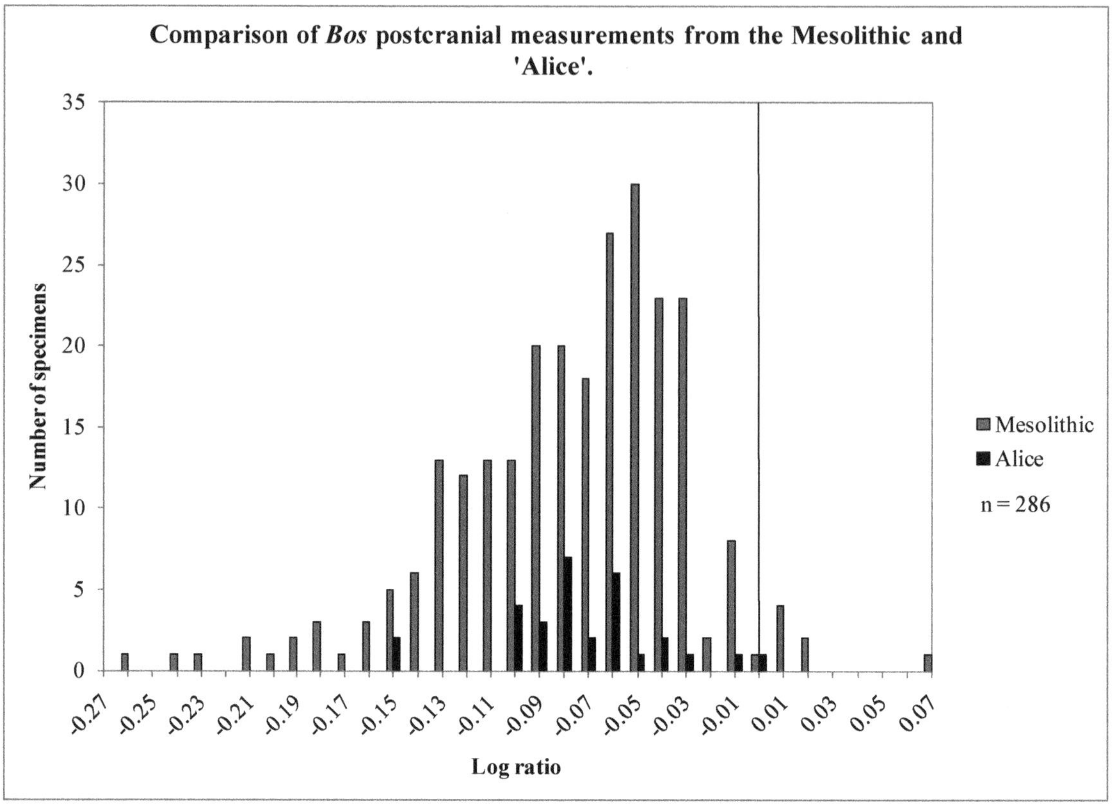

Figure 4.18 Comparison of *Bos* postcranial measurements from Mesolithic sites and 'Alice'. Fully fused specimens have been included. The standard measurement is shown by the vertical black line at 0.

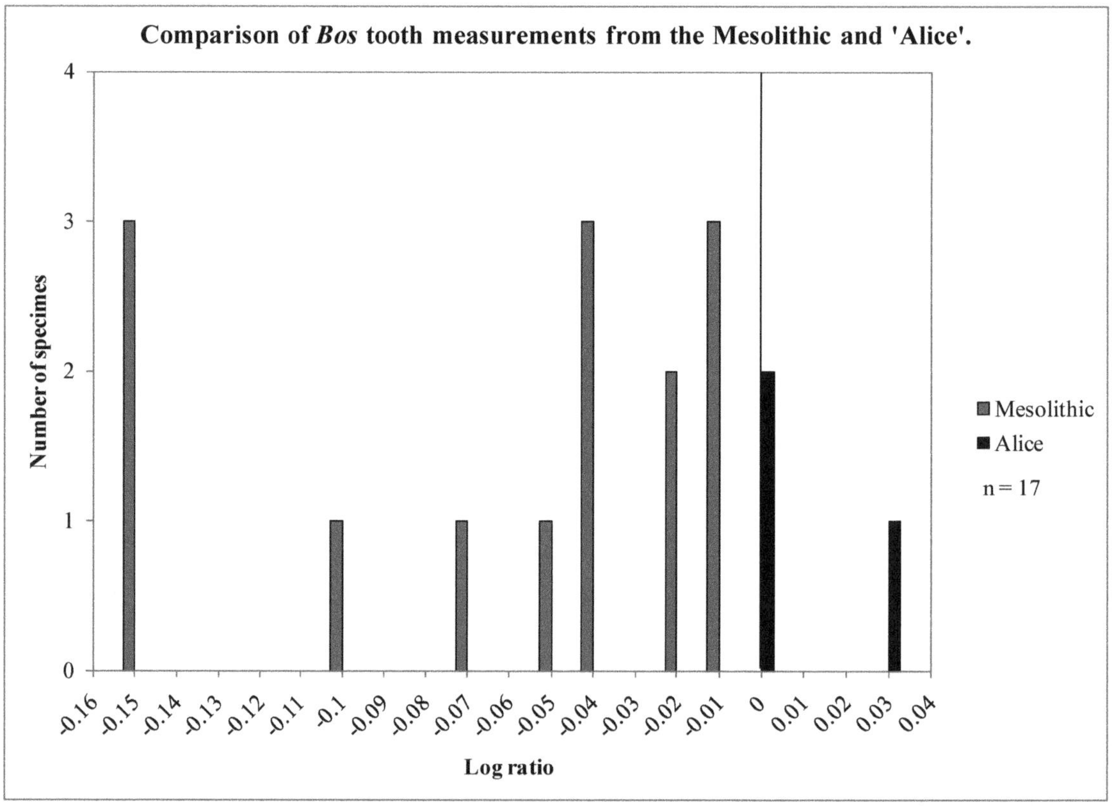

Figure 4.19 Comparison of *Bos* tooth measurements from Mesolithic sites and 'Alice'. Heavily worn specimens have been excluded. The standard measurement is shown by the vertical black line at 0.

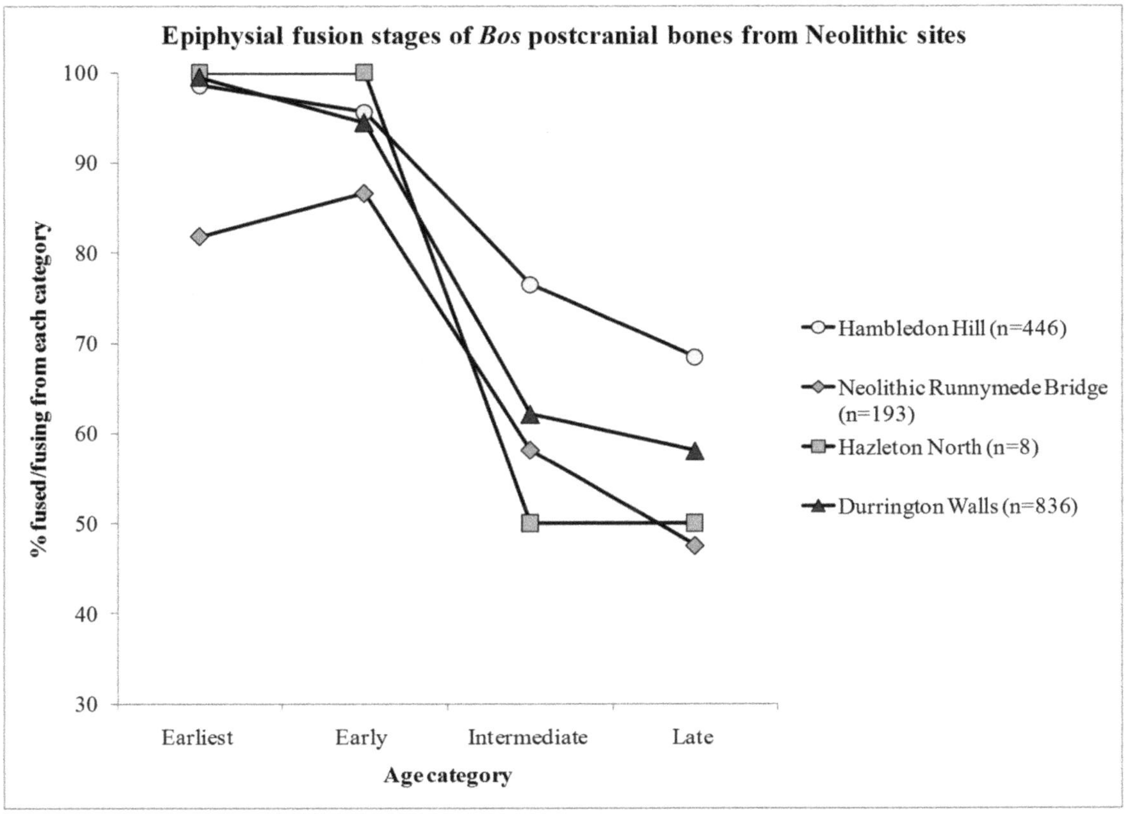

Figure 4.20 *Bos* epiphyseal fusion from Early Neolithic Hambledon Hill, Runnymede Bridge and Hazleton North, and Late Neolithic Durrington Walls.

Mandibles from Runnymede Bridge and Hambledon Hill reinforce some of the trends that were identified from mandibles in Section 4.2.1 (Figure 4.2). Both Runnymede Bridge and Hambledon Hill are dominated by mandibles from Adult animals. The very small sample from Runnymede Bridge is made up of 2 Adult mandibles, one Immature and one Subadult. No Elderly animals were present according to the results of mandibular wear. This pattern supports of the idea that there was greater emphasis on young animals than was found at other sites of the same period.

At Hambledon Hill the assemblage was also dominated by Adult animals, and a significant proportion of mandibles came from Elderly cattle (Figure 4.22). In contrast, no mandibles from the three youngest categories (Juvenile, Immature and Subadult) could be included from Hambledon Hill. Therefore the main emphasis at Hambledon Hill was on the use of adult animals, and the extended coverage of an animal's life that is afforded by mandibular wear (when compared with the limited epiphyseal fusion timescale) allowed the identification of several mandibles from Elderly animals. A greater emphasis on Adult cattle existed at Hambledon Hill when compared to other Neolithic sites, and the system employed made use of cattle until quite advanced in age. This evidence does not vary significantly from the epiphyseal fusion evidence presented in the previous section. The emphasis is on adult cattle, with no evidence of young animals (although they were clearly present on the site in small numbers evidenced by the unfused bones discussed above).

Mandibles from Coneybury Anomaly (Figure 4.23) were too few in number to provide a detailed picture of the cattle population at the site. However, the results do show that both young animals and those that had reached adulthood were present. The presence of Elderly mandibles also indicates that animals survived to advanced age. A similar system persisted into the Late Neolithic (see Figure 4.4 above). Evidence from Durrington Walls suggests that while young animals were killed and deposited at the site, the majority of animals reached adulthood, and some survived to be Elderly.

Sex

Evidence presented above has indicated that both female and male cattle bones were deposited on sites during the Neolithic, and perhaps that the sex balance was skewed in favour of females. One of the elements that suggested the possible prevalence of female animals compared to male animals was the distal humerus (Figure 4.9). Too few specimens from Runnymede Bridge were available for a sound interpretation of the evidence, but the two specimens that could be plotted in Figure 4.24 were relatively small and are more likely to be females than males. The differential presence of females and males is apparent in both the Hambledon Hill and the Late Neolithic Durrington Walls results. At Hambledon Hill the separation is less pronounced than at Durrington Walls. Those specimens from the Late Neolithic were clearly distinct from the main distribution. Unfortunately, the Mesolithic sample was too small to allow comparison with the later Neolithic sample, and it is therefore impossible to be sure that these larger specimens are not aurochsen. These plots suggest that both Hambledon Hill and Durrington Walls had a group of female and a group of male animals, but other sites did not provide enough specimens to allow an interpretation of the sex distribution.

When measurements of the astragalus from aurochsen of the Mesolithic period were compared to those from Danish aurochsen of known sex it became apparent that the element was a useful indicator of sex groups within a population (Figure 4.5), but the distinction was not continued into the Neolithic. In the analysis above all measurements from Early Neolithic sites were combined and presented as a single sample, therefore it was necessary to separate the sites so that pooling of samples did not mask variation. In Figure 4.25, 4.26 and 4.27 the astragalus measurements from Runnymede Bridge, Hambledon Hill and Coneybury Anomaly have been plotted. In all three of the plots the distribution of different Neolithic sites was very similar. No sex groups can be distinguished with any certainty in any of the astragalus measurements. The single Coneybury Anomaly specimen plots well within the range of the other Early Neolithic sites.

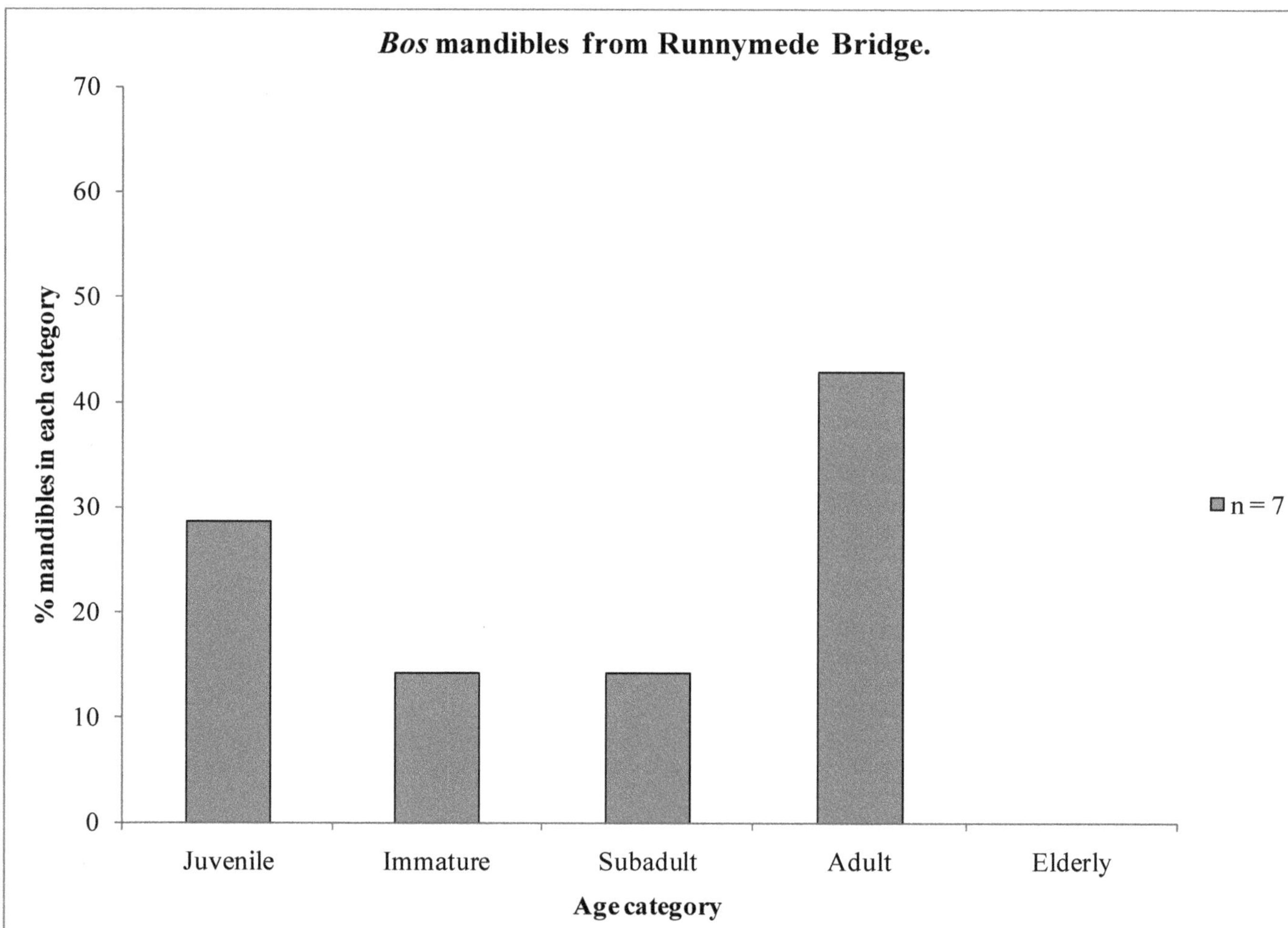

Figure 4.21 Wear of *Bos* mandibles from Early Neolithic Runnymede Bridge. Age categories are after O'Connor 1988.

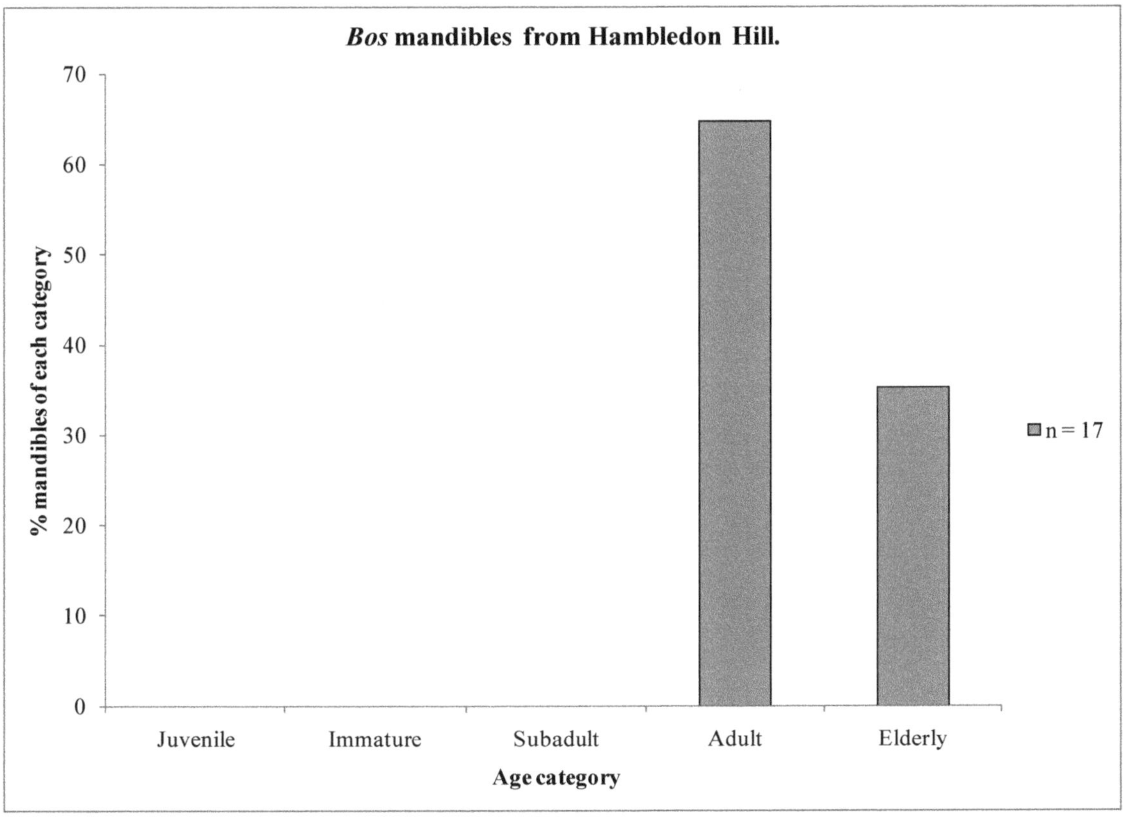

Figure 4.22 Wear of *Bos* mandibles from Early Neolithic Hambledon Hill. Age categories are after O'Connor 1988.

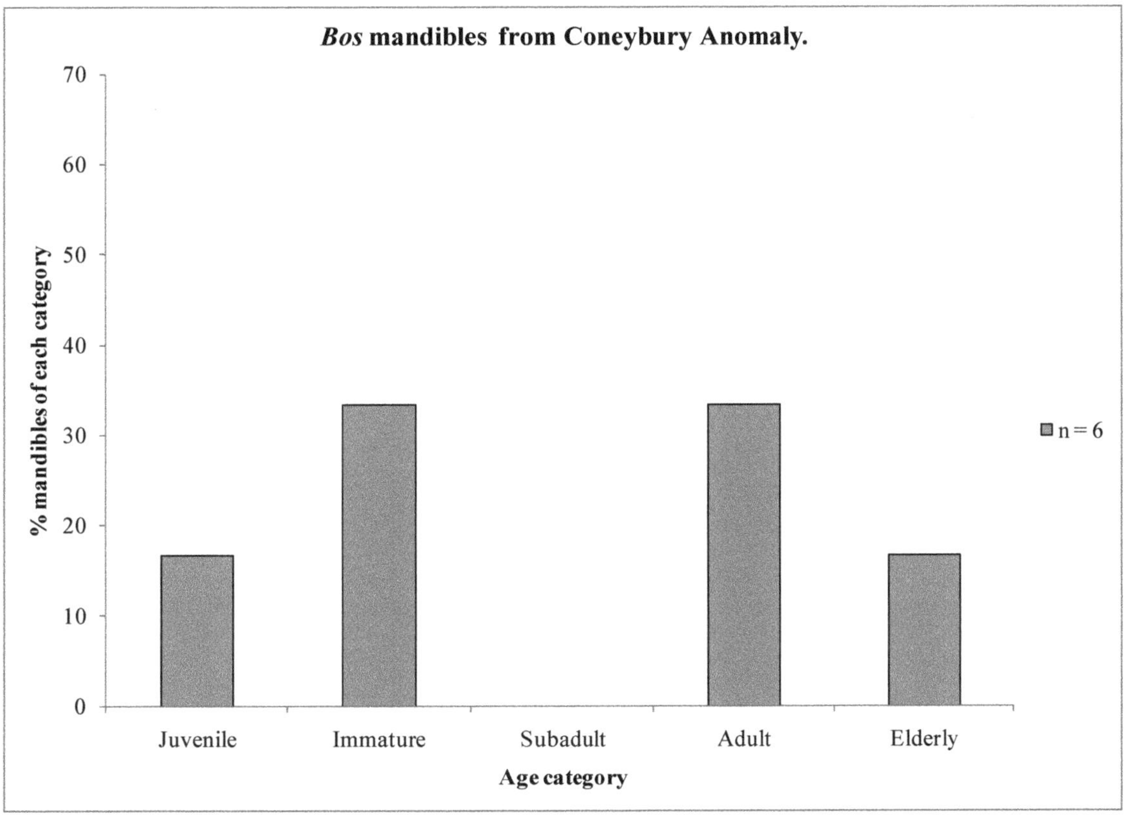

Figure 4.23 Wear of *Bos* mandibles from Early Neolithic Coneybury Anomaly. Age categories are after O'Connor 1988.

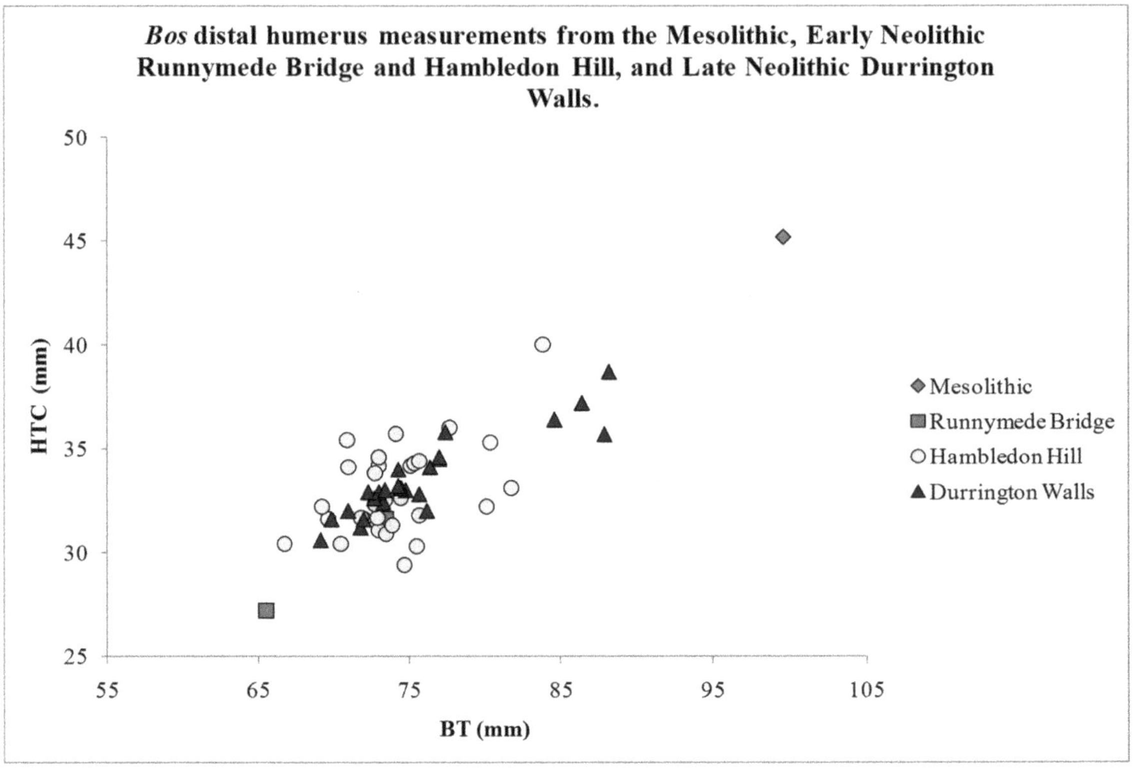

Figure 4.24 *Bos* distal humerus measurements (HTC = height of the trochlea, BT = breadth of the trochlea, von den Driesch 1976) from the Mesolithic, Early Neolithic Runnymede Bridge and Hambledon Hill, and Late Neolithic Durrington Walls. Only fully fused specimens have been included.

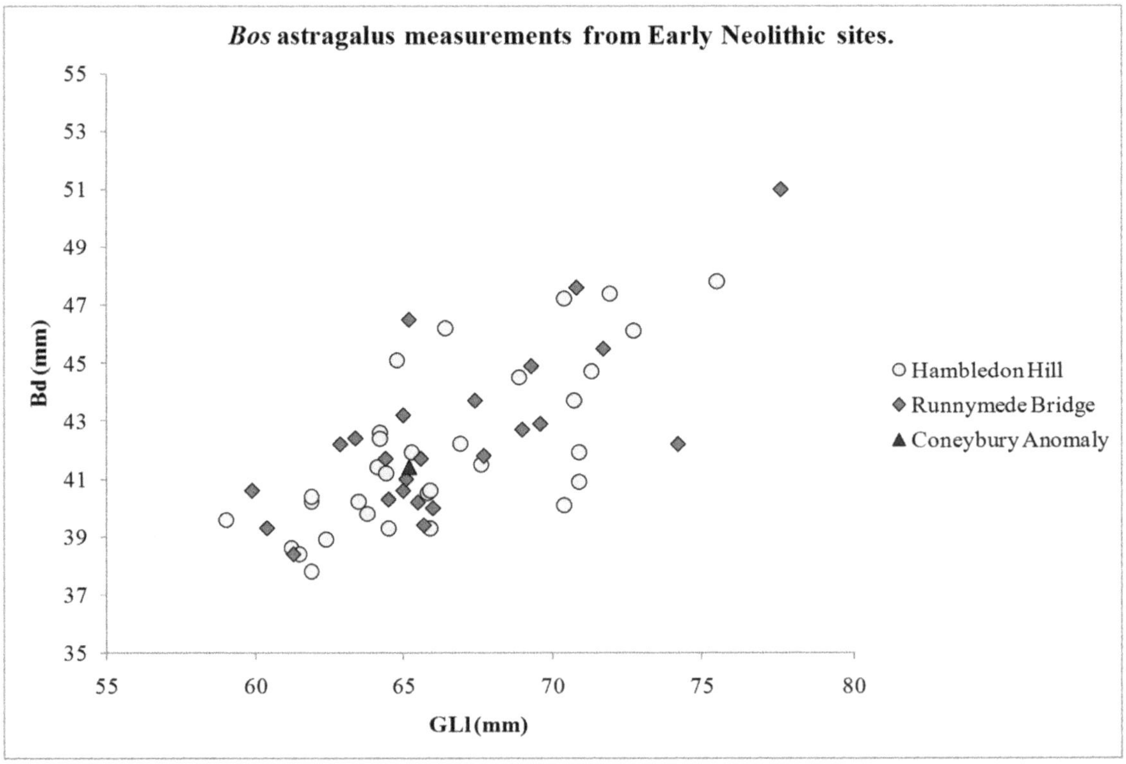

Figure 4.25 Comparison of *Bos* astragalus measurements from the Early Neolithic Runnymede Bridge, Hambledon Hill and Coneybury Anomaly (Bd = breadth of the distal end, GLl = greatest lateral length, after von den Driesch (1976).

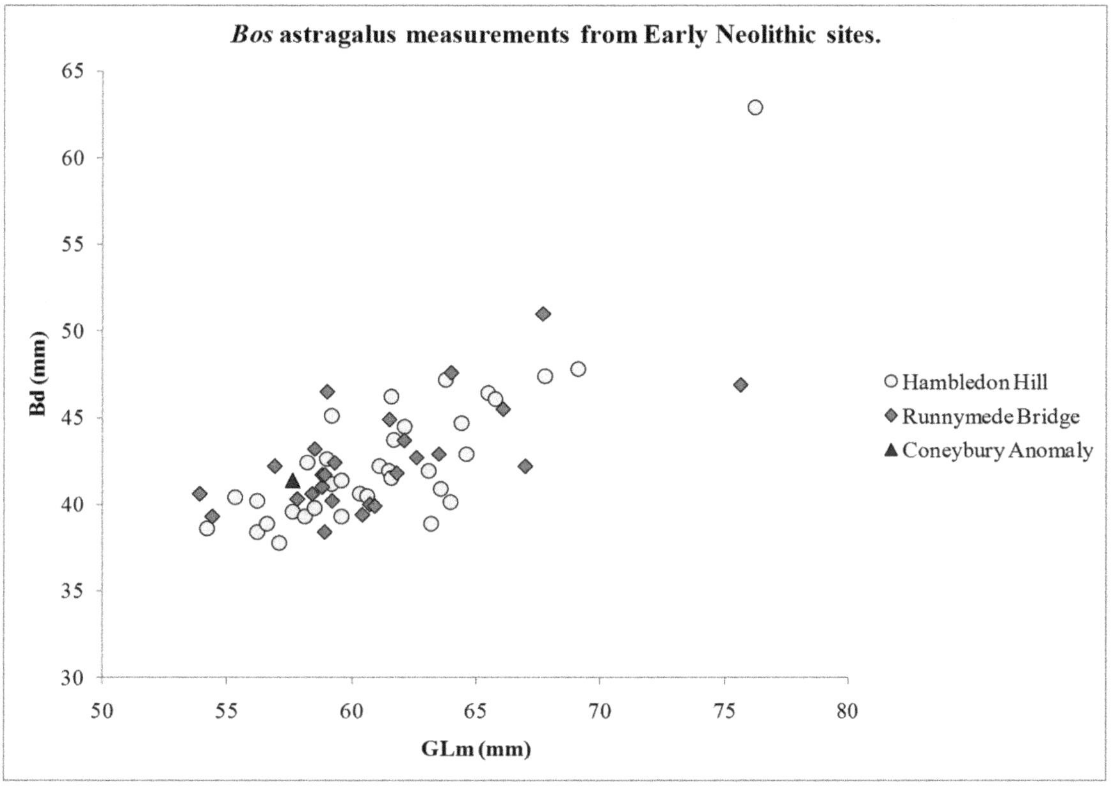

Figure 4.26 Comparison of *Bos* astragalus measurements from the Early Neolithic Runnymede Bridge, Hambledon Hill and Coneybury Anomaly (Bd = breadth of the distal end, GLm = greatest medial length, after von den Driesch (1976).

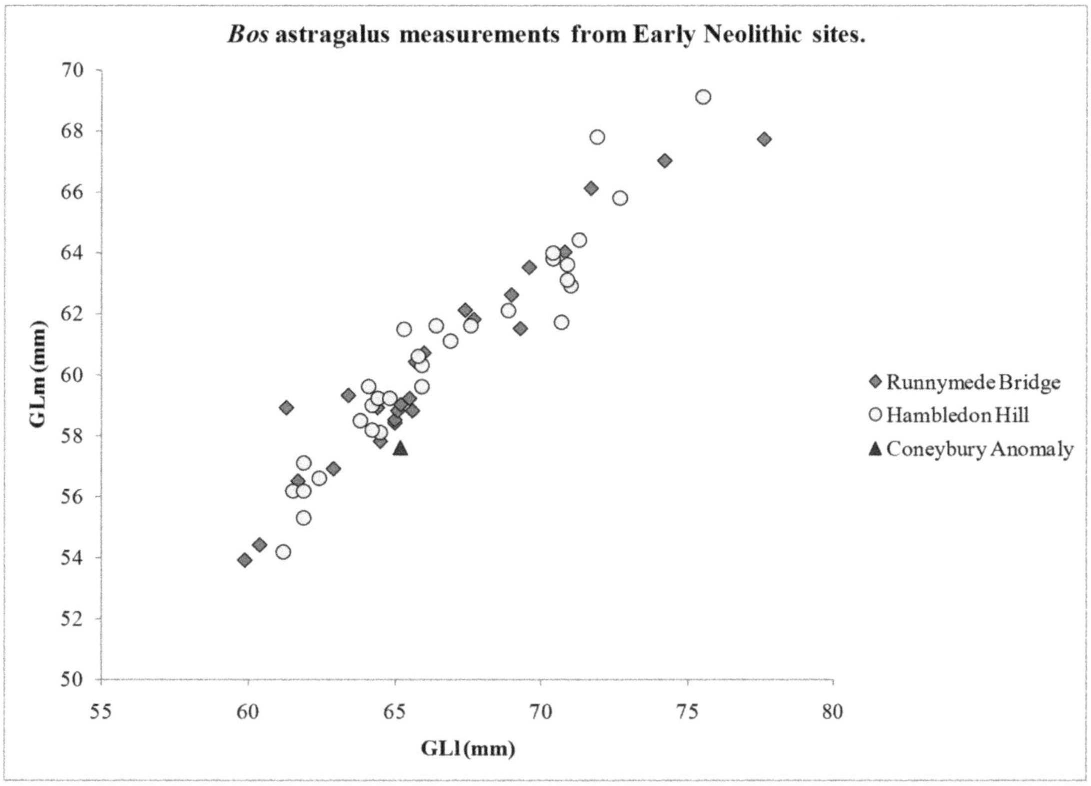

Figure 4.27 Comparison of *Bos* astragalus measurements from the Early Neolithic Runnymede Bridge, Hambledon Hill and Coneybury Anomaly (GLl = greatest lateral length, GLm = greatest medial length, after von den Driesch (1976).

Size and shape
Biometrical analysis of cattle remains at the different sites of the Early Neolithic revealed a high degree of uniformity. Measurements of the astragalus (usually a good indicator of population differences) provide no hint of difference between the populations during the Early Neolithic (Figure 4.25, Figure 4.26 and Figure 4.27). T-test results (Table 4.1) indicate no significant difference between the astragalus measurements from Hambledon Hill and Runnymede Bridge. The same pattern is repeated in the plot of distal tibia measurements (as in Figure 4.28, Table 4.1). The distribution of measurements at Hambledon Hill and Runnymede Bridge are almost identical, and single points from Coneybury Anomaly and Hazleton North are situated well within the ranges of the two.

Log ratio comparisons of postcranial measurements reinforce the overall impression from the Neolithic data that cattle at different sites were very similar to one another (Figure 4.29). The distribution of measurements from Runnymede Bridge, Hambledon Hill and Coneybury Anomaly are all some distance to the left of the standard and have means of -0.19 (Runnymede Bridge and Hambledon Hill), and -0.20 (Coneybury Anomaly). This pattern of similar range and distribution is not unexpected given the very limited variation in the individual skeletal elements explored so far. The results reinforce the idea that animals, and therefore husbandry practices were common to the Early Neolithic sites being explored.

Tooth measurements continued the trend (Figure 4.30). The Hambledon Hill comparison provided a mean that was marginally larger than the Runnymede plot. Hambledon Hill has a single very large outlier that is too large to be a domestic specimen and must therefore come from an aurochs. Measurements from Coneybury Anomaly were too few to be very informative (and the higher mean is unsound given the small sample size; the addition of a few more specimens could alter the plot significantly), but the presence of aurochs within the sample is illustrated by one very large outlier to the right of the standard. The patterns of the Early Neolithic are generally continued into the Late Neolithic (compare with Figure 4.15). As well as the similarity of domestic cattle size overall in the Early Neolithic and Late Neolithic, a small number of aurochs teeth were identified in the log ratio comparison of Late Neolithic tooth measurements.

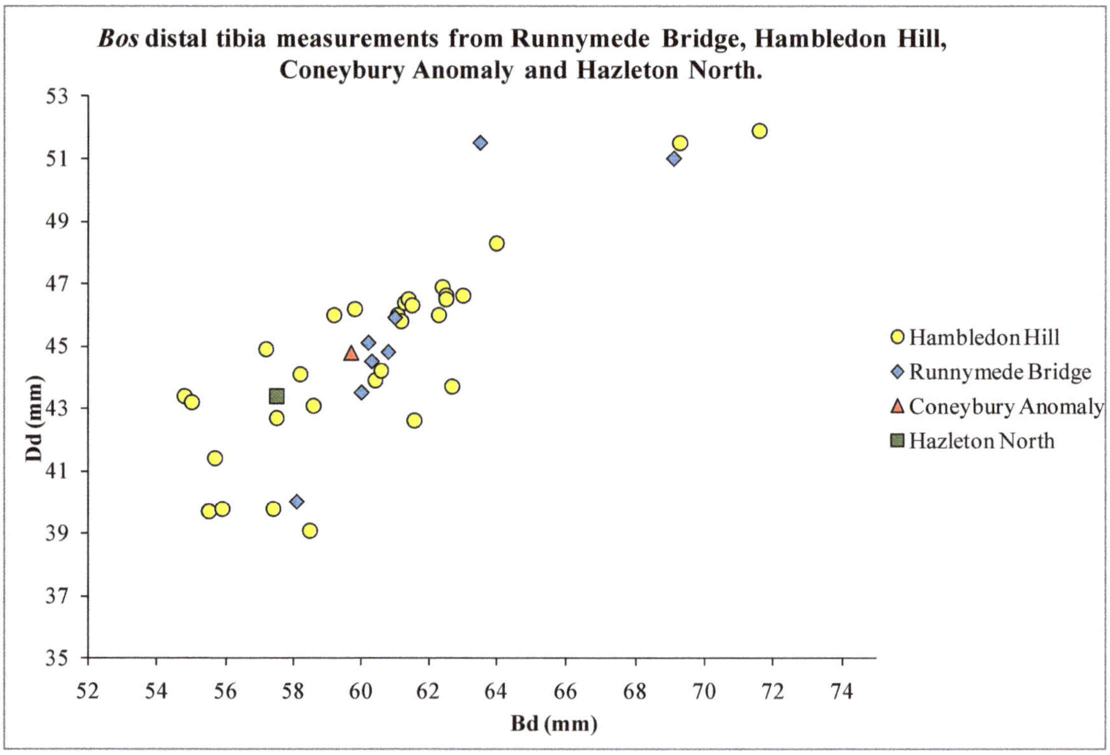

Figure 4.28 Comparison of *Bos* distal tibia measurements (Dd = depth of the distal end, Bd = breadth of the distal end, von den Driesch 1976) from Early Neolithic sites. Only fully fused specimens have been included.

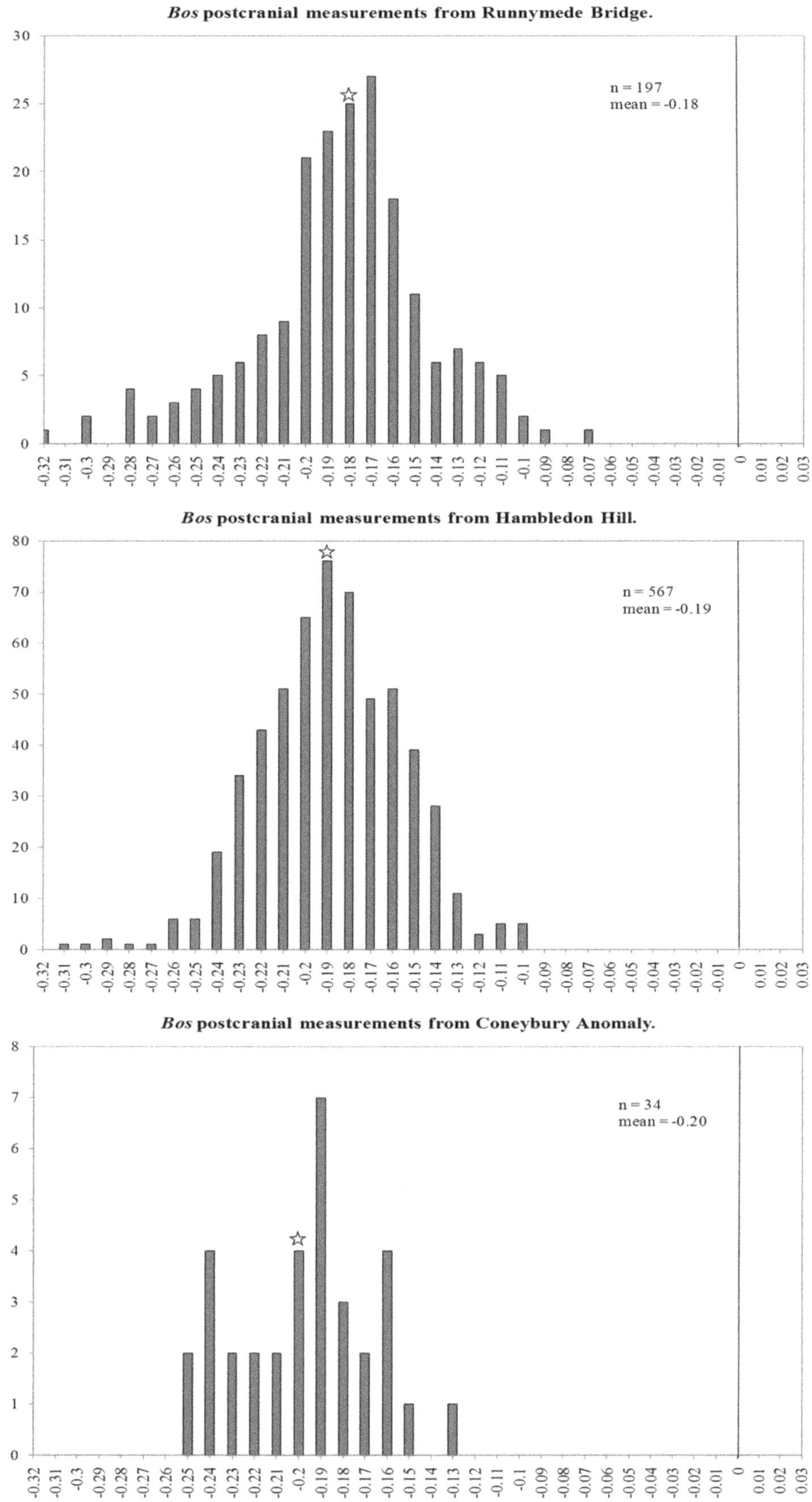

Figure 4.29 Log ratio comparison of *Bos* postcranial measurements from Early Neolithic Runnymede Bridge (upper diagram), Early Neolithic Hambledon Hill (middle diagram) and Early Neolithic Coneybury Anomaly (lower diagram). Fully fused specimens (excluding the scapula and horn cores) have been included. The mean is marked by a star, and the standard measurement by the black vertical line at 0.

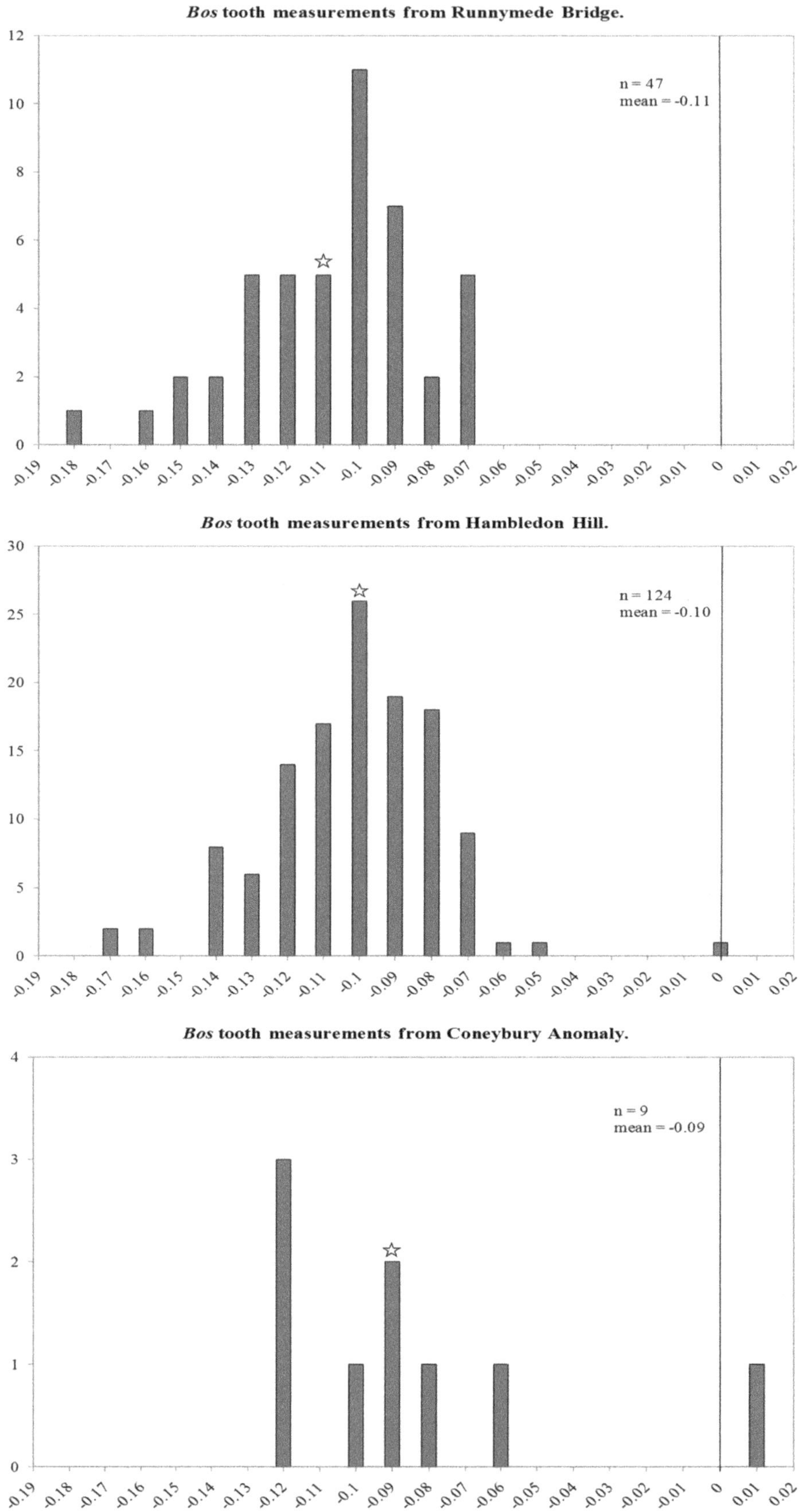

Figure 4.30 Log ratio comparison of *Bos* postcranial measurements from Early Neolithic Runnymede Bridge (upper diagram), Early Neolithic Hambledon Hill (middle diagram) and Early Neolithic Coneybury Anomaly (lower diagram). Heavily worn specimens have been excluded. The mean is marked by a star, and the standard measurement by the black vertical line at 0.

4.4 Conclusion

The data presented here was gathered in an attempt to answer a number of specific research questions about the way that domesticated cattle arrived in Britain and how they were exploited during the periods being studied. One of the main aims of this chapter was to identify changes in the way that *Bos* was exploited through time. Central to the understanding of *Bos* exploitation is a consideration of the age of cattle when they died. The age of animals deposited on prehistoric sites was investigated through the use of epiphyseal fusion and mandibular tooth wear. Overall, the results from the Mesolithic found that the exploitation focused on the use of Adult animals, with evidence of very few young animals on any of the sites investigated. There is a possibility that this is partially a taphonomic issue and that small sample size and poor preservation on Mesolithic sites influenced the age distribution, but an overall predominance of adults still remains the most likely scenario. Concentration on Adult and Elderly animals continued into the Neolithic, but the focus of exploitation had changed, suggesting a significant shift in the motivation for using cattle. This new motivation led to the death of a substantial proportion of the population before they reached the intermediate category, in a system that otherwise focussed on the use of Adult and Elderly cattle.

The identification of strongly distinguished sex groups in the Mesolithic (with the aid of comparative material from Denmark (Degerbøl and Fredskild 1970) highlights the real change in *Bos* use through time. The shift with the beginning of the Neolithic cannot be explained as a result of changing exploitation of male/female aurochs. Therefore the evidence presented here supports the view that domesticated cattle were introduced during the Early Neolithic.

No such strongly differentiated sex groups could be identified in the Neolithic itself, but there is limited evidence of a distinction based on sex from specific skeletal elements (in particular the humerus). Plots of this element led to the suggestion that female cattle outnumbered male cattle on Early and Late Neolithic sites. If male animals of all ages were largely absent from the Early Neolithic sites that were investigated, their site of deposition must be considered. It is probable that if male animals were killed early in their lives and female animals kept into adulthood that taphonomic influences would have destroyed some of the male remains. But this does not account for the entire pattern, and the possibility that male animals were not brought to the sites must also be considered.

Biometrical analysis was employed to address the central themes of local domestication or external cattle origins, intermediate types of cattle, and the possibility of mixing between domestic and wild *Bos*. There was a sharp contrast between the Mesolithic and Neolithic animals in terms of their physical characteristics that was apparent in both postcranial and dental elements. No evidence for a gradual reduction in *Bos* size was found (although the large time gap must be considered) and no group of intermediate animals was discovered that would support the idea of local domestication processes (of the type suggested by Bökönyi 1969 and other researchers). In addition, differences between Mesolithic and Neolithic populations suggest that there was little visible mixing of wild and domestic types. Whether this was deliberate or inadvertent is impossible to determine.

In the Late Neolithic, the comparability of age, sex and biometrical information with that from the Early Neolithic suggests significant continuity in exploitation and motivation through time.

In an attempt to exclude non-anthropogenic size reduction as the main influencing factor in the changes seen in the *Bos* skeleton through the time period in question, the remains of a Neolithic aurochs was compared to the Mesolithic specimens studied. These found no substantial difference in the size of animals from the two periods. This evidence has three main implications; the first is that it can be expected that Neolithic aurochsen were generally significantly larger than the domesticated animals being investigated; the second is that the long stretch of time for which no zooarchaeological material is available does not mask a gradual, non-anthropogenic change in the size of aurochsen; and finally that the Mesolithic material can provide an appropriate baseline for wild/domestic comparison in Neolithic Britain.

Aurochsen were used infrequently in the Neolithic, but they did occur in small numbers at a number of sites (both Early and Late Neolithic), thus supporting the suggestion that hunting of these animals continued after the end of the Mesolithic.

As a result of the large changes in exploitation practices, and in the animals themselves with the onset of the Neolithic, it was necessary to determine whether these patterns were common to all sites of the period or whether there was variability between sites. In general, the observed patterns were the same at most of the sites investigated. An exception was the age of animals from Early Neolithic Runnymede Bridge, in which assemblage there were considerably more young animals than other sites in the study. This difference might be a consequence of different site use as Runnymede Bridge was the only clearly domestic settlement included in the study. Conversely, Hambledon Hill had a relatively greater proportion of Adult and Elderly cattle, a pattern that has further implications for husbandry practices. The interpretation of these results will be explored further in Chapters 5 and 7.

Distinctions between different Early Neolithic sites in terms of sex and biometrical information were few. The implications of this evidence is that the systems of exploitation, and the animals that were exploited, were homogenous over large parts of southern Britain during the Early Neolithic, and homogenous over long time periods from the Early Neolithic to the Late Neolithic.

In the next chapter the results gathered from pigs and cattle from British sites will be compared with other published evidence of animals from Britain, Ireland and continental Europe. In Chapter 7 the results of both pigs and cattle will be discussed in greater detail.

Table 4.1 Results of t-tests conducted on measurement groups from sites used in the study. Significant results are marked with a *.

Measurement	Group/n		Group/n		Significance
Tibia Bd	Fused (Neo)	69	Fusing (Neo)	13	0.581
	Fused (Neo)	69	Unfused (Neo)	7	0.904
	Fusing (Neo)	13	Unfused (Neo)	7	0.631
	Ilford	4	Mesolithic	4	0.062*
	Ilford	4	Early Neolithic	43	0.000**
	Ilford	4	Late Neolithic	26	0.000**
	Mesolithic	4	Early Neolithic	43	0.026*
	Mesolithic	4	Late Neolithic	26	0.027*
	Early Neolithic	43	Late Neolithic	26	0.933
	Runnymede Bridge (EN)	11	Hambledon Hill (EN)	30	0.281
Tibia Dd	Fused (Neo)	69	Fusing (Neo)	13	0.187
	Fused (Neo)	69	Unfused (Neo)	7	0.455
	Fusing (Neo)	13	Unfused (Neo)	7	0.583
	Ilford	4	Mesolithic	4	0.022*
	Ilford	4	Early Neolithic	43	0.000**
	Ilford	4	Late Neolithic	26	0.000**
	Mesolithic	4	Early Neolithic	43	0.006*
	Mesolithic	4	Late Neolithic	26	0.006*
	Early Neolithic	43	Late Neolithic	26	0.755
	Runnymede Bridge (EN)	11	Hambledon Hill (EN)	30	0.416
Astragalus GLl	Ilford	7	Mesolithic	11	0.000**
	Ilford	7	Early Neolithic	57	0.000**
	Ilford	7	Late Neolithic	16	0.000**
	Mesolithic	11	Early Neolithic	57	0.000**
	Mesolithic	11	Late Neolithic	16	0.000**
	Early Neolithic	57	Late Neolithic	16	0.528
	Runnymede Bridge (EN)	25	Hambledon Hill (EN)	31	0.721
Astragalus GLm	Ilford	7	Mesolithic	11	0.000**
	Ilford	7	Early Neolithic	57	0.000**
	Ilford	7	Late Neolithic	16	0.000**
	Mesolithic	11	Early Neolithic	57	0.000**
	Mesolithic	11	Late Neolithic	16	0.000**
	Early Neolithic	57	Late Neolithic	16	0.806
	Runnymede Bridge (EN)	25	Hambledon Hill (EN)	31	0.702
Astragalus Bd	Ilford	5	Mesolithic	12	0.000**
	Ilford	5	Early Neolithic	57	0.000**
	Ilford	5	Late Neolithic	16	0.000**
	Mesolithic	12	Early Neolithic	57	0.000**
	Mesolithic	12	Late Neolithic	16	0.000**

Table 4 continued

	Early Neolithic	57	Late Neolithic	16	0.454
	Runnymede Bridge (EN)	24	Hambledon Hill (EN)	33	0.643
M_3 length	Ilford	4	Early Neolithic	57	0.001*
	Ilford	4	Late Neolithic	30	0.000**
	Early Neolithic	57	Late Neolithic	30	0.979
M_3 width	Ilford	4	Early Neolithic	57	0.015*
	Ilford	4	Late Neolithic	30	0.016*
	Early Neolithic	57	Late Neolithic	30	0.392

Table 4.2 Coefficient of variation of postcranial and tooth measurements from sites used in the study.

Ast GLm	**Number of specimens**	**Standard deviation**	**Mean**	**Coefficient of variation**
Ilford	5	4.16	88.71	4.69
Mesolithic	11	2.69	79.38	3.39
Early Neolithic	57	3.49	60.46	5.78
Late Neolithic	16	8.89	62.20	14.29
Ast Bd				
Ilford	5	3.16	66.74	4.73
Mesolithic	12	4.58	54.10	8.47
Early Neolithic	57	5.32	42.78	12.43
Late Neolithic	16	7.49	43.28	17.31
Tibia Bd				
Ilford	4	2.66	91.23	2.91
Mesolithic	5	9.65	80.58	11.98
Early Neolithic	43	4.80	59.70	8.04
Late Neolithic	26	3.11	59.78	5.20
Tibia Dd				
Ilford	4	1.86	69.48	2.67
Mesolithic	5	5.89	61.56	9.56
Early Neolithic	43	3.68	44.38	8.30
Late Neolithic	26	2.44	44.15	5.52
M_3 length				
Ilford	4	1.82	46.65	3.90
Early Neolithic	57	1.93	37.98	5.08
Late Neolithic	30	2.51	37.96	6.62
M_3 width				
Ilford	4	1.64	20.40	8.03
Early Neolithic	57	0.64	16.33	3.93
Late Neolithic	30	1.08	16.64	6.52

Chapter 5
Pigs and cattle in context

5.1 Introduction

So far discussion has concentrated on the analysis of primary data collected during the course of the research. It is therefore important that the results expressed in Chapters 4 and 5 be compared to information from other contemporaneous assemblages to determine whether observed patterns can be applied to southern Britain as a whole. A key debate concerning the Mesolithic/Neolithic transition has been whether domestic species were allochthonous or autochthonous, and in an attempt to add to this debate a broader comparison between British and European material will also be conducted. The location of British sites used for comparison are shown in Figure 5.1.

The comparison of British material means that the sites used in the first part of this chapter ordinarily provide information on both cattle and pigs. But the often cited shift from a concentration on cattle in the Early Neolithic to a greater use of pigs in the Later Neolithic also means that the amount of useable data from each site is different for the two species.

After presenting information regarding the background of the British and continental European sites used for comparison, this chapter will discuss the wider context of pigs and their use in Britain and further afield, in the Mesolithic, Early and Late Neolithic periods and, to a more limited extent, also later times. The second part of the chapter will be concerned with British cattle from the Early Neolithic and Late Neolithic, and their broader European position.

The main problem facing this kind of comparison of biometrical information is consistency of measurement. Choice of measurements is often based on the specific research questions being applied, and therefore many biometrical investigations use incomparable measurements. For instance, Bartosiewicz (1999), Chaix and Arbogast (1999) and Street (1999) all provide large numbers of aurochs measurements, but the biometrical information cannot be compared to data from this study as different measurements were used. In addition, inconsistency in the way different researchers may have taken apparently equivalent measurements could be a cause of discrepancies between results.

Pre-Neolithic material

Three areas will form the focus of comparison with British cattle and pigs. These areas have been chosen because of geographical and archaeological relevance, and, by necessity, the availability of data. Southern Scandinavia will be investigated because of its proximity to Britain, abundant data and comparability with the British situation. The Netherlands and northern France have been chosen because of their close geographical relationship, and the premise that the British Neolithic may originate from this part of continental Europe. Finally, the Iberian Peninsula will also be investigated. Portugal in particular occupies a similar position in Europe (albeit further south) to Britain on the Atlantic seaboard.

Pre-Neolithic wild boar and aurochs remains are severely limited in Britain, and the largest collections have been studied as part of this study. Especially if Neolithic animals are to be compared with the Mesolithic wild animals it is essential to understand how British animals relate to their counterparts in other parts of Europe and other periods. Denmark has provided an abundance of material with which the British wild fauna can be compared. Wild boar in prehistoric southern Scandinavia has been intensively investigated and has been the focus of biometrical analysis (for instance by Rowley-Conwy 1995, 2003; Rowley-Conwy and Dobney 2007). The data used here was collected by Peter Rowley-Conwy and Keith Dobney, and is discussed in Albarella *et al.* (2009). The measurements come from the Early Mesolithic sites of Holmegaard, Lundby, Mullerup, and Svaerdborg; Kongemose and Bloksbjerg that both date to the Middle Mesolithic; and from the Late Mesolithic sites of Agernæs, Flynderhage, Nivaa, Norslund and Sludegaard (Albarella *et al.* 2009).

Denmark also provides one of the largest biometrical studies of aurochs remains conducted to date. Degerbøl and Fredskild (1970) studied aurochs remains that date from the pre-Boreal to the Boreal, and the research included material from excavations and stray finds from Danish bogs. In addition to the detailed study of aurochs remains from Denmark the same publication provided biometrical data from cattle recovered from the Danish Køkkenmøddinger (kitchen middens) that date to the Ertebølle period (5300-3950 cal BC). Comparisons between the process of Neolithisation in Denmark and Britain are often made due to the similarity of the situation in the two areas (*i.e.* both are located on the northern fringe of the European landmass, both potentially had long periods of contact between Mesolithic and Neolithic groups before Neolithic practices were fully adopted, *etc.*). It will be interesting to determine whether the same patterns of change found within British faunal assemblages are shared with the Danish transition.

The Netherlands are one of the closest parts of continental Europe to the British mainland, and as such are an important area for comparison with British wild and domestic fauna. Wild boar biometrical data from the Dutch Mesolithic that is included in this chapter were collected by Umberto Albarella and Roel Lauwerier (Albarella *et al.* (2009), and come from a single site: Hardinxfeld-Giessendam Polderweg.

Portuguese wild boar measurements were collected by Simon Davis. They came from the sites of Cabeço da Arruda, Cabeço do Pez, Moita do Sebastião and Poças de Sao Bento and are discussed in Albarella *et al.* (2005). They all date to the Mesolithic and will be included in the comparison.

Pre-Neolithic Spanish material was collected by Angelos Hadjikoumis as part of his PhD research at the University of Sheffield (Hadjikoumis 2010). Pre-Neolithic material came from Cueva de Cubío Redondo, Cueva de Zatoya, Cueva de Covalanas, and Cueva de Zatoya.

Finally, measurements collected by Umberto Albarella and Keith Dobney from the Italian Mesolithic sites of Grotta Dell'Uzzo, Grotta della Madonna and Grotta delle Mura (Albarella *et al.* 2009) will also be used for comparison.

Early Neolithic material

The number of Early Neolithic sites with large numbers of cattle and pig remains from southern Britain is relatively small. This is particularly the case for pig, which tends to be rarer than cattle in this period (a trend that is seen to reverse during the Late Neolithic). Windmill Hill has, however, provided a large assemblage of cattle and, to a lesser extent, pig remains. The site was utilised during multiple periods in prehistory, with the Early Neolithic contexts providing bones from cattle, fewer numbers of pigs, and sheep/goat. Very few wild animal remains were found (Grigson 1999). Windmill Hill is located in Wiltshire and was excavated initially by Alexander Keiller (Keiller 1934; Smith 1958; 1959; 1965), subsequently by Isobel Smith between 1957 and 1958 (Smith 1958; 1959; 1965), and most recently by a team headed by Alasdair Whittle in 1988. The faunal remains were investigated by Caroline Grigson, and the information used for comparison in this chapter has been taken from her published work (Grigson 1965; 1999). Bones included here came from a variety of contexts including an old land surface, and the primary and secondary fills of three concentric ditches. The old land surface provided the earliest date from the site (BM-73 4910±150BP, 3775-3755 cal BC or 3705-3695 cal BC; (Ambers and Housley 1999, 119). The youngest ^{14}C date (BM-2673 4310±60BP, 3040-2960 cal BC or 2940-2880 cal BC) was obtained from cattle bone retrieved from the upper secondary fill of the outer ditch (Ambers and Housley 1999, 119).

In addition to the Windmill Hill remains, further direct comparison can be made between the primary data gathered as part of this research and the site of Ascott-under-Wychwood, Oxfordshire. Ascott-under-Wychwood is an Early Neolithic Cotswold-Severn Chambered Long Barrow that was excavated by Don Benson between 1965 and 1969 (Benson and Whittle 2007). The primary construction of the monument has been dated to 3760-3700 cal BC (Bayliss *et al.* 2007). It is therefore contemporary with Windmill Hill. Faunal remains from the site, which included a relatively small number of cattle bones, were investigated and published by Mulville and Grigson (2007). Both aurochs and cattle bones were identified. The bones that are included for comparison here came from a pre-monument Neolithic phase and the period of the barrow construction.

West Kennet Long Barrow is a tomb of the Cotswold-Severn type located in Wiltshire (Thomas and Whittle 1986). The absolute date of this monument is unclear, but Thomas and Whittle (1986) have suggested a sequence that began in the later part of the Early Neolithic and that may have continued until the Late Neolithic. Analysis of the animal remains was undertaken originally by Clarke and Lisowski (Piggott 1962, 53) and a reassessment was made for an MSc thesis by Liddle (1998). Faunal remains from the site were not abundant, but did include both cattle and pig bones (Piggott 1962; Liddle 1998). Measurements from the two species were included in Liddle's 1998 dissertation but were not linked to the different phases of activity at the site and potentially come from an extended time period. Despite the limitations of the available data, the site has been included for comparison. It is important due to its relationship with other monuments in the area (for instance Windmill Hill) and its similarity to sites such as Hazleton North and Ascott-under-Wychwood.

The Early Neolithic in Britain will also be compared with contemporary European material. This will include biometrical information from Cerny, France, that dates to the late 5th millennium BC (Tresset 1997). Tresset (2003) has postulated a Paris Basin origin for domesticated animals in the British Early Neolithic, and a comparison of these two areas might help to support or refute this idea. In addition, comparison between British *Sus* and material from Middle Neolithic Rocadour (Lot, France) (Lesur cited in Boyle 2006) will be made.

The Dutch site of Schipluiden will also be used. This site has been dated to 3630-3380 cal BC (Louwe-Kooijmans and Jongste 2006). Activity at Schipluiden is chronologically similar to the Early Neolithic in Britain. The economic system, however, was quite different and inhabitants continued to rely heavily on wild resources (Zeiler 2006).

Finally, the work of Hadjikoumis has provided some biometrical information from Early Neolithic *Sus* remains in Iberia and will be used for comparison here (Hadjikoumis 2010). Measurements from the sites of Cova Fosca, Cueva Chaves, La Draga, Los Castillejos and San Pau are included.

Late Neolithic material

The comparison between The Late Neolithic sample used in Chapter 3 and Chapter 4 with contemporary material is of particular importance as the Late Neolithic sample used for the study was provided by a single site, Durrington Walls. Any broad interpretation of the Late Neolithic is problematic as Durrington Walls may prove to be unique, and should not necessarily be considered representative of the Late Neolithic in general. Available faunal assemblages from the period tend to be small and again, it is expected that Late Neolithic sites in Britain will provide more significant assemblages of pig remains than cattle bones. Three main sites have been used for comparison with Durrington Walls (Barrow Hills, West Kennet palisade enclosures and Puddlehill, locations given on Figure 5.1), and the remains from a number of other Late Neolithic sites will be commented on briefly.

Barrow Hills is a monument complex near Radley in Oxfordshire, which has a history of investigation that began with its discovery by OGS Crawford in the early 20th century and was most recently excavated by Alistair Barclay and Claire Halpin between 1983 and 1985 (Barclay *et al.* 1999). The limited Late Neolithic faunal assemblage was investigated by Levitan and Serjeantson (1999). Late Neolithic bones from this site come from the first phase of Ring Ditch 611 and a number of pits (Levitan and Serjeantson 1999).

The West Kennet palisade enclosures are part of the Avebury/Silbury Hill complex in Wiltshire. The palisade enclosures have been dated to 2600/2500 cal BC to 2200-2100 cal BC, dates that place it within the final phases of the Neolithic in the area (Whittle 1997, 139). Faunal remains were the focus of two MSc theses at the University of Sheffield (Edwards 1994; Horne 1994; Edwards and Horne 1994) from which measurements used in this comparison were taken.

Numerous Late Neolithic pits were excavated at Puddlehill in Bedfordshire and produced a small amount of faunal material that is a useful comparison with Durrington Walls. The animal bones from Pits 1, 2 and 3 at Puddlehill were investigated by Ewbank, who identified pig along with domestic cattle and aurochs remains (Field *et al.* 1964, 364-366). A subsequent investigation by Caroline Grigson concentrated on the animal bones from Pit 6, work that provided evidence of an assemblage containing wild boar, pig, aurochs and possibly domestic cattle (Grigson 1976). Pottery of the Late Neolithic Clacton, Durrington Walls and Woodlands Grooved Ware styles have been identified from pits at the site (ceramic material from Pits 1, 2 and 3 was identified as Rinyo-Clacton ware; Field *et al.* 1964, 362-363), which is now known as Grooved Ware (Smith 1976).

Clearly the sites that have been chosen from the British Neolithic are not exhaustive and a number of other sites will also be discussed. These, however, have provided only very few specimens, or very few published measurements that were taken in a way that is consistent with the current study. These sites include Mount Pleasant henge located on chalkland in Dorset (Harcourt 1979; Wainwright 1979), Maiden Castle hillfort (Armour-Chelu 1991; Sharples 1991), and the enclosure at Marden, Wiltshire (Harcourt 1971; Wainwright *et al.* 1971). All of these sites are also included in Figure 5.1.

The site of Newgrange in County Meath, Ireland, provides a good comparison with the Late Neolithic contexts at Durrington Walls. The assemblage from Newgrange is the only one of significant size from the Irish Late Neolithic (Waddell 1998, 30). The site itself has a series of ^{14}C dates that suggest it was constructed around 2500 cal BC (van Wijngaarden-Bakker 1986, 19). Animal bones were studied originally by van Wijngaarden-Bakker (1986) and all of the cattle data included in this chapter came from her published report.

The Late Neolithic of the Netherlands is roughly contemporary with the dates from Durrington Walls. The sites of Vlaardingen, Zandwerven and Hekelingen in the Netherlands have been used for comparison with the British Late Neolithic, these sites were included in the extensive zooarchaeological study conducted by Clason (Vlaardingen and Zandwerven Clason 1967) and Prummel (Hekelingen, Prummel 1987). Vlaardingen and Zandwerven are described as Eneolithic, while Hekelingen is referred to as Late Neolithic and has been dated to 3000-2600 cal BC. The reliance on hunting in the Early Neolithic of Holland was mentioned above, but in the Late Neolithic this seems to be variable and largely dependent on local conditions.

The comparison with Iberia will be continued into the Late Neolithic period. The Spanish Chalcolithic period is contemporary with the British Late Neolithic. Durrington Walls pig remains will be compared with those collected by Hadjikoumis (2010).

Roman Wroxeter

The final site with which comparison will be made is Roman Wroxeter (*Viroconium Cornoviorum*). This site provided a large amount of wild boar measurements that were collected by A. Hammon (2005) Hammon 2005, Albarella *et al.* 2009), and it will be used to investigate changes in the size of wild boar through time in Britain.

5.2 A comparison between British wild boar and those from Europe

Introduction

Recent work has highlighted the great diversity of wild boar across their modern distribution. In Europe two main trends have been isolated in wild boar: 1) a size increase from south to north; 2) a size increase from west to east (Albarella *et al.* 2009). The west/east and south/north clines can be explained by a number of factors. In the case of the west/east variation continentality (which causes more extreme temperature variation in areas removed from the moderating effects of the sea) and temperature (western Europe is warmed by the Gulf Stream) have been cited as possible causes of smaller size in the west (Albarella *et al.* 2009). The south/north cline is probably also linked to temperature (though, possibly indirectly) and its occurrence supports Bergmann's rule (Albarella *et al.* 2009; Davis 1981). Britain, therefore, being located as it is on the central western periphery of Europe provides a rather complex situation regarding the size of mammals and the factors that influence their body size. In theory, wild boar from Britain should be smaller than those found further to the east on the European mainland. But in contrast the more northerly latitude of Britain should also influence the body size of *Sus scrofa* and it would be expected that British wild boar will be larger than their counterparts to the south.

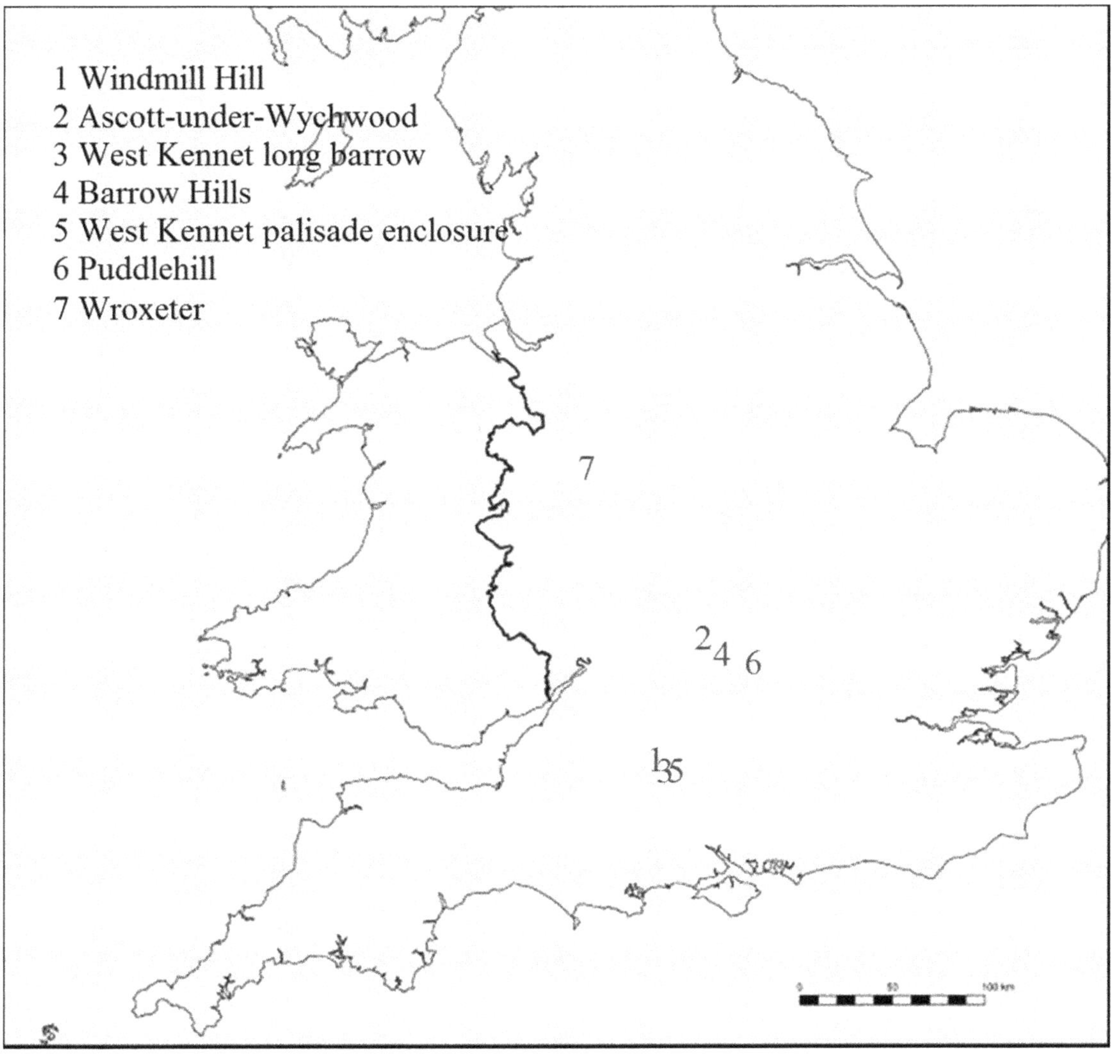

Figure 5.1 The location of British Neolithic and Roman sites used for comparison.

It is important to understand the position of British wild boar in relation to other populations in Europe. Establishing a clear picture of wild boar size in Britain is particularly important when addressing the issue of the appearance of domesticated types, as domestic status is most commonly established based on reduced skeletal size compared to wild counterparts. The following section will compare postcranial and tooth measurements gathered from Mesolithic assemblages in Britain, with data from Mesolithic Netherlands, Denmark and Italy (Albarella *et al.* 2009), Portugal (Albarella *et al.* 2005; 2009) and Spain (Hadjikoumis 2010).

Biometrical analysis

Figure 5.2 and Figure 5.3 compare the mandibular second molar length and width measurements collected by me from Britain, to those collected by Albarella *et al.* (2009) from the Netherlands and Denmark (appropriate measurements were not available from the Iberian peninsula or Italy).

According to the results in Figure 5.2 (which compares the length and anterior width of the second molar) the British second molar measurements are considerably smaller than those from Denmark. A small area of overlap between the British and Danish Mesolithic measurements is apparent, but the Danish specimens are clearly in the larger part of the British plot. The few specimens from the Dutch Mesolithic overlap with the British and Danish sample, being towards the largest extent of the former and the smallest end of the latter. The same pattern is clear in the comparison of second molar length and posterior width (Figure 5.3), but there seems to be more variability between the measurements within different regions. A similar finding was discussed in Chapter 3, where the anterior width of the second molar was shown to vary more through time than the posterior width.

Length measurements were not available for all of the areas under investigation (specifically the Italian and Portuguese Mesolithic and the Spanish pre-Neolithic), so the comparison has been made using the two second molar width measurements (Figure 5.4). The results of the Danish, Dutch and British material have much the same relationship as in the previous plots, with large Danish, small British and the Dutch material falling almost exclusively within the area of overlap between the two.

Further than acknowledging that all of the Spanish pre-Neolithic and the Italian Mesolithic specimens tend to fall in the lower half of the distribution of teeth from other areas of Europe relatively little information can be gathered due to the small sample size. The Portuguese specimens on the other hand are comparatively numerous. Portuguese wild boar tend towards the smaller end of the British distribution. This would suggest a generally smaller size in the Portuguese animals.

In Figures 5.2 and 5.4 there is a single large British outlier. This outlier (from Thatcham, Berkshire) was identified in Chapter 3, but its comparison with material from other European sites makes clear that the animal is of a size comparable with Danish wild boar. Two scenarios could have led to the inclusion of this animal on a British Mesolithic site: either British wild boar had the potential to reach very large size equivalent to the largest Danish individuals or this is an introduced animal.

Plots of the second molar cusp widths compared to length allow the identification of differences in the variation of the cusps. In the second molar measurements plotted here between group variation is gradually less pronounced towards the back of the tooth (*i.e.* there is more difference between the anterior widths, than the posterior widths). This geographical variation in tooth cusps is interesting due to some of the results discussed in Chapter 3 that found that the anterior width was less different between groups of the Mesolithic and Neolithic than the posterior width was. The relationship between changing cusp size in pig teeth is extremely complex, but it would appear that the geographical variation between pig groups and the temporal variation within Britain have different effects on teeth, and, therefore, different causes. Perhaps then the change from wild to domestic (*i.e.* from the Mesolithic to the Neolithic in Britain) causes changes in the posterior width of teeth, while environmental/climatic differences are more prone to cause change in the anterior width.

In general mandibular third molar measurements provide a similar story (Figures 5.5, 5.6 and 5.7). Again, Danish and Dutch wild boar are larger than the ones recorded from Britain. But in contrast with the second molar evidence there is no overlap between the Dutch and British specimens. This may purely be a result of the small Dutch sample size. Portuguese specimens are smaller than those from Britain, as are the majority of Italian specimens. The small size of Italian specimens has been explained as a result of partial insular dwarfism (Albarella *et al.* 2009) as the data derive from a site in Sicily, and reduction in size of mammals is known to occur as a result of confinement on islands.

Tooth measurements from British wild boar fit well with the trends identified for Europe by Albarella *et al.* (2009). In terms of the west/east size increase expected to influence the wild boar population British teeth tend to be smaller than those from Holland. The difference is small, and this reflects the similarity between the two countries in terms of climate and average temperatures. The south/north cline is also clear. British wild boar are larger than the Portuguese ones, most likely a result of the higher temperatures enjoyed at more southerly latitudes. Finally, the British wild boar teeth are smaller than those from Denmark and again, this can most readily be explained by different temperatures in the two areas. The higher temperatures in Britain because of Gulf Stream warming and a more southerly latitude are conducive with smaller wild boar size when compared to southern Scandinavia.

Discussion will now turn to postcranial remains from wild boar in those regions we have already discussed. Distal humerus measurements from pre-Neolithic sites in Britain and Europe have been plotted in Figure 5.8 below. In some respects the postcranial bones mirror the mandibular teeth. Danish Mesolithic wild boar are substantially larger than those from the British Mesolithic, although the smallest Danish measurements are comparable to the largest British ones. Mesolithic wild boar from Holland overlap with the British Mesolithic animals, but they tend to be in the upper part of the British range and a number of the specimens are much larger. The main difference between teeth and postcranial bones can be seen in the Portuguese material. Portuguese wild boar teeth were noticeably smaller than the British, but the humerus measurements from the two areas exhibit distributions that are almost identical. The single pre-Neolithic specimen from Spain is at the extreme lower left of the plot, a pattern that was also hinted at from tooth measurements in Figure 5.4.

Astragalus measurements were available from the Spanish pre-Neolithic period and have been compared to British Mesolithic specimens in Figure 5.9 in order to further explore the relationship between these two areas. In contrast to the second molar measurements (Figure 5.4) and distal humerus measurements (Figure 5.8) from pre-Neolithic Spain, the astragali are larger than their British Mesolithic counterparts. The Spanish specimens are towards the upper distribution of the British and a number are larger than those from the British Mesolithic. Tentatively, there are two groups in the Spanish material, and it can be suggested that those that overlap with the British measurements are from smaller female wild boar.

It is unclear whether the differences seen in the postcranial elements from Britain and those from Europe contribute to general trends or whether they apply only to specific measurements or elements. In Figure 5.10 the postcranial measurements from British and Portuguese Mesolithic and Spanish pre-Neolithic sites have been compared using the log ratio technique. Portuguese and British Mesolithic postcranial remains are larger than the standard (*i.e.* to the right of the vertical line) but similar to each other. The mean of the British material is marginally higher (0.04 compared to 0.03), but in general the postcranial remains suggest animals were very similar biometrically from the two areas. The log ratio comparison provides the same results as the distal humerus measurements. In contrast the Spanish material suggests a generally larger population. The mean (0.06) is higher than both the British and Portuguese postcranial elements. However, the sample size from pre-Neolithic Spain is small and the addition of a few samples could change the results significantly. Unfortunately, comparable tooth measurements are not abundant from pre-Neolithic Spain and therefore cannot be used to compare with the British wild boar.

The large size of postcranial bones relative to teeth in the Portuguese wild boar has been identified previously and also suggested for other wild boar groups in Europe (Albarella *et al.* 2005, 43). The pattern is not reproduced in the British material in which there is no difference in the relative size of bones and teeth found on Mesolithic sites.

Roman wild boar

A large body of data from Britain is available from excavations at Roman Wroxeter (Hammon 2005). Roman Britain is outside the main period with which this research is concerned, but a comparison with the material will help to identify trends in the size of wild boar within the British Isles. Wild boar were extant in Britain for some time after the Mesolithic, and various changes in the size of the animals have been envisaged.

Distal tibia measurements from the Mesolithic and Roman period have been plotted in Figure 5.11, and astragalus measurements in Figure 5.12. It is very likely that at Wroxeter the bulk of the population was domestic, whereas the spread of large outliers – more clearly distinct in the astragalus plot – is represented by wild boar. In both figures the Mesolithic wild boar tend to plot in between the domestic and wild Roman clusters. This on the one hand indicates the clear reduction in average pig size due to domestication and, on the other, the likely increase in wild boar size after the Mesolithic, a trend previously identified for Britain and other European regions (Albarella *et al.* 2009, 129). An alternative explanation for the large size of the Wroxeter wild boar is that they represent the result of an introduction, but this does not seem to have been the case as wild boar of similarly large size have been identified, though in small numbers, also in later prehistoric sites in Britain (Albarella 2010).

Third and second molar plots (Figures 5.13 to 5.15) show no occurrence of large outlier in the Wroxeter assemblage. Wild boar therefore do not seem to be represented in these plots, probably as a consequence of the fact that they were killed off-site and that the heads may have been disposed of at the killing site. As a consequence, these diagrams do not allow any direct comparison between British Mesolithic and Roman wild boar.

Conclusion

Researchers interested in the geographical variation of wild boar have noted a west/east cline in their body size, but the antiquity of this variation is still uncertain (for instance Albarella *et al.* 2009 cannot rule out a modern influence for the change in size with more easterly location). By comparing the measurements from Mesolithic Britain with those from other parts of Europe it is, however, clear this variation has a time length, and that it confirms the expected variation in wild boar size. However, it must also be noted that while the teeth of Portuguese animals are smaller than their more northerly counterparts in Britain, their bones are not. The situation therefore is more complex than merely larger or smaller size, and therefore skeletal elements need to be assessed in isolation as well as in combination in order to reconstruct correctly the factor affecting morphometric patterns.

The size of the British wild boar did not remain stable in the course of its history and my data confirm the previously detected increase in the size of the wild boar after the Mesolithic.

5.3 Early Neolithic pigs in Britain

Large collections of pig remains from Early Neolithic Britain are rare, and as a result there is little additional data available. Nevertheless pig remains from Windmill Hill (Wiltshire, Grigson 1999) and Ascott-Under-Wychwood (Oxfordshire, Mulville and Grigson 2007) will be compared with results from the early Neolithic assemblages re-analysed as part of this research: Hambledon Hill, Runnymede Bridge, Hazleton North and Coneybury Anomaly. In Chapter 3, postcranial remains from the main Early Neolithic sites were compared and found to be extremely similar to one another. Here the results from all sites will be combined.

In terms of their postcranial remains, the pigs from Windmill Hill plot well within the distribution of the Early Neolithic sites studied as part of this research (Figure 5.16). The Windmill Hill mean is substantially lower but this may be the result of the combined effect of small sample size, the absence of large outliers and the occurrence of a very small specimen at this particular site. We cannot therefore be confident of a genuine smaller size of the Windmill Hill pigs compared to those from contemporary sites.

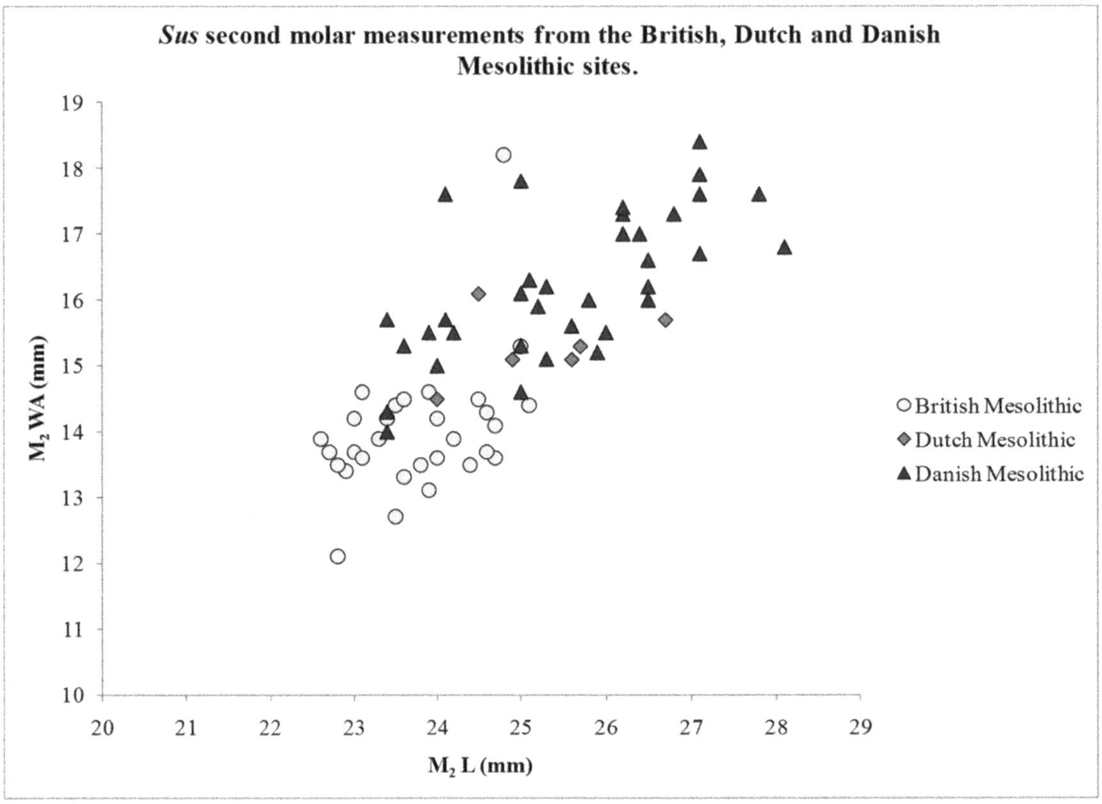

Figure 5.2 ***Sus*** **second molar measurements (L = length, WA = width of the anterior cusp) from British, Dutch and Danish Mesolithic sites. Dutch and Danish material supplied by Albarella *et al.* 2009. Measurements follow Payne and Bull 1988.**

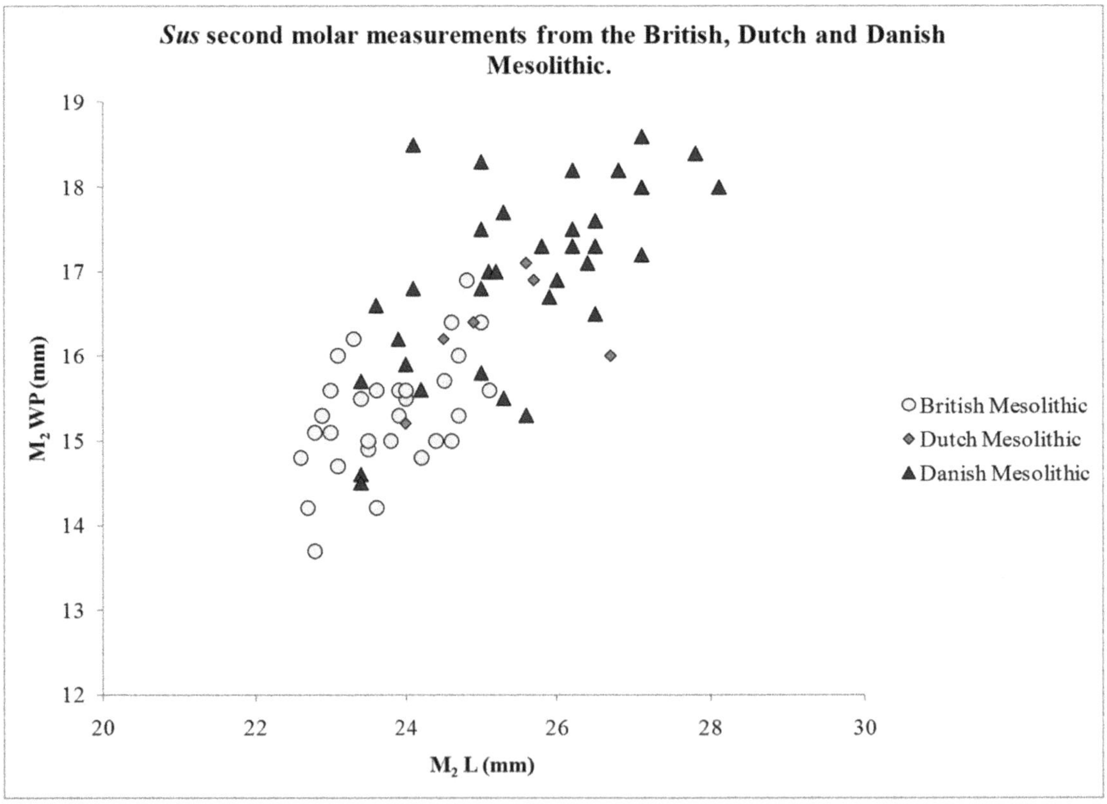

Figure 5.3 ***Sus*** **second molar measurements (L = length, WP = width of the posterior cusp) from British, Dutch and Danish Mesolithic sites. Dutch and Danish material supplied by Albarella *et al.* 2009. Measurements follow Payne and Bull 1988.**

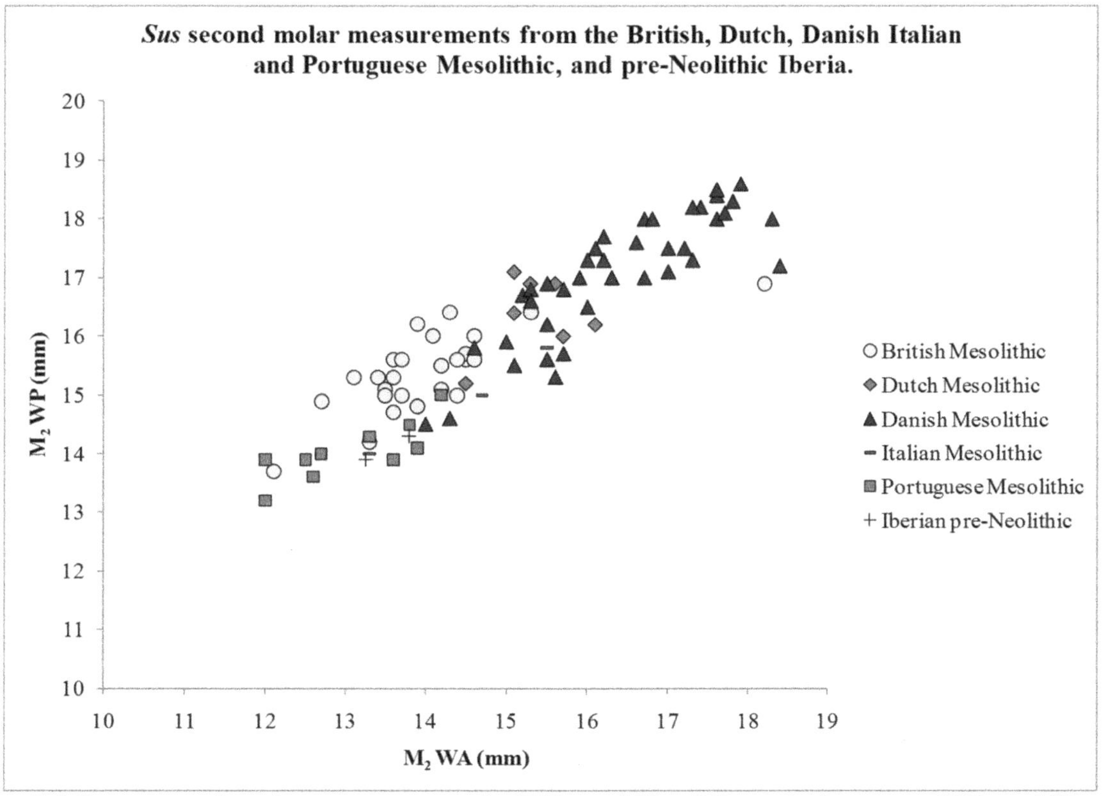

Figure 5.4 ***Sus*** **second molar measurements (WA = width of the anterior cusp, WP = width of the posterior cusp after Payne and Bull 1988) from British, Dutch, Danish, Italian and Portuguese and pre-Neolithic Iberia. British data was collected by the author, Dutch, Danish, Italian and Portuguese provided by Albarella *et al.* (2009), and data labelled as pre-Neolithic Iberia from Hadjikoumis (2010).**

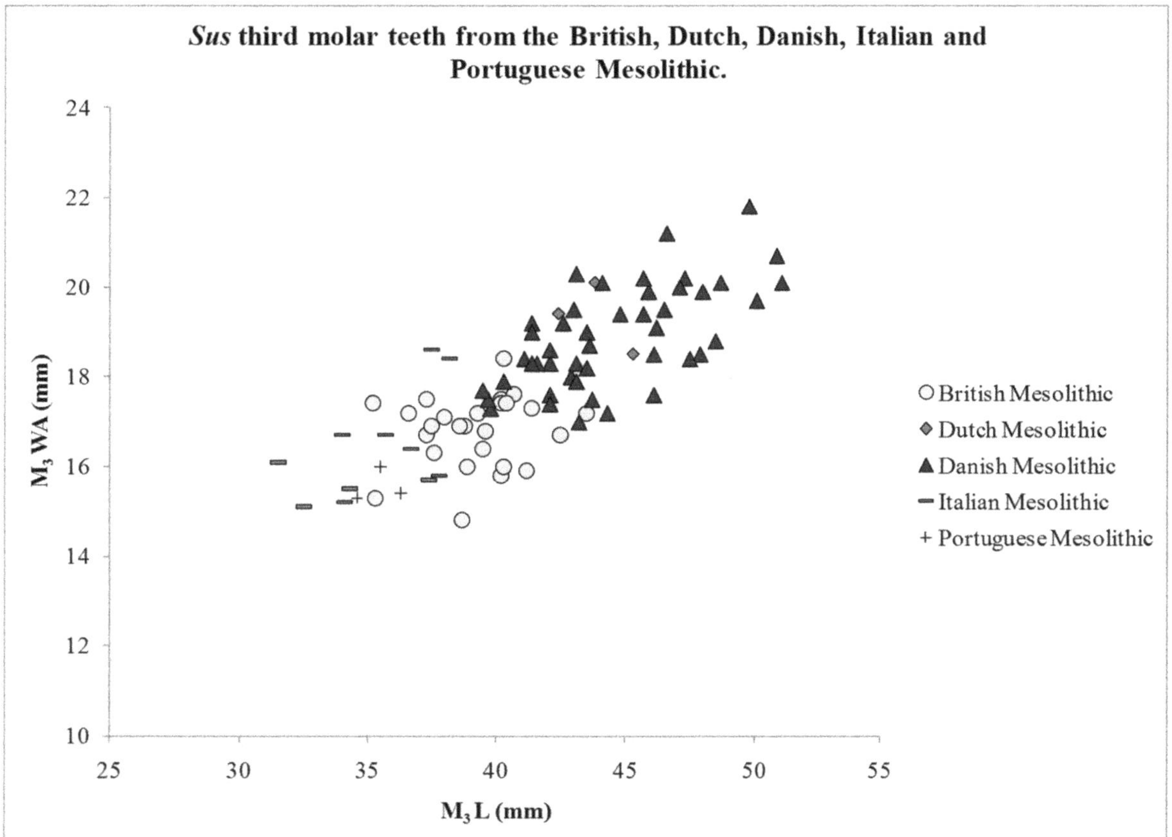

Figure 5.5 ***Sus*** **third molar measurements (L = length, WA = width of the anterior cusp following Payne and Bull 1988) from the British, Dutch, Danish, Italian and Portuguese Mesolithic.**

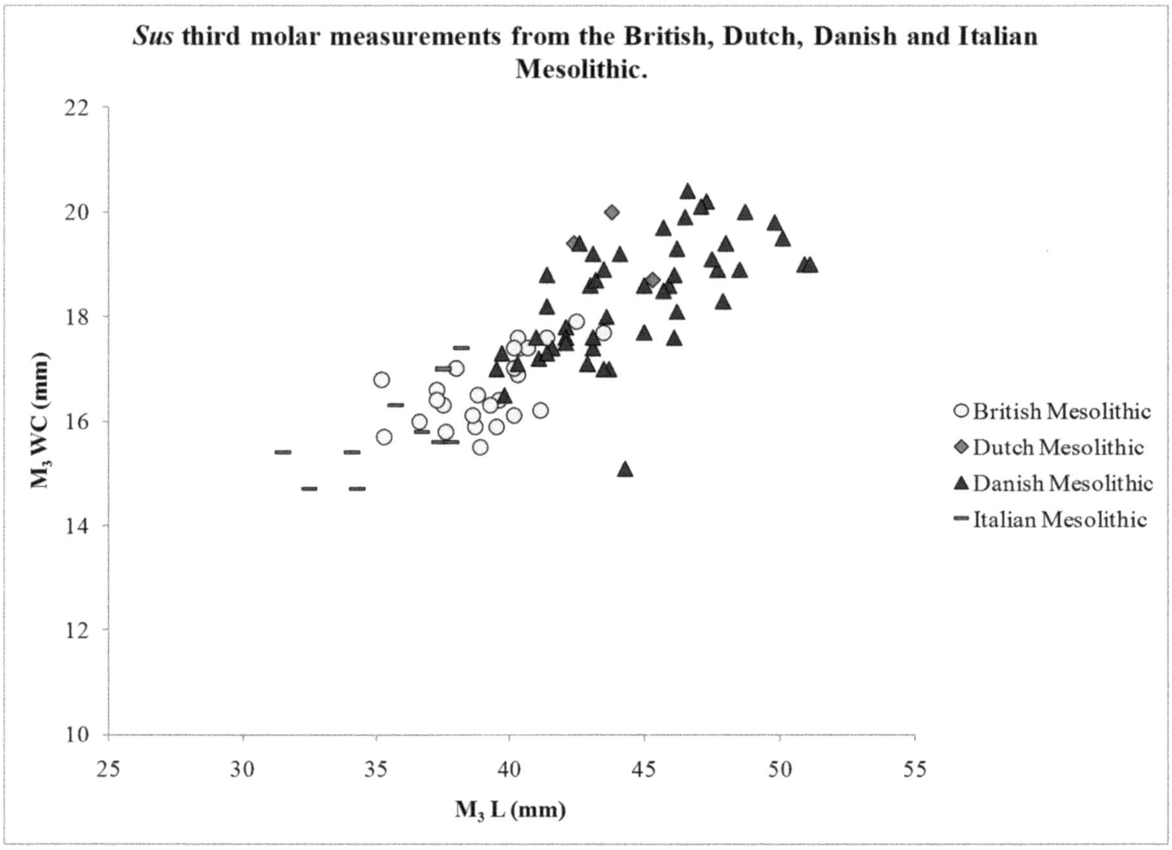

Figure 5.6 ***Sus*** **third molar measurements (L = length, WC = width of the central cusp following Payne and Bull 1988) from the British, Dutch, Danish, Italian and Portuguese Mesolithic.**

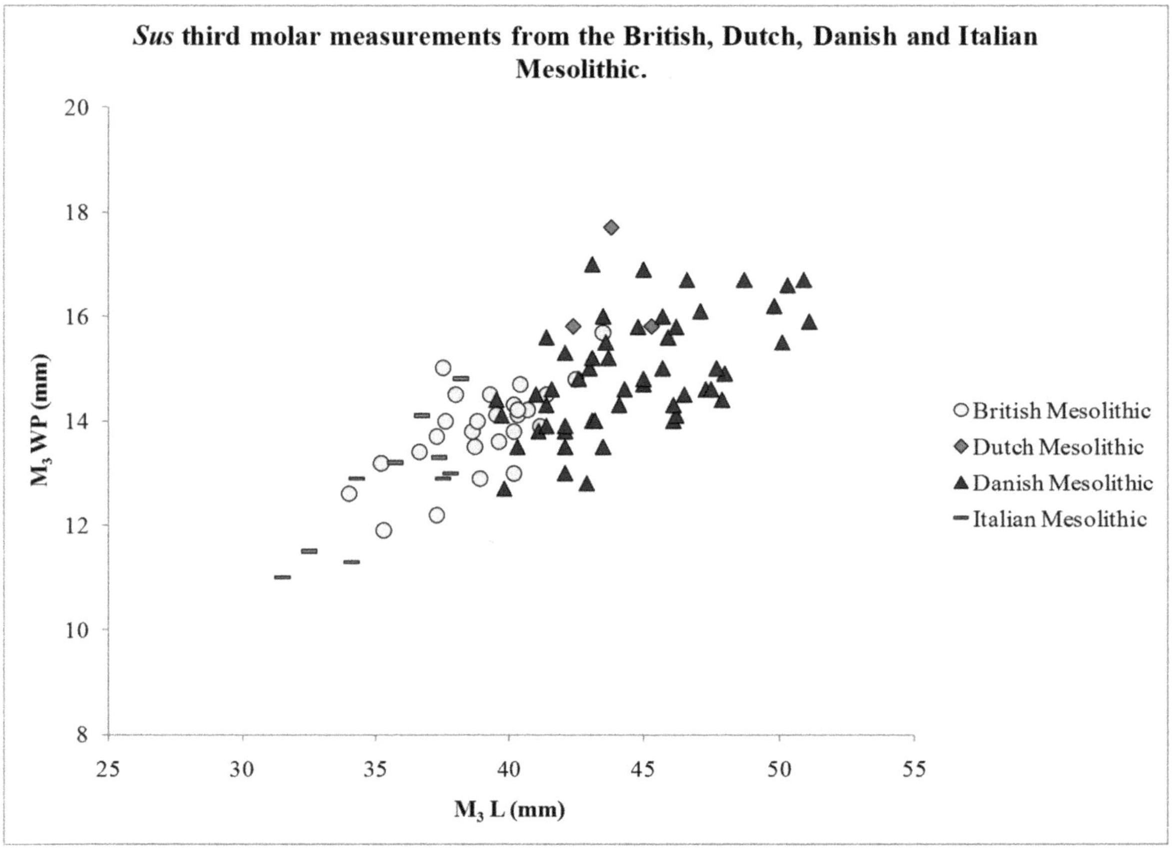

Figure 5.7 ***Sus*** **third molar measurements (L = length, WP = width of the posterior cusp following Payne and Bull 1988) from the British, Dutch, Danish, Italian and Portuguese Mesolithic.**

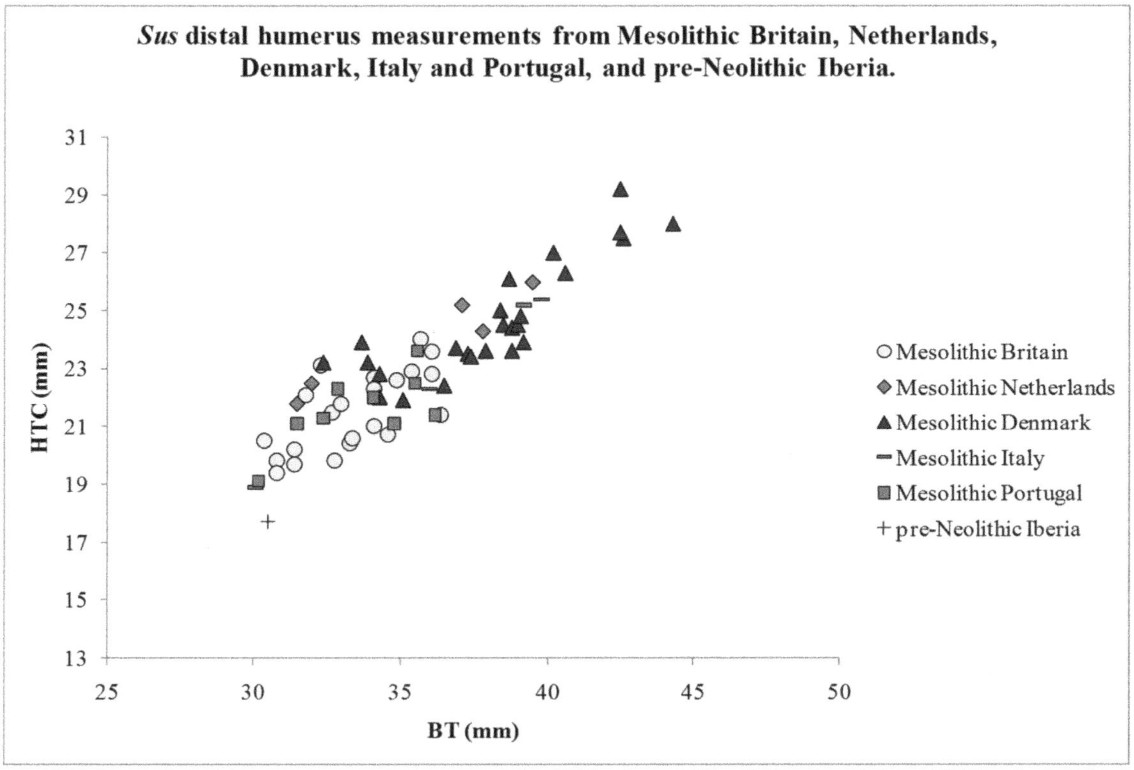

Figure 5.8 ***Sus*** **distal humerus measurements (BT = breadth of the trochlea, HTC = height of the trochlea after Payne and Bull 1988) from Mesolithic Britain, Netherlands, Denmark, Italy and Portugal, and pre-Neolithic Iberia. Only fully fused specimens have been included.**

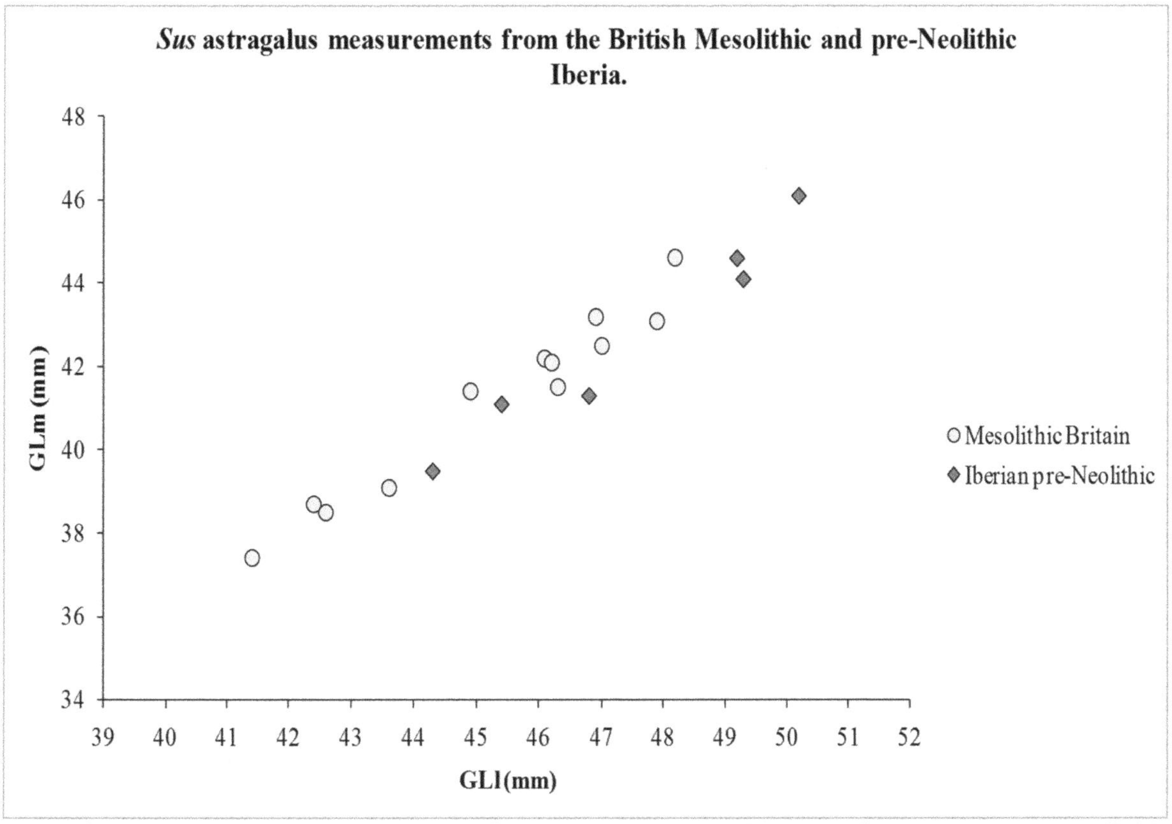

Figure 5.9 ***Sus*** **astragalus measurements (GLl = greatest lateral length, GLm = greatest medial length after Payne and Bull 1988) from the British Mesolithic and pre-Neolithic Iberia.**

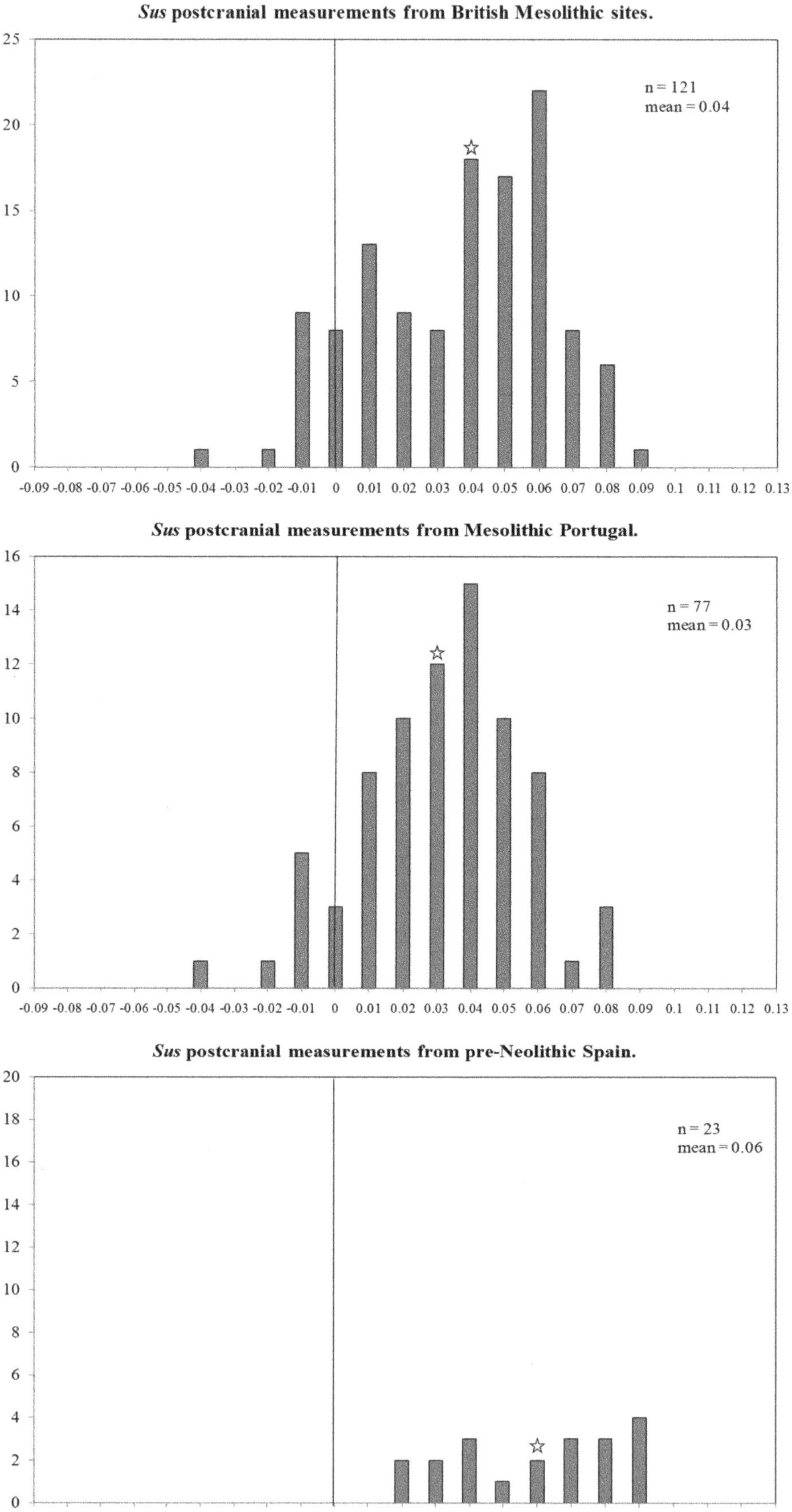

Figure 5.10 *Sus* postcranial measurements from Mesolithic Britain (upper diagram), Mesolithic Portugal (middle diagram, data from Albarella *et al.* 2009) and pre-Neolithic Spain (lower diagram, data from Hadjikoumis 2010). Fully fused postcranial bones have been included (excluding the scapula). The mean is marked by a star, and the standard is marked by the vertical black line at 0.

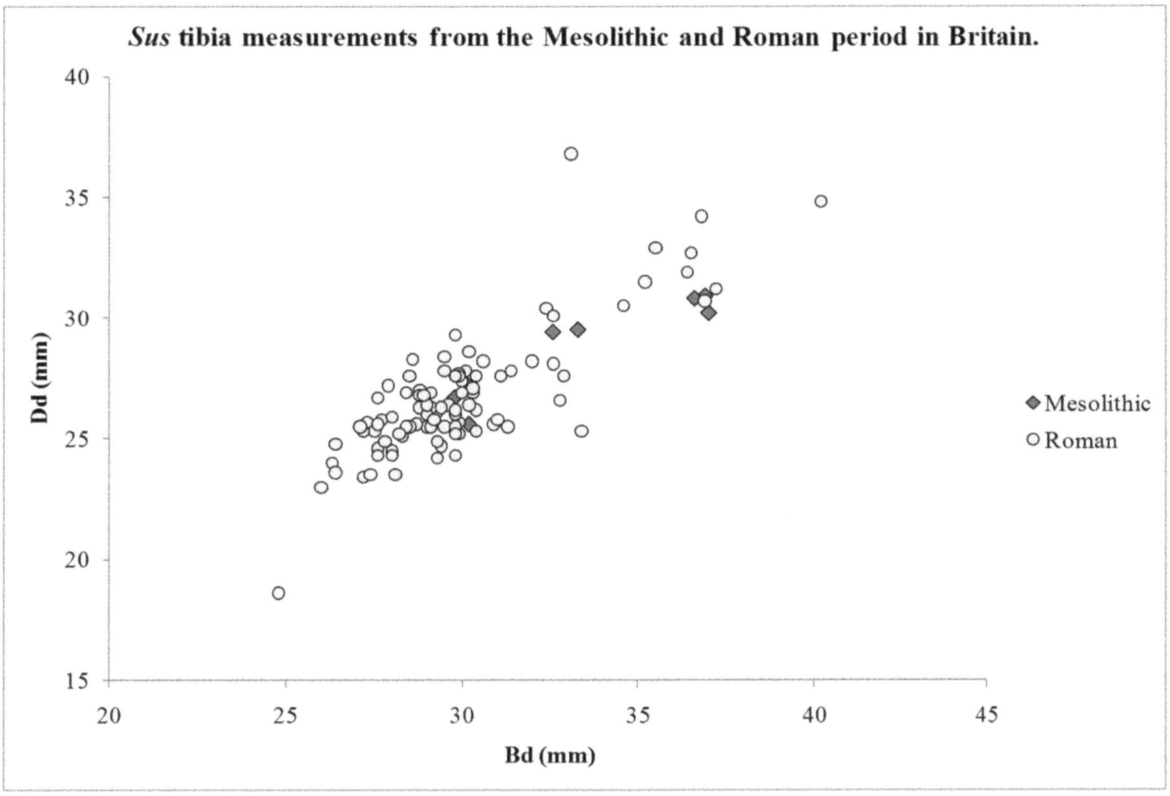

Figure 5.11 ***Sus*** **tibia measurements (Dd = depth distal end, Bd = breadth of the distal end after Payne and Bull 1988) from the Mesolithic and Roman period (Wroxeter). Roman measurements are assumed to be from wild boar.**

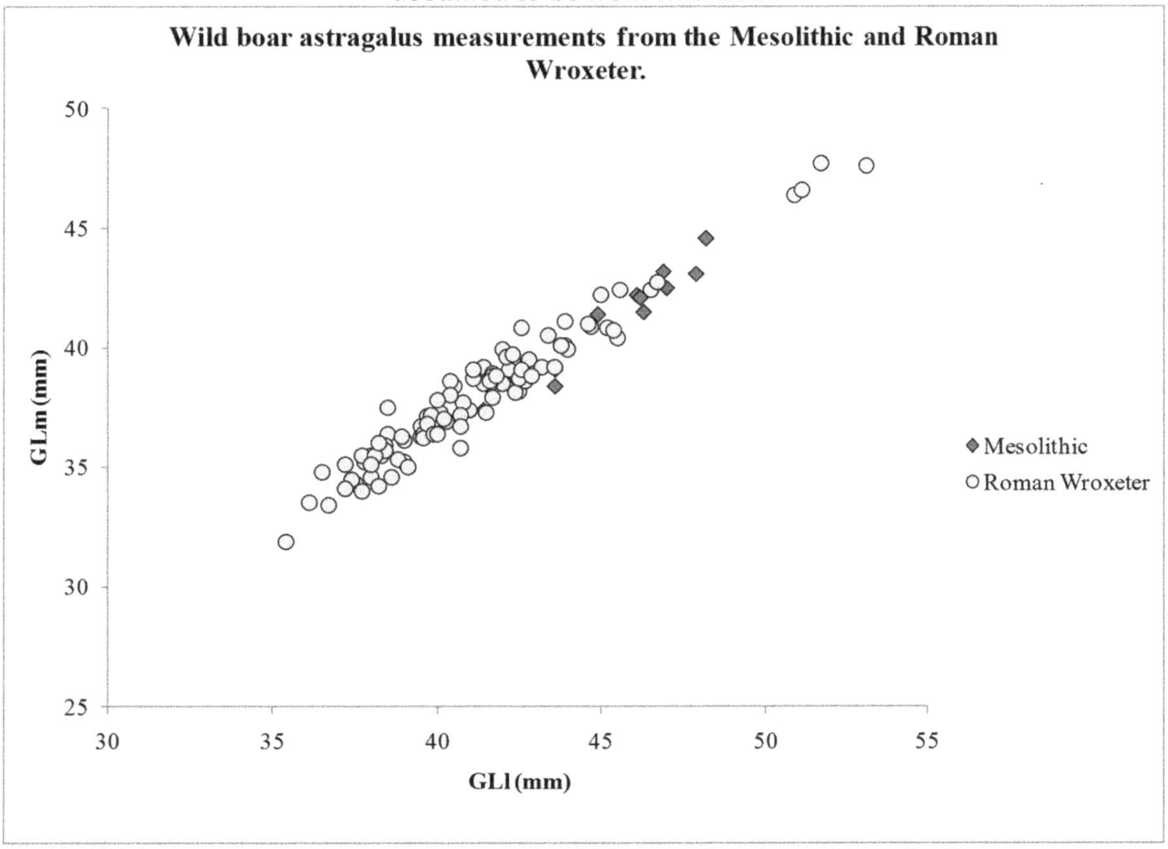

Figure 5.12 Wild boar astragalus measurements (GLl = greatest lateral length, GLm = greatest medial length von den Driesch 1976) from the Mesolithic and Roman (Wroxeter) periods.

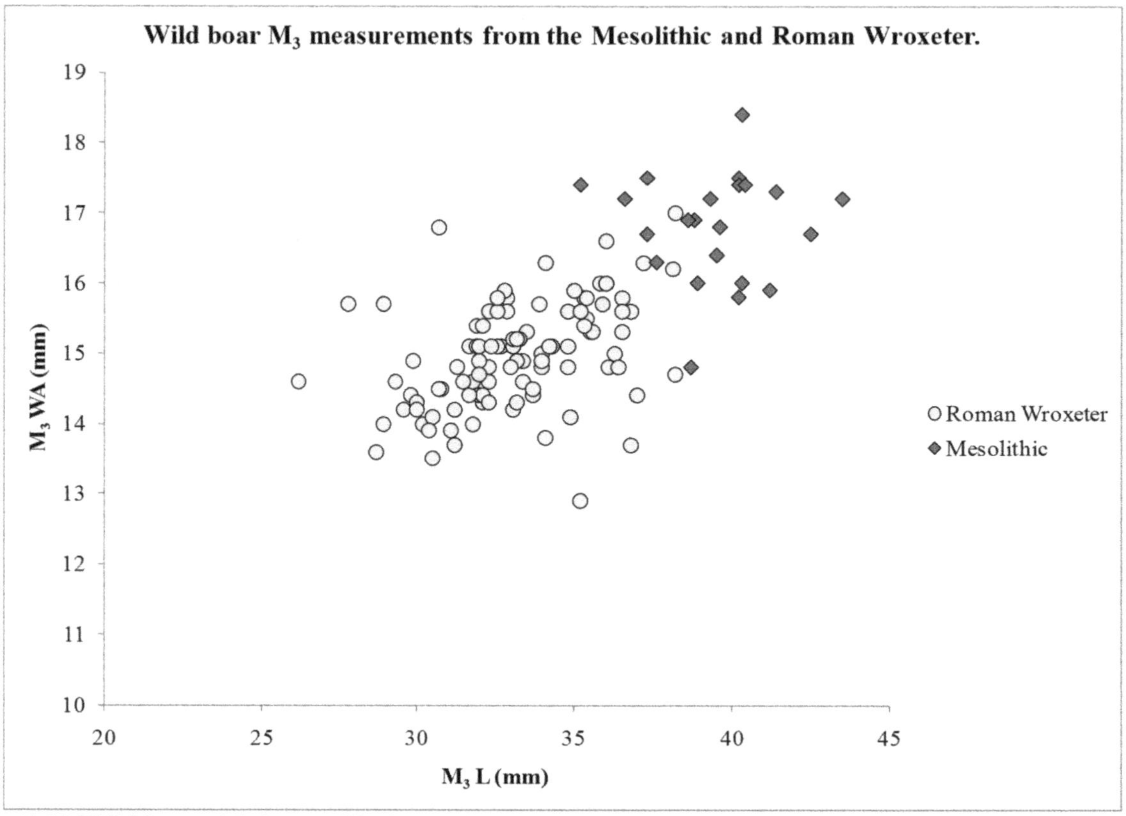

Figure 5.13 Wild boar third mandibular molar measurements (L = length, WA = width of the anterior cusp after Payne and Bull 1988) from the Mesolithic and Roman (Wroxeter) periods. Both loose teeth and jaws are included.

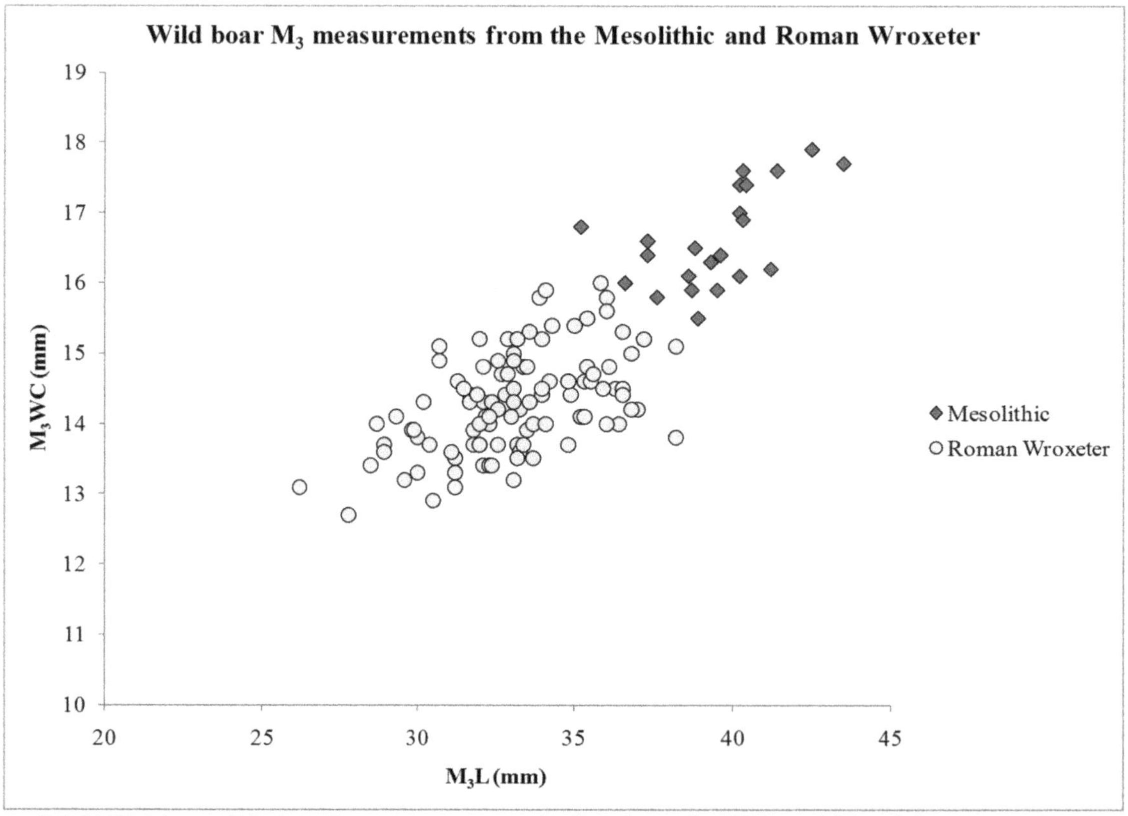

Figure 5.14 Wild boar third molar measurements (L = length, WC = width of the central cusp after Payne and Bull 1988) from the Mesolithic and Roman (Wroxeter) periods. Both loose teeth and jaws are included.

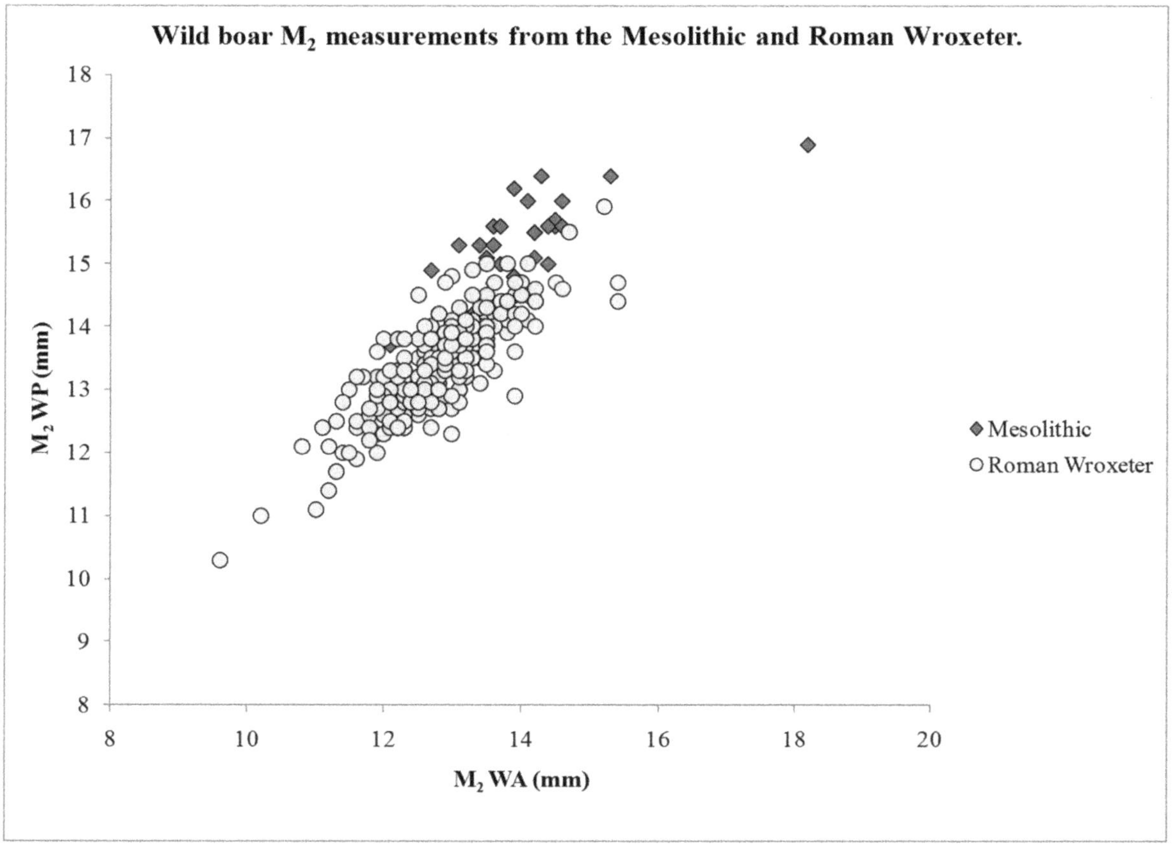

Figure 5.15 Wild boar second molar measurements (WP = width of the posterior cusp, WA = width of the posterior cusp after Payne and Bull 1988) from the Mesolithic and Roman (Wroxeter) periods. Both loose teeth and jaws are included.

In contrast, the Ascott-Under-Wychwood postcranial remains (shown in Figure 5.17) are all at the larger end of the Early Neolithic plot. This large size led to a mean value of 0.07, much higher than the Early Neolithic mean from sites studied here. All of the specimens were recorded as wild boar by the original specialists (Mulville and Grigson 2007). If the measurements are compared to Mesolithic wild boar specimens they are certainly large enough to be wild animals. However, the substantial overlap between wild boar of the Mesolithic and domesticates of the Early Neolithic make it impossible to securely identify these animals as wild or domestic. No very large specimens comparable to the outliers on the Early Neolithic plot were present.

Tooth measurements were uncommon from both of these sites and cannot be presented graphically. Worth mentioning is a large mandibular third molar from Ascott-Under-Wychwood (Mulville and Grigson 2007), which had a length of 40.5mm. This is larger than the majority of Early Neolithic specimens and only two specimens (from Hambledon Hill) are larger. All three of these should probably be considered as wild boar.

A comparison between British and European Neolithic pigs

Evidence of Early Neolithic pigs is more abundant from the European mainland. Astragalus measurements for the British Early Neolithic, the Early and Late/Final Neolithic in Spain and the Early Neolithic of Schipluiden (Netherlands) are plotted (Figure 5.18) Early Neolithic sites in Spain include domestic, wild and possibly crossings (Hadjikoumis 2010), which explains the wide range of variation of pigs from this region and period. The British range overlaps greatly with that from Spain though the very largest specimens are missing and most measurements tend to cluster in the lower two thirds of the distribution, probably indicating a less diversified scenario. Late Neolithic samples from Spain and the Early Neolithic of the Netherlands are consistent with the existence of well distinct small (domestic) and large (wild) populations, but the samples are tiny and require cautious interpretation.

As a single element the astragalus cannot exhaust our analysis of patterns of variation in the Early Neolithic of western Europe. Therefore postcranial measurements from Holland, France, Spain and Britain have been plotted using the log ratio technique in Figure 5.19. Early Neolithic *Sus* remains from Britain, Holland and Spain areas are largely consistent in size, but the British material represents the only group that plots clearly unimodally with large outliers. This presumably indicates a largely domestic population with a few genetically distinct wild boar, while in Spain and the Netherlands more mixing seems to occur. Such mixing may concern population management but in the Spanish case also reflects the combination of sites that experienced very different patterns of animal exploitation.

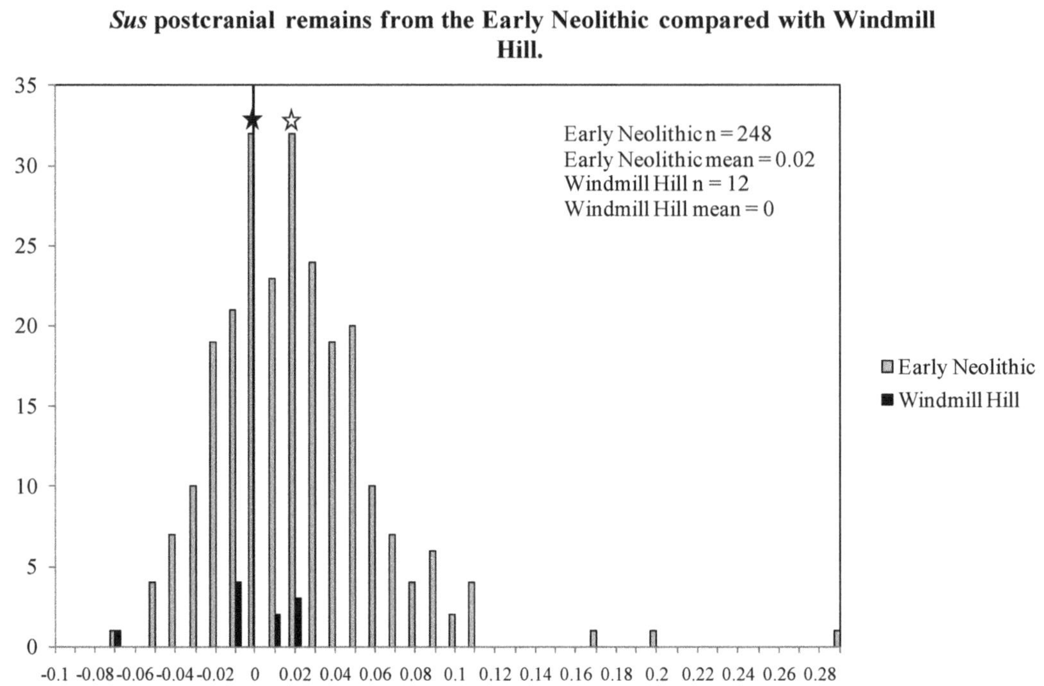

Figure 5.16 ***Sus*** **postcranial remains from combined Early Neolithic sites used in this study compared to Windmill Hill. Fused postcranial remains (excluding the scapula) have been included. The standard is marked by the vertical black line at 0. The Windmill Hill mean is marked by the black star, and Early Neolithic mean is marked by the white star.**

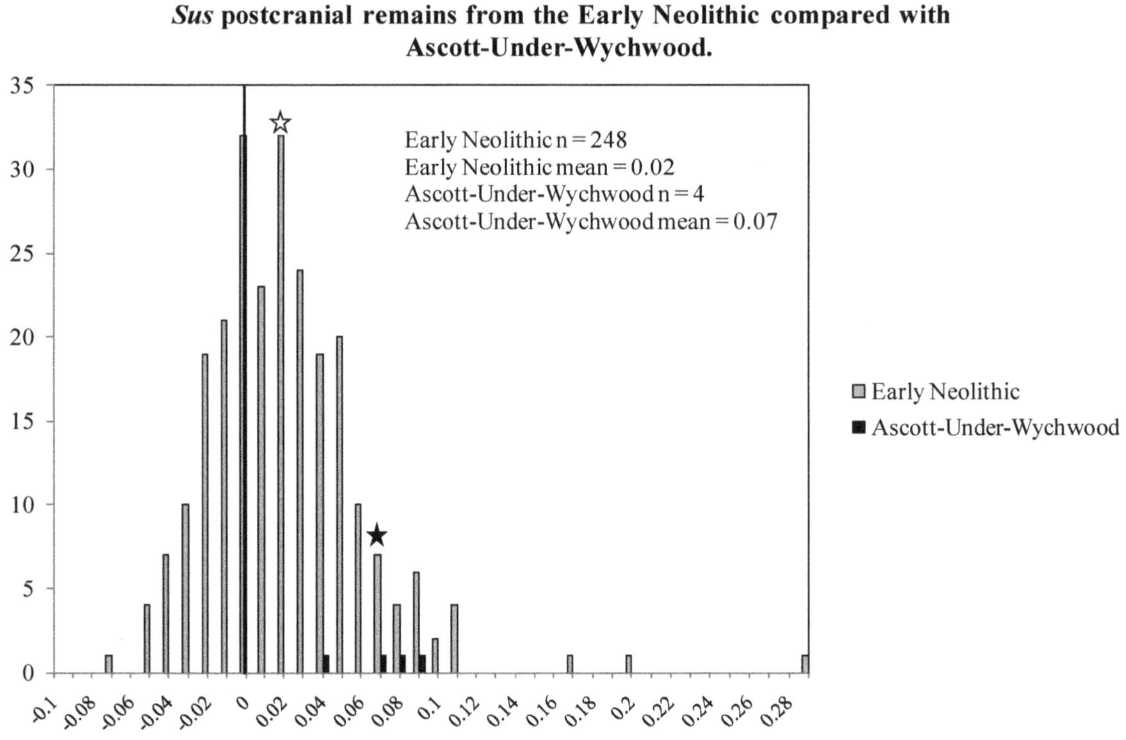

Figure 5.17 ***Sus*** **postcranial remains from combined Early Neolithic sites used in this study compared to Ascott-Under-Wychwood (Mulville and Grigson 2007). Fused postcranial remains (excluding the scapula) have been included. The standard is marked by the vertical black line at 0. The Ascott-Under-Wychwood mean is marked by the black star, and Early Neolithic mean is marked by the white star.**

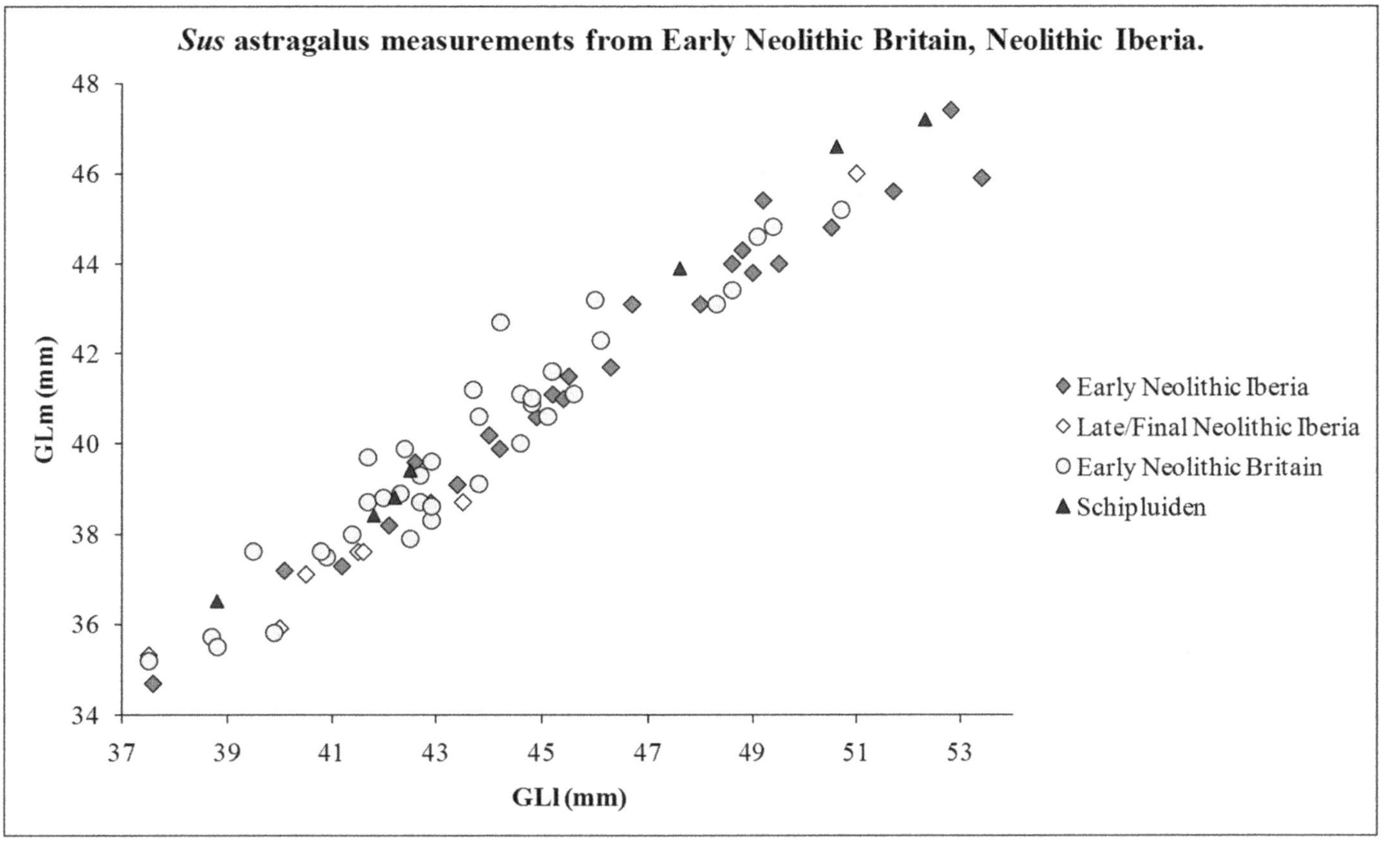

Figure 5.18 *Sus* astragalus measurements (GLl = greatest lateral length, GLm = greatest medial length Payne and Bull 1988) from the Early Neolithic in Britain, Neolithic Spain, and Early Neolithic Schipluiden.

The *Sus* remains from Middle Neolithic Rocadour Cave, France, are markedly different from those of the Early Neolithic of Britain, Holland and Spain. The animals from Rocadour are significantly larger than the standard (the mean value is 0.10). Although there is some overlap between these pigs and those from other areas, it is the largest British, Dutch and Spanish specimens that are comparable in size to the smallest French specimens. The large size of the pigs from Rocadour, compared to pigs from other parts of Europe, supports the wild status of animals from this site.

Despite its geographical distance from the British study area the Rocadour Cave wild boar are a useful comparative assemblage, especially given the extreme paucity of our evidence for the presence and characteristics of this species in Britain during the Early Neolithic. Some of the Early Neolithic British specimens fall into the range of wild boar measurements from Rocadour, including those from Ascott-Under-Wychwood that were identified as wild boar by the original investigators. The few large British outliers are even larger than the specimens from Rocadour and may represent animals that were especially sought after for their exceptionally large size. As such they may only be indicative of the upper end of the wild boar distribution in Early Neolithic Britain. It is thus possible that the slight right tail in the British distribution also includes some animals that were wild and are in fact compatible in size with those from Rocadour. Despite the obvious geographic differences, the size of the French Middle Neolithic wild boar, combined with the larger size of the British Mesolithic specimens, leaves, however, little doubt that the bulk of the British Early Neolithic pigs is represented by domestic animals.

Tooth measurements were unavailable from Holland and Portugal in the Early Neolithic, and therefore only British and Spanish tooth widths have been compared in Figure 5.20. Tooth measurements from the two areas have the same mean, both of which are identical with the standard. Contrary to the bones the Spanish material has a more restricted range compared with the British, which indicates that tooth size is less affected by the population variation mentioned for Spain, than postcranial bones.

Conclusion

Despite potential differences in domestication patterns and husbandry strategies there is no striking morphometric difference between British Early Neolithic pigs and those from other European countries. Windmill Hill pigs are all well within the range from other important sites in southern Britain. The large size of the Ascott-Under-Wychwood pigs by and large confirms the original interpretation of them being wild, though we have seen that wild boar of even larger size were indeed present in the same period in Britain. Since no measurements from smaller domestic pigs are available at

Ascott-Under-Wychwood it is likely that at this site wild boar hunting played an important role. However, the generally low frequency with which it is found on archaeological sites of the period provides evidence that wild boar was not of central importance in the provision of food.

By comparison with Middle Neolithic French wild boar it is clear that some of the British Neolithic pigs are far too small to be wild boar, but also that British wild boar of the period could potentially be larger than those from France.

In general an overall similarity rather than difference seems to characterise the morphometry of Early Neolithic domestic pigs across different European areas. Small variations, however, also exist and we must also not forget that, however useful, large scale analysis tends to be crude and will mask more nuanced patterns of variation in pig exploitation, which more detailed, smaller scale analysis, has proved to have occurred.

5.4 Late Neolithic pigs in Britain and Europe

Introduction

Despite the commonly cited dominance of cattle in the Early Neolithic and pigs in the Late Neolithic of southern Britain there is a surprising paucity of large well preserved collections of pig remains from the period. As a result, the comparison between Durrington Walls and other Late Neolithic British sites will be brief, and will be followed by a more substantial comparison between British and European material. The main European sites that will be used for the Late Neolithic are Hekelingen III (Holland, Prummel 1987), Late Neolithic/Chalcolithic Leceia and Zambujal (Portugal, Albarella *et al.* 2005) and a collection of Late Neolithic/Chalcolithic sites from Spain (data collected by Angelos Hadjikoumis).

Late Neolithic pigs in Britain

Very few measurements were available from other Late Neolithic sites in Britain. Figure 5.21 plots the only measurements available from Barrow Hills along with the same measurements from Durrington Walls. Despite the tiny sample the Barrow Hills specimens are clearly comparable with the largest pigs from Durrington Walls. Two potential explanations can be proposed. The first is that these are the remains of wild boar, and if that is the case it provides good evidence that some wild boar were included in the Durrington Walls sample too. Such a situation is not unique in the British Neolithic, since at Ascott-Under-Wychwood (discussed above) the only *Sus* bones found appear to be from wild boar. In addition, the large specimens from Durrington Walls and Barrow Hills (along with Ascott-Under-Wychwood) are of a similar size to the animals from Rocadour. The other possibility is that there was very substantial difference between the animals at different Late Neolithic sites in southern Britain. Although this cannot be ruled out until more Late Neolithic *Sus* remains are found, it seems to be a much less likely possibility.

Late Neolithic pigs in Europe

So far, the pig remains from Neolithic Britain have been comparable in size to their counterparts in Europe. In Figure 5.22, below, measurements of postcranial remains from Late Neolithic/Chalcolithic Spain and Portugal have been plotted using the log ratio technique. Here, for the first time, the postcranial remains of pigs from the Iberian Peninsula are smaller than those from Britain. For the first time there is a very stark difference in the postcranial remains of the pigs in Britain and those found in Portugal. Portuguese and Spanish pigs are both, on average, much smaller than their British counterparts. In all of the samples the small number of outliers towards the right of the plot suggest that wild boar were a component of the Late Neolithic assemblages.

From the Mesolithic specimens that were discussed above, it is clear that the relationship between British and Portuguese wild boar and pigs is not straightforward, and that in some respects they are very similar whereas other aspects of the skeleton vary. Therefore, a more detailed consideration of the British Late Neolithic, and the Portuguese Late Neolithic/Chalcolithic data will follow. In Figure 5.23, 5.24 and 5.25 measurements of the mandibular third molar indicate that, unlike the postcranial bones, there is little to distinguish the British from Portuguese third molars. Durrington Walls teeth all fall well within the distribution from both Leceia and Zambujal, but from both sites there are a significant number of much smaller and much larger specimens. Large specimens from Leceia and Zambujal have been identified as wild boar (Albarella *et al.* 2005). In both cases the specimens towards the upper right of the plots are likely to be wild boar, and at Leceia the separation between wild boar and domestic pigs that was first identified by the original investigators is also apparent here (most clearly in Figure 5.23 and 5.25). Thus, some of the largest specimens from Durrington Walls, by comparison with Leceia and Zambujal, may well belong to wild boar. By comparison, the number of wild boar at Durrington Walls is small, and the tail in the Portuguese sample is more significant.

Unfortunately, published *Sus* remains from Holland are rather limited, and raw data suitable for comparative analysis is difficult to obtain in English publications. Hekelingen is an important Vlaardingen culture site and tooth lengths from this site (Prummel 1987) have been used in Figure 5.26 in comparison with tooth lengths from Durrington Walls. The range of measurements is similar from both sites, recreating the results from the Early Neolithic and in contrast to the slightly larger size exhibited in postcranial remains from the Mesolithic of the country.

5.5 Variation through time

This chapter has found that the wild boar and pigs that inhabited Britain during prehistory were different, to varying degrees, from their continental neighbours. In pre-Neolithic Europe Danish wild boar had the largest teeth and postcranial remains. These were followed in size by Dutch and British wild boar, although the Dutch animals were apparently slightly larger than their British counterparts. Finally, the animals from Iberia and Italy were the smallest investigated. But the small size of the Portuguese animals was not uniform; their teeth were smaller than contemporary animals, but their postcranial bones were akin to the British animals in size.

In the Early Neolithic, evidence from both postcranial remains and teeth suggest great homogeneity in the animals from Holland, Spain, Portugal and Britain. Iberian and British animals exhibit size reduction compared to the earlier Mesolithic period. The degree of change apparent from the Mesolithic to the Neolithic is of a similar degree in all of the areas included.

Finally, in the Late Neolithic the evidence from Britain suggests a small change in size (as discussed in Chapter 3), but this contrasts sharply with the evidence from Portugal and Spain. In Iberia, a clear reduction in the postcranial bones of pigs is evident, and results in a population that is much smaller than the one collected from Durrington Walls. The evidence is limited, but the pigs from Holland appear similar, at least in terms of their postcranial remains, to those from Durrington Walls.

5.6 British pre-Neolithic aurochsen

Introduction

This section will explore the comparability of British pre-Neolithic aurochs remains to European *Bos primigenius*. Unlike pigs, there are few studies of Mesolithic aurochs remains with which British material can be compared. By far the largest study of aurochsen was made by Degerbøl and Fredskild (1970), and this data will provide the basis for comparison with the English *Bos primigenius* remains.

Grigson (1969) argued that there was no significant difference in the size of aurochs from northern Europe, but that those from northern Europe (specifically Denmark) were larger than those from more southerly parts of Europe (specifically Hungary). This assertion has been challenged by results from the Czech Republic (Kyselý 2008), though it is possible that the effects of temperature and climate could have affected aurochs and caused variation in their size in different geographic areas.

Biometrical analysis

Figure 5.27 provides a comparison of *Bos primigenius* measurements from Ilford, the Mesolithic in Britain and aurochs from Denmark collected by Degerbøl and Fredskild (1970). The astragalus measurements illustrate both the change in size of aurochsen through time, and also potential differences between different geographical areas. Measurements from Ilford (that date to MIS 7, 186-242kya, Andy Currant pers. comm. 2008) are the largest, with only a single specimen overlapping with the next largest group - the males from Denmark. The Mesolithic animals from Britain overlap with the Danish material, though the males tend to plot in the lower half of the Danish male distribution. The Danish sexed specimens help us to discern two sex groups in the British material too (as was discussed in Chapter 4). The female groups from Britain and Denmark are particularly analogous, with even the outlying specimens being similar from the two areas. As mentioned, a slight difference can be seen in the male distribution – the British Mesolithic specimens are all at the smaller end of the Danish male sample. This could suggest that the British animals tended to have a lesser degree of sexual dimorphism, but the small size of the British sample is problematic and could have biased the results somewhat (only three specimens plot with the 'male' group). Grigson (1969) published a mean value of astragalus measurements (greatest length) from Hungary as 84.1mm (based on 96 specimens), along with the mean of values from Star Carr, which she regarded to be much larger. With the inclusion of post-glacial aurochs astragalus measurements from other British sites it is necessary to adjust Grigson's mean value to 84.7mm, which is not dramatically larger than their Hungarian counterparts. The mean value itself, however, is of limited value as it is much affected by the proportion of females and males that contribute to it. In the British Mesolithic material there is female to male ratio of 8:3, if the interpretation of the clusters is correct, but this ratio is unknown for the Hungarian material.

In Figure 5.28 the breadth of the humerus trochlea from British and Danish sites has been plotted (HTC measurements are not available from Denmark). All of the Ilford specimens are much larger than the Danish or British Mesolithic humeri, and are clearly separated from the main Danish distribution. In this diagram the largest of the Danish specimens was female according to Degerbøl and Fredskild's identification (1970). This is surprising given the significant sexual dimorphism of aurochs which results in males that are generally much larger than females. It is likely that this represents either a misidentification on the part of the original researchers, or else the absence of the largest males from the sample. Only one British Mesolithic specimen was available, and it falls, well within the Danish sample.

The results from tooth measurements are not so clear as for postcranial bones, due in part to the small sample size from the Mesolithic. Figure 5.29 compares length and width measurements of the lower third molar from Ilford, Denmark (after Degerbøl and Fredskild 1970), and the British Mesolithic and Neolithic periods. The specimens from Ilford are, again, in the upper part of the plot. But unlike the astragalus measurements in which the Ilford specimens were clearly larger than the Danish Holocene animals, measurements of mandibular third molars from Ilford are comparable to the Danish males. They also overlap with Danish females to a degree. Only one Mesolithic specimen was available which is similar in size to the Danish aurochs measurements, but it is impossible to determine whether this was a female or male animal. Two other British Mesolithic specimens with lengths of 46.6mm and 47.7mm respectively reinforce the impression that aurochs third molars from the time were like those from Denmark.

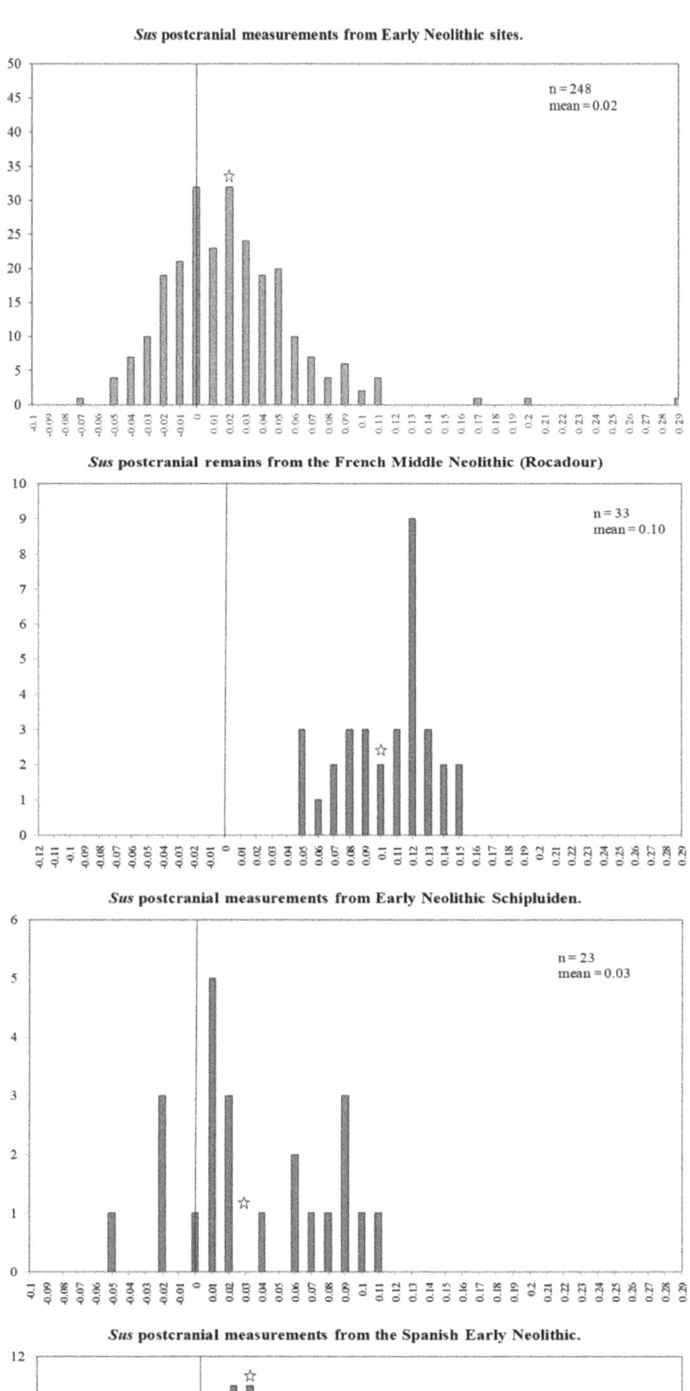

Figure 5.19 *Sus* postcranial measurements from the British Early Neolithic (upper diagram), French Middle Neolithic (second diagram), Early Neolithic Schipluiden (third diagram) and Spanish Early Neolithic (lower diagram). Fused postcranial remains (excluding the scapula) are included. The standard is shown by the vertical black line at 0, and the mean by the star

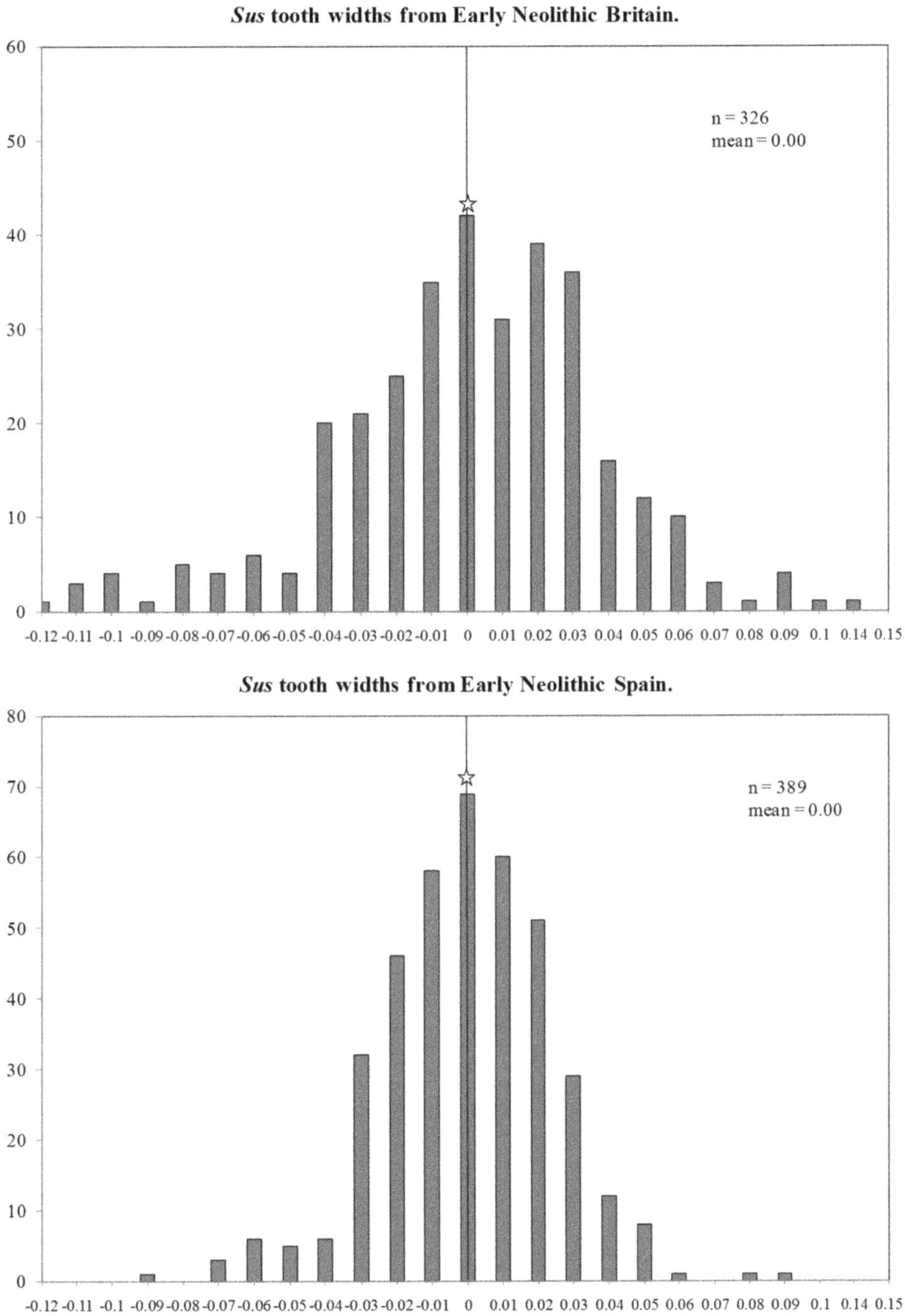

Figure 5.20 *Sus* tooth widths from Early Neolithic Britain (upper diagram) and Early Neolithic Spain (lower diagram). The standard is shown by the vertical line at 0, and the mean identified by a star.

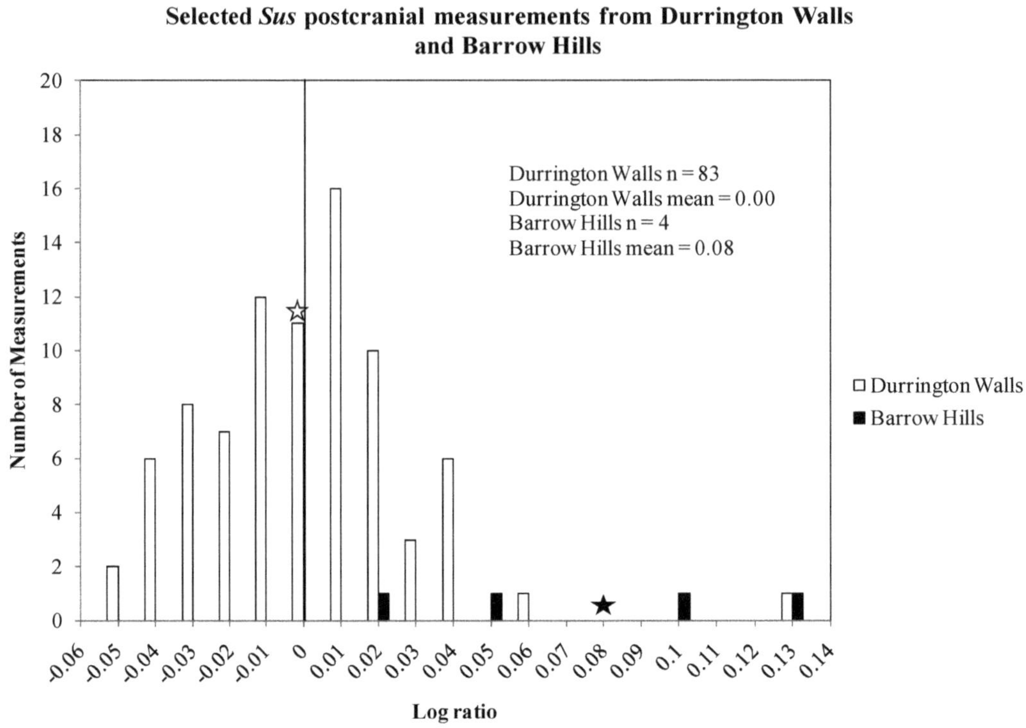

Figure 5.21 Comparison of *Sus* postcranial remains from Durrington Walls and Barrow Hills. The standard is marked by a line at 0, and the means by a black star (Barrow Hills) and a white star (Durrington Walls).

These results have a number of implications for our understanding of aurochs from the Pleistocene onwards. Postcranial bones are, from this limited evidence, sexually dimorphic, and in some cases specimens can be securely attributed to female or male. This does not appear to be the case for teeth, and according to the results of third molar analysis, sexual dimorphism is not pronounced in aurochs teeth. Other teeth and larger samples would be needed to test this assertion adequately. Secondly, a significant size reduction in aurochs has been postulated between the Pleistocene and the Holocene associated with rapid warming (*e.g.* Kyselý 2008; van Vuure 2005). This reduction is clear in the postcranial bones that have been studied, but is not apparent in the (admittedly limited) tooth sample. Size reduction may not have affected the entire skeleton to the same degree. Finally, there is no clear evidence to suggest a difference between British and Danish aurochs size during the Mesolithic. The expectation that, like in pigs, smaller animals might exist in Britain than Denmark because of a warmer climate is not supported by the evidence presented here, but larger samples are needed to investigate this question further.

5.7 Early Neolithic cattle

Introduction

This section will begin by comparing the data collected by the author from Early Neolithic sites in Britain to other published information. The aim of this section is to determine whether cattle were treated the same in different parts of Britain, and whether they are skeletally comparable. Chapter 4 highlighted the characteristics that typified cattle in the Early Neolithic period of Britain, and identified a high degree of homogeneity in both the husbandry practices employed, and the physical characteristics of the cattle deposited at different Early Neolithic sites. Homogeneity was exhibited in both the postcranial and dental elements, and both will be investigated where possible in this chapter. As with pigs, the main Early Neolithic sites to be used on a comparative basis are Windmill Hill and Ascott-Under-Wychwood.

In addition, a comparison will be made with Early Neolithic material from Holland and France. This is an important undertaking given the conclusion drawn from Chapter 4 that domesticated cattle were introduced into Britain from an external source. Incomparability of measurements and a lack of published raw data have all limited the available appropriate faunal assemblages. Two sites will be used in detail here: Cerny, France (Tresset 2005) and Schipluiden, Holland (Zeiler 2007). The Dutch site Schipluiden was chosen because of its geographical location on the coast of Holland, an area that would have been easily within reach of Britain during the Early Neolithic, and its contemporaneity with the Early Neolithic in Britain. Cerny was chosen for its location, and although slightly earlier than the British material investigated for this research it still provides important information (see section 5.1 for details of these two sites). According to the research conducted by Tresset (2003) the Paris Basin is a good contender for the likely origin of domestic cattle in Britain.

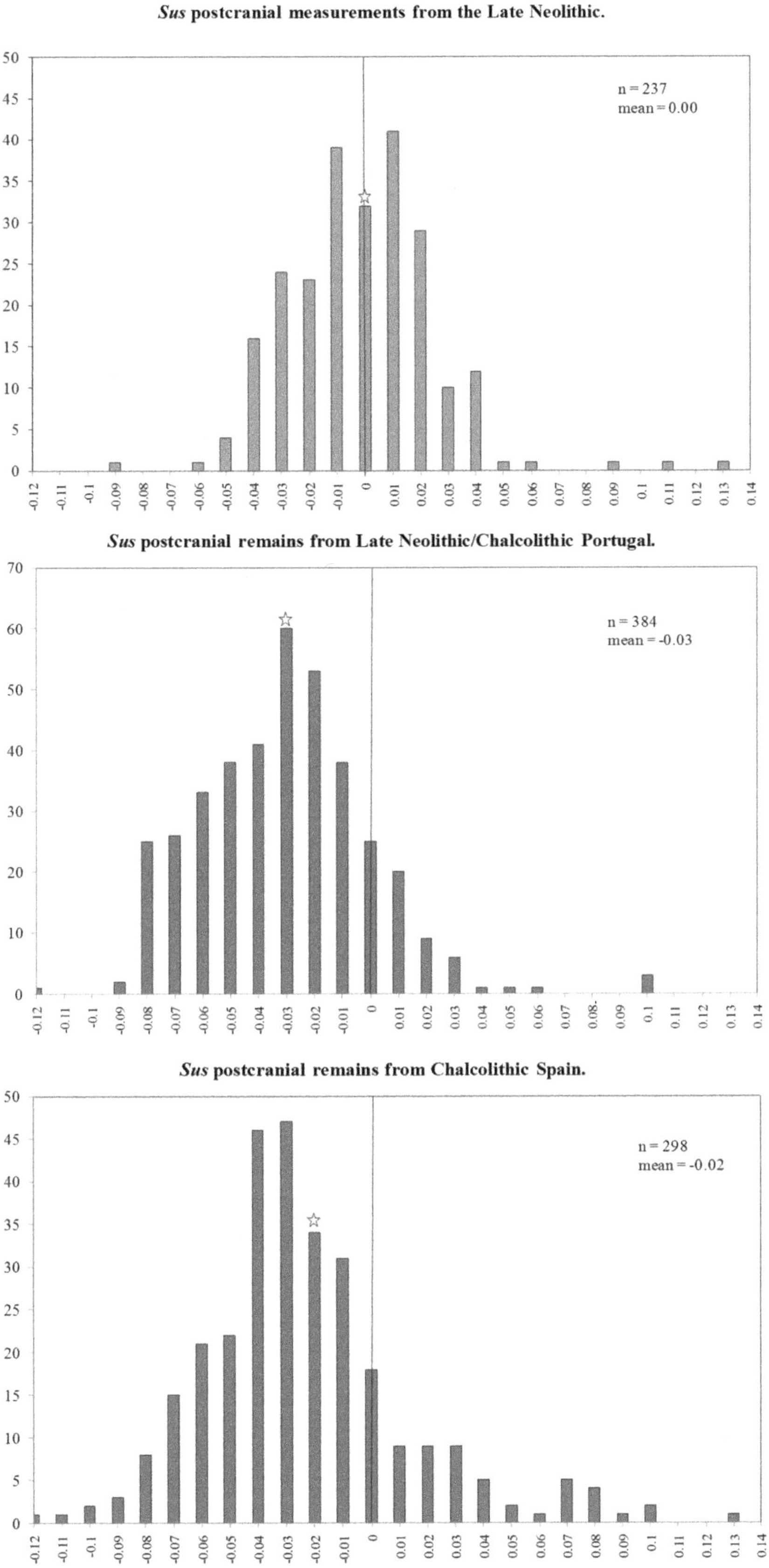

Figure 5.22 *Sus* postcranial measurements from Late Neolithic Britain (upper diagram), Late Neolithic/Chalcolithic Portugal (middle diagram) and Chalcolithic Spain (lower diagram). Measurements from fused postcranial remains (excluding the scapula) have been included. The standard is shown by the black vertical line at 0, and the mean is indicated by a star.

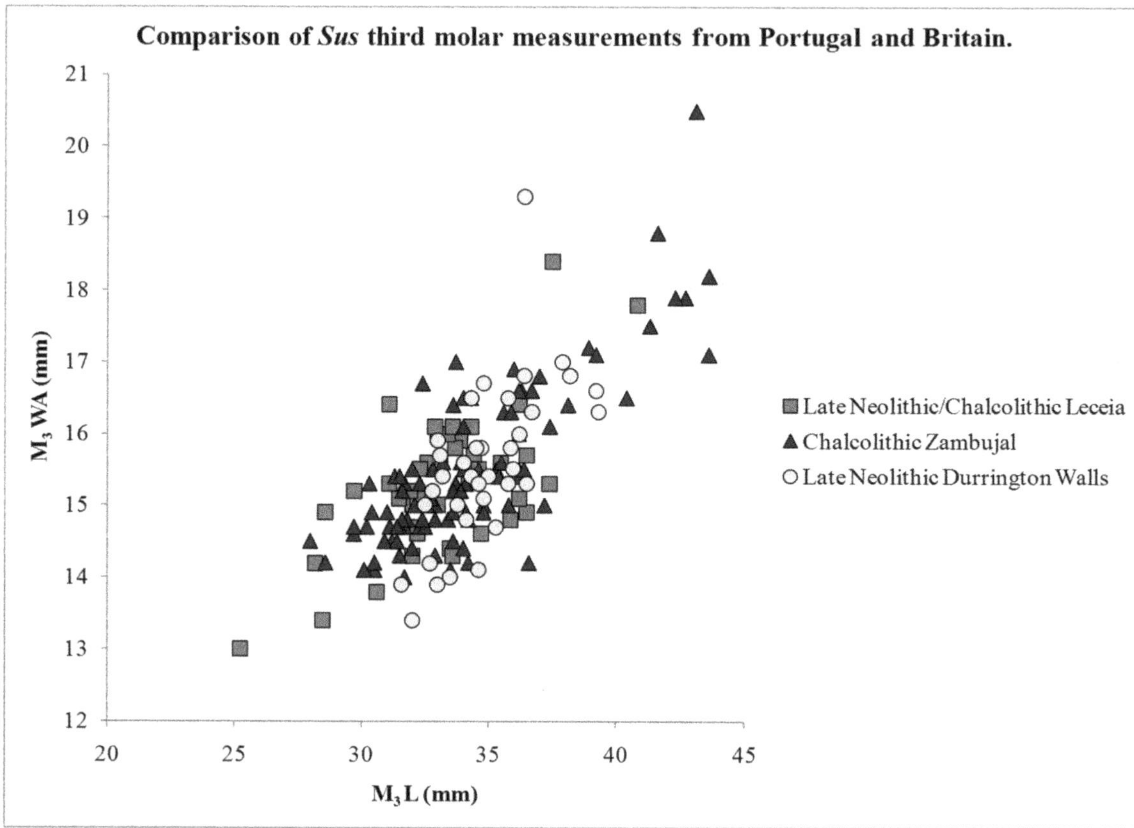

Figure 5.23 ***Sus*** **third mandibular molar measurements from Leceia and Zambujal (Portugal, data collected by U. Albarella), and Durrington Walls (L = length, WA = anterior width, following Payne and Bull 1988). Heavily worn teeth have been excluded.**

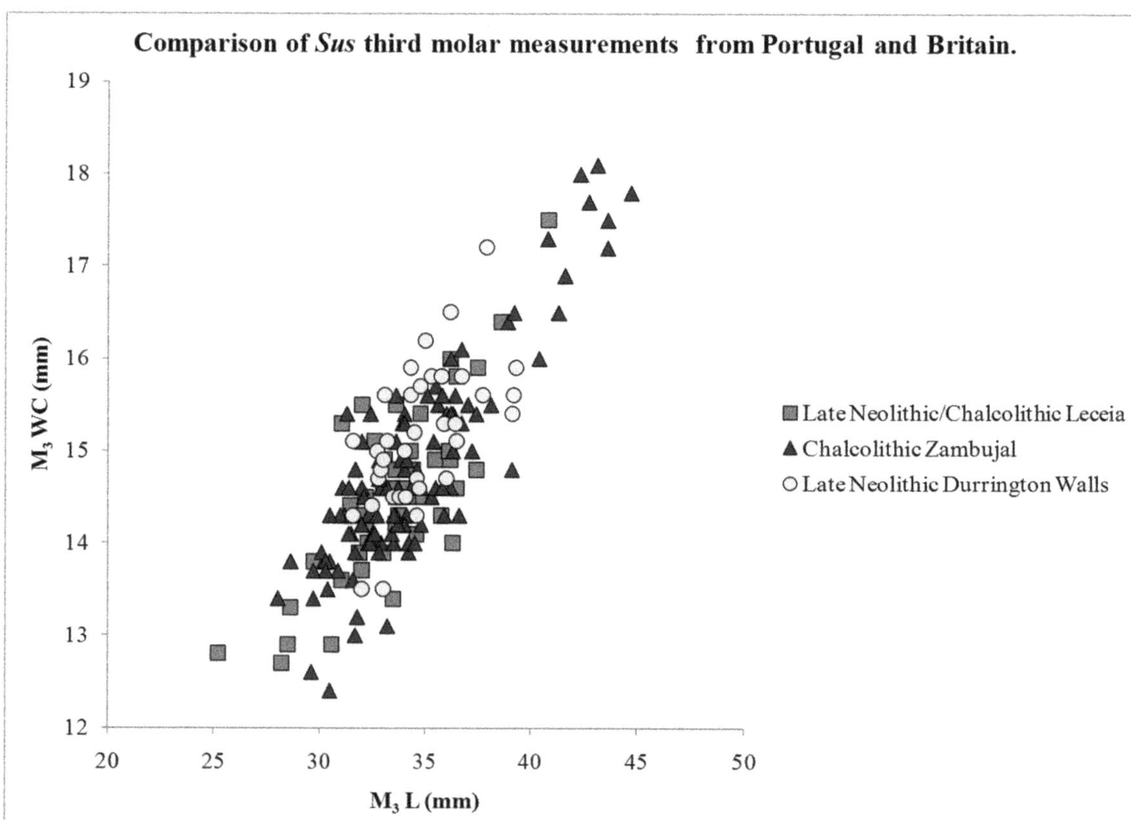

Figure 5.24 ***Sus*** **third mandibular molar teeth from Leceia and Zambujal (Portugal, data collected by U. Albarella), and Durrington Walls (L = length, WC = central width, following Payne and Bull 1988). Heavily worn teeth have been excluded.**

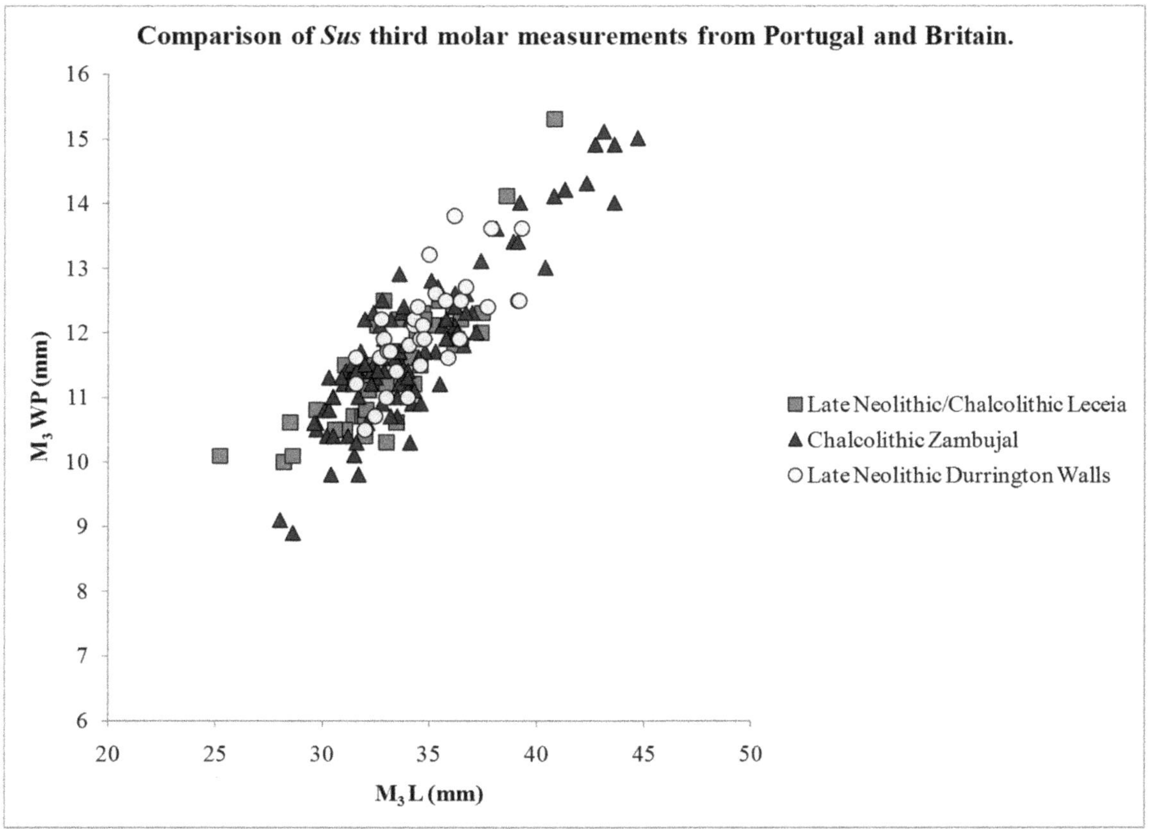

Figure 5.25 *Sus* third mandibular molar measurements from Leceia and Zambujal (Portugal, data supplied by U. Albarella) and Durrington Walls (L = length, WP = posterior width, following Payne and Bull 1988). Heavily worn teeth have been excluded.

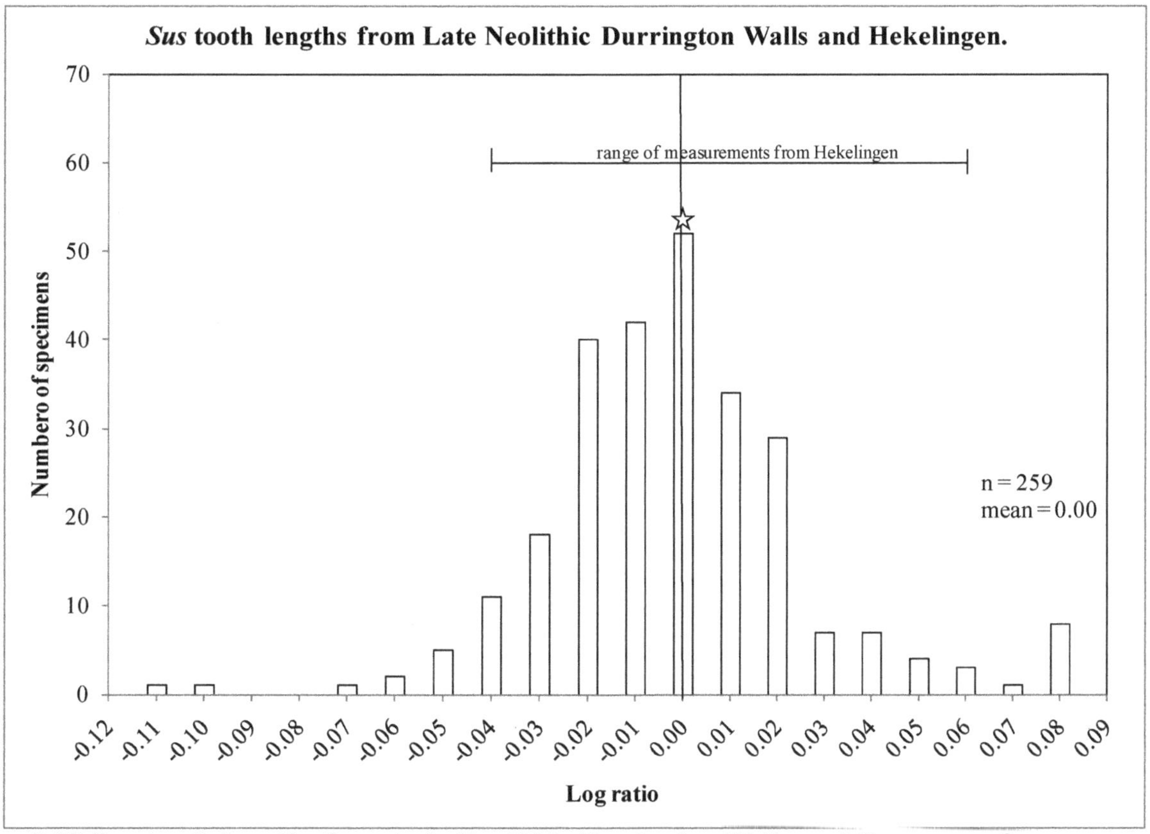

Figure 5.26 *Sus* tooth lengths from Late Neolithic Durrington Walls and the range of measurements from Hekelingen. Heavily worn teeth were excluded.

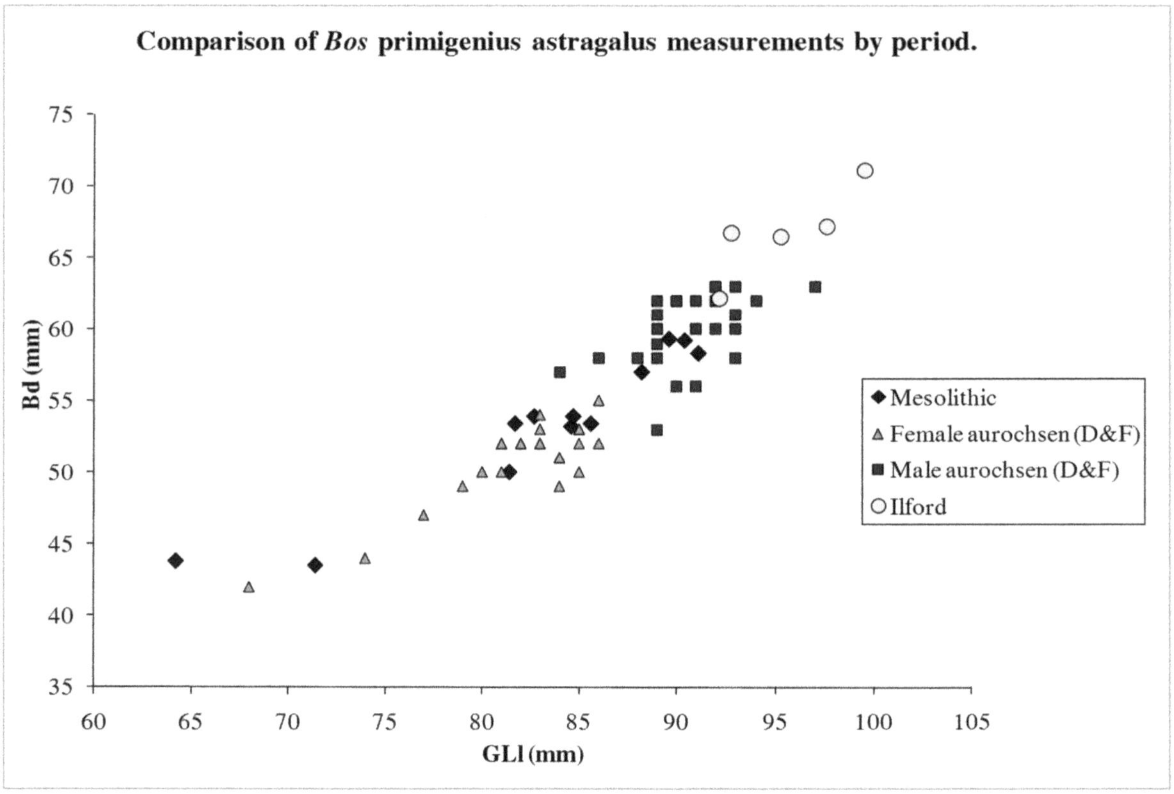

Figure 5.27 ***Bos*** **astragalus measurements (GLl = greatest length, Bd = breadth of the distal end, after von Den Driesch 1976) from the British site of Ilford, Mesolithic sites used in the study, and female and male aurochsen data presented by Degerbøl and Fredskild (1970).**

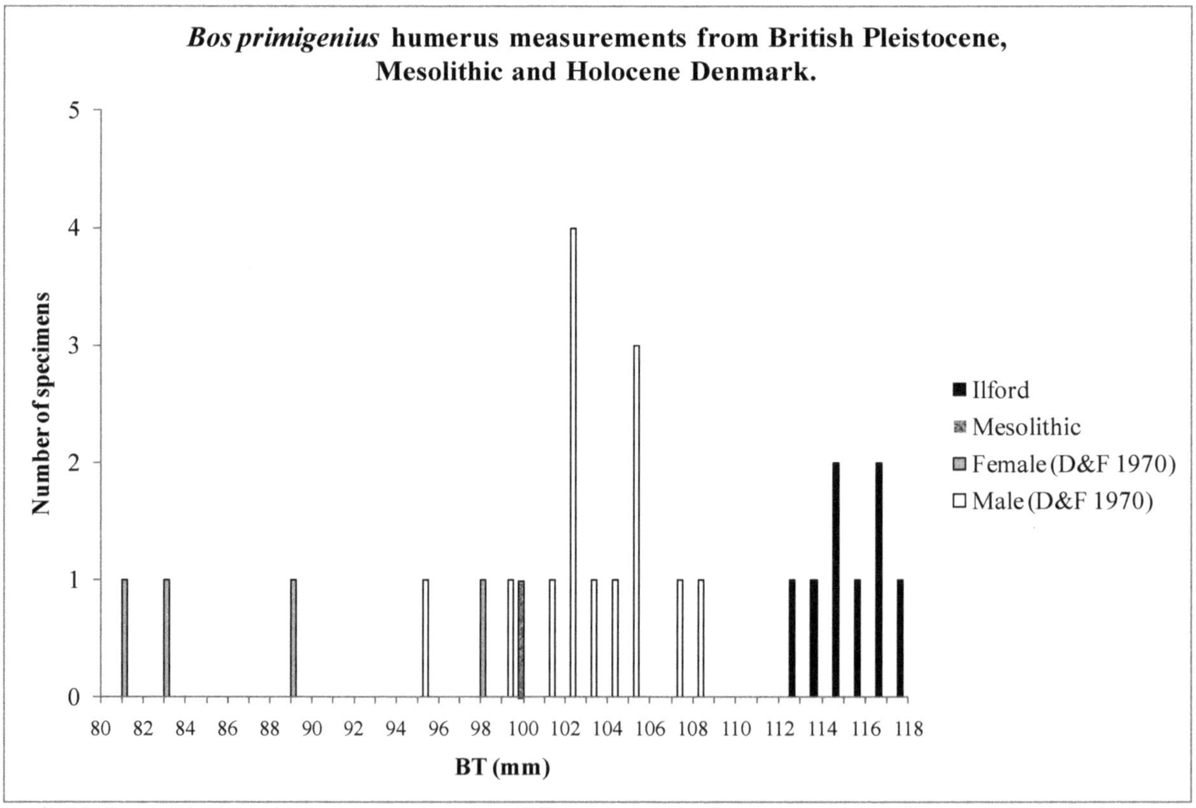

Figure 5.28 ***Bos primigenius*** **measurements (BT = breadth of the trochlea after von Den Driesch 1976) from the British Pleistocene, Mesolithic and Danish Holocene (data from Degerbøl and Fredskild 1970). Fused specimens only.**

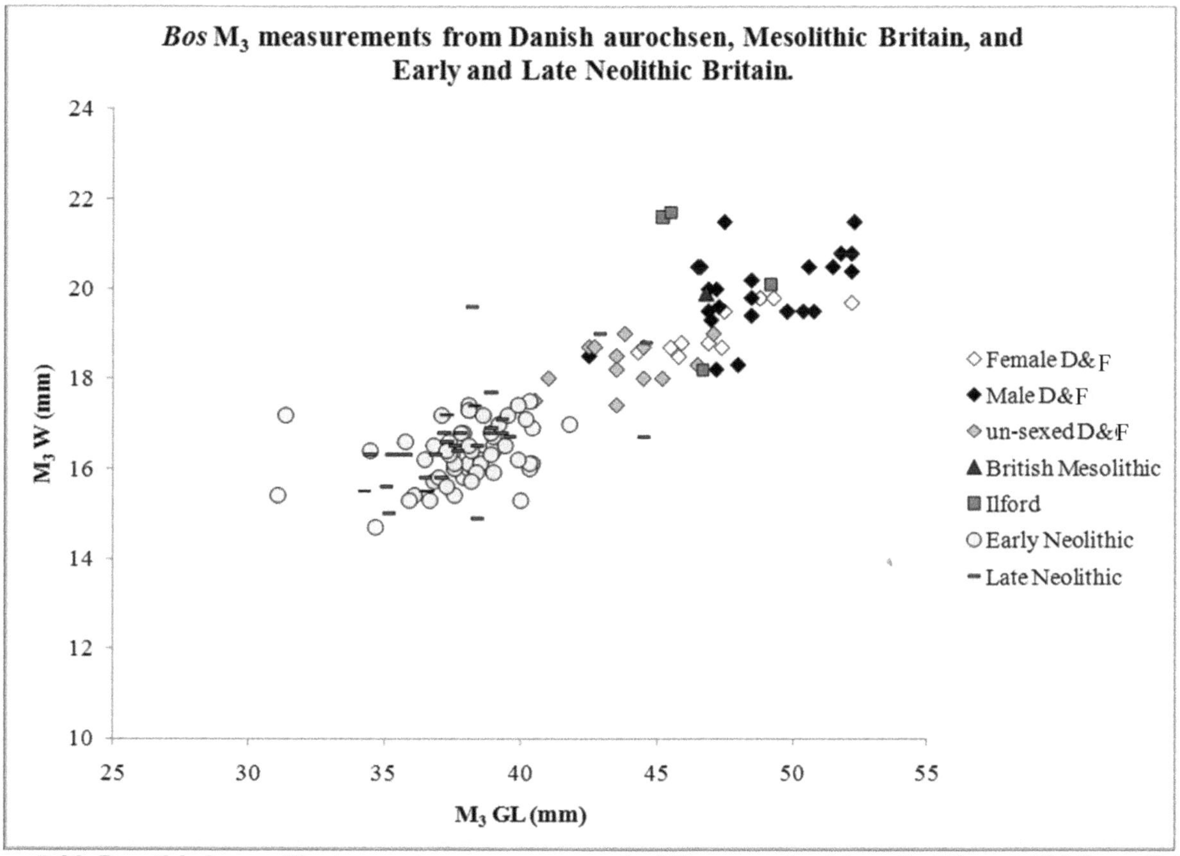

Figure 5.29 *Bos* third mandibular molar measurements (GL = greatest length, W = width) from British aurochs remains from Ilford, Mesolithic Star Carr, Early Neolithic and Late Neolithic British sites, and aurochsen from Danish sites recorded by Degerbøl and Fredskild (1970).

Age

Although direct comparisons between the ages of animals at different archaeological sites can be problematic, it still has potential to inform us about varying husbandry practices. Especially in a small, well defined research area such as southern Britain, it would be tempting to assume that the treatment of animals would vary little between different sites, and in fact, the information presented in Chapter 4 did provide evidence of uniformity in the age at which animals were killed in the Early Neolithic.

Of the two main sites used for comparison (Windmill Hill and Ascott-under-Wychwood), only Windmill Hill provided sufficient data to allow detailed comparison of the material. By comparing the proportion of fused and unfused bones that were recovered during the two main periods of excavation at the site, it can be seen that those from the most recent excavations indicate a situation very similar in nature to that seen at Hambledon Hill, Runnymede Bridge and Hazleton North (see Figure 5.30 for Windmill Hill and Figure 5.31 for others). Like the other sites used in this study, Windmill Hill provides evidence of relatively few young animals in Early Neolithic contexts. Those Early Neolithic sites analysed in Figure 5.31 (Hambledon Hill, Runnymede Bridge and Hazleton North) provide a pattern in which between 80% and 100% of the earliest fusing postcranial elements were fused. Windmill Hill varies from a little less than 80% to 90% fused/fusing in the first two categories. Results from the earlier excavation by Alexander Keiller are different. There is no evidence of an early cull, and less dependence on the use of younger animals in general. Instead, a gradual reduction in animals is apparent, which results in a greater number of adult animals occurring in each age category. There are perhaps two main reasons for this pattern: the first is that the sample size from Keiller's excavation is small; and the second is that excavation bias might have resulted in the preferential recovery of bones from adult animals.

In Chapter 4 the sharp contrast between fused/fusing bones of the Early and Intermediate categories was interpreted as the result of a significant cull of animals. This cull appears to have taken place, to a greater or lesser extent, at all of the sites studied. The faunal assemblage from the main Windmill Hill excavations is no exception, and as Figure 5.30 shows, there was an abrupt change in the proportion of fused/fusing bones between the Early and Intermediate categories. Less than 50% of the bones were fused in the Intermediate category compared with more than 80% in the Early category, suggesting that less than half of the animals at Windmill Hill survived until the Intermediate stage.

Too few bones that provided fusion data were available from Ascott-under-Wychwood to undertake analysis of the epiphyseal fusion pattern. However, a few mandibles provided information about wear stages. According to mandibular tooth wear analysis a variety of ages were present at the site (Mulville and Grigson 2007, 245). Two mandibles from adult cattle were allocated to the adult/senile category, and two younger mandibles were aged to 18-30 months and 30-36 months respectively (Mulville and Grigson 2007, 245).

The age profiles provided by the new primary data collected as part of this research, and the information from Grigson's work at Windmill Hill suggest that in general mortality at the sites was very similar. The evidence suggests that there was very little variation in the motivation for cattle husbandry at different sites and in different areas of southern Britain. At all sites use was made of very young animals to only a small degree. There was a cull of animals between 2-3 years, but many animals survived into adulthood, and a small number reached old age.

Sex

A disparity between the number of female cattle and male cattle was identified at a number of Early Neolithic sites in Chapter 4, as well as at Windmill Hill (Grigson 1999). Grigson based sex determination in cattle on two skeletal elements: pelves and horn cores. Using morphological distinctions 19 female and four male pelves were identified, while larger more robust male horn cores numbered only six compared to the shorter horn cores of females that totalled 57. Horncore and pelvis data were not collected as part of this study, but other skeletal elements provided evidence of the greater abundance of females at some of the Early Neolithic sites studied.

The data collected from Hambledon Hill and Runnymede Bridge suggest a greater abundance of female cattle in Early Neolithic assemblages, despite evidence to suggest decreasing sexual dimorphism in the Early Neolithic (for instance the substantial difference between female and male astragalus measurements of aurochsen from the Mesolithic that was not repeated in measurements from Early Neolithic cattle). In Chapter 4 biometrical analysis based on distal humerus measurements identified two distinct groups of animals that were interpreted as female and male (Chapter 4, Figure 4.10 and 4.11).

Biometry

Biometrical analysis presented in Chapter 4 provided information about the size of animals from different Early Neolithic sites. Further biometrical comparisons will be made here between data published from Windmill Hill and Ascott-Under-Wychwood and that collected by the author from Hambledon Hill, Runnymede Bridge, Coneybury Anomaly and Hazleton North (the latter two sites being presented only when biometrical data was sufficient). The homogeneity/difference in the size and shape of Early Neolithic cattle will be assessed. To prevent flawed results only those measurements that were taken in the same way in both the published zooarchaeological reports and in this study have been included in the comparison.

Measurements of the astragalus have been used in this study to distinguish between different cattle populations. A comparison of astragalus Bd and GLl measurements highlighted the strong differences between Mesolithic and Neolithic populations, but failed to identify any distinction between animals that all came from the Early Neolithic. Further evidence of homogeneity in the cattle from the Early Neolithic in southern Britain was provided by comparing astragalus measurements from Hambledon Hill and Runnymede Bridge with published information from Windmill Hill (Figure 5.32). No distinction could be drawn between the astragali from different Early Neolithic sites. Measurements from Hambledon Hill, Runnymede Bridge and Windmill Hill share the same distribution and range. However, the two outliers in the top right of the plot should be discussed. The specimens (one from Hambledon Hill and one from Windmill Hill) form the upper limit of *Bos* astragali from the main sites investigated. The astragalus from Windmill Hill was not identified as belonging to aurochs by Grigson (1999). These two specimens, however, fall into the smallest end of the *Bos* astragali from Mesolithic sites. The specimens could therefore represent either large domestic bulls, or small female aurochs.

The postcranial measurements gathered from Hambledon Hill, Runnymede Bridge and Coneybury Anomaly, when compared using the log ratio method to those from Windmill Hill, reveal animals that were dimensionally alike (Figure 5.33). The animals from Windmill Hill, in common with the majority of Early Neolithic *Bos*, are far smaller than the standard established from Ilford (shown by the vertical line on Figure 5.33). They are also generally smaller than most of the Mesolithic measurements presented in Chapter 4. It is clear from the log ratio diagrams that bones collected during different periods of excavation at the site of Windmill Hill were biometrically comparable and came from the same type of animals. Ascott-Under-Wychwood provided a much smaller sample (the bottom diagram in Figure 5.33).The measurements from this site share a similar range to those from the other Early Neolithic sites studied. Despite the general comparability of the Ascott-Under-Wychwood remains to other Early Neolithic sites, the mean is substantially higher (-0.13 compared to between -0.20 and -0.18 for the other sites). This high mean value is best explained by the presence of the two large outliers that pull the mean of the small sample to the right. These specimens at -0.05 and -0.04 were identified as aurochsen by the original investigators (Mulville and Grigson 2007, 249) and will be discussed in more detail below.

The very similar postcranial measurements from the sites investigated for this study when compared with the measurements from other Early Neolithic sites reinforce the characteristic homogeneity of the physical dimensions of Early Neolithic cattle. No differences between the postcranial elements and dental elements were found as a result of log ratio comparisons in Chapter 4. Tooth measurements will be compared here to ensure that this is a common characteristic of the Early Neolithic.

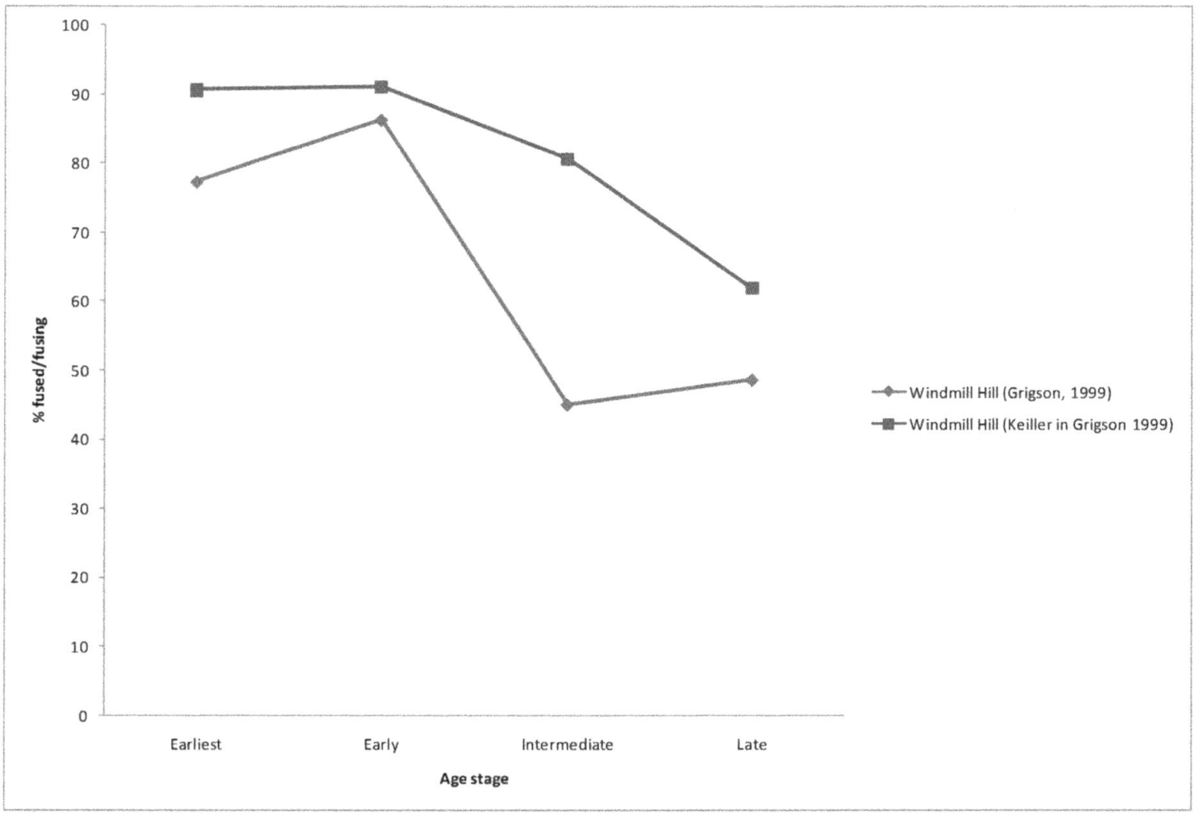

Figure 5.30 Epiphyseal fusion of cattle remains from excavations at Windmill Hill between 1925-34 and 1988 (Grigson 1999).

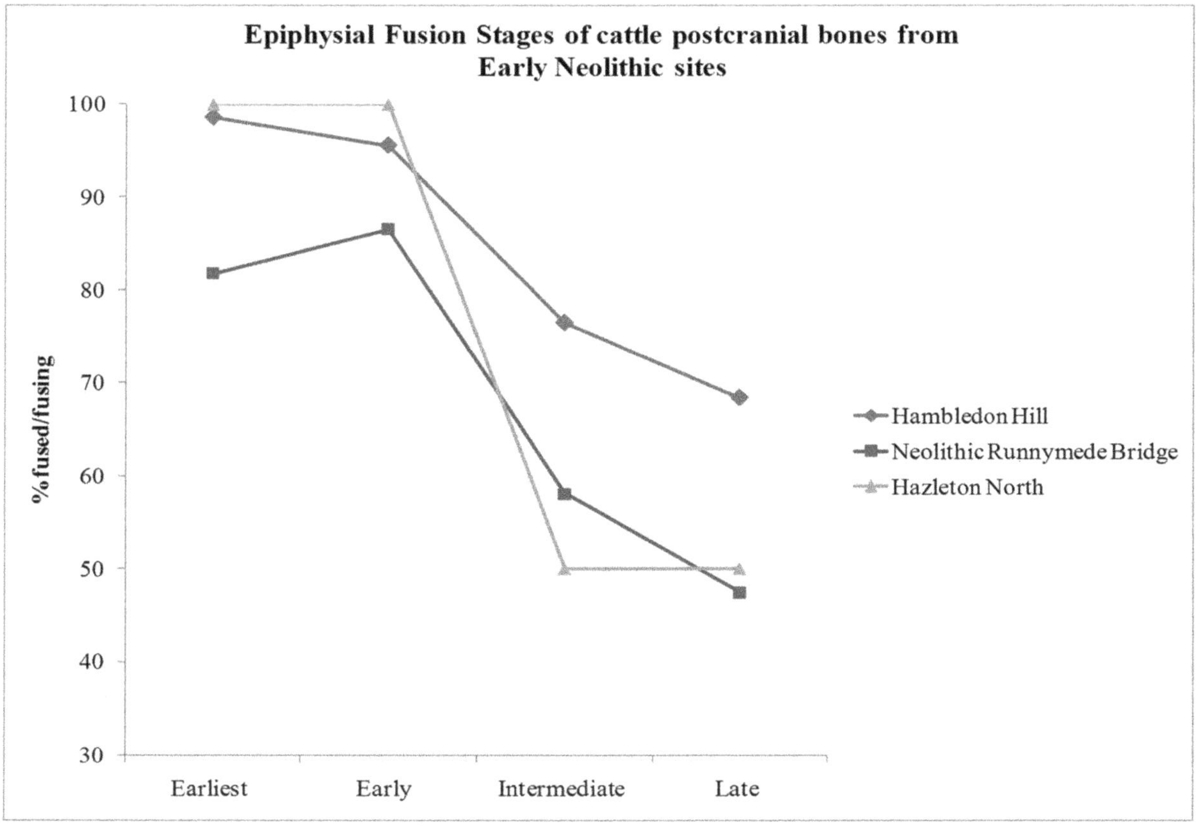

Figure 5.31 Epiphyseal fusion of cattle postcranial elements from Hambledon Hill, Runnymede Bridge and Hazleton North.

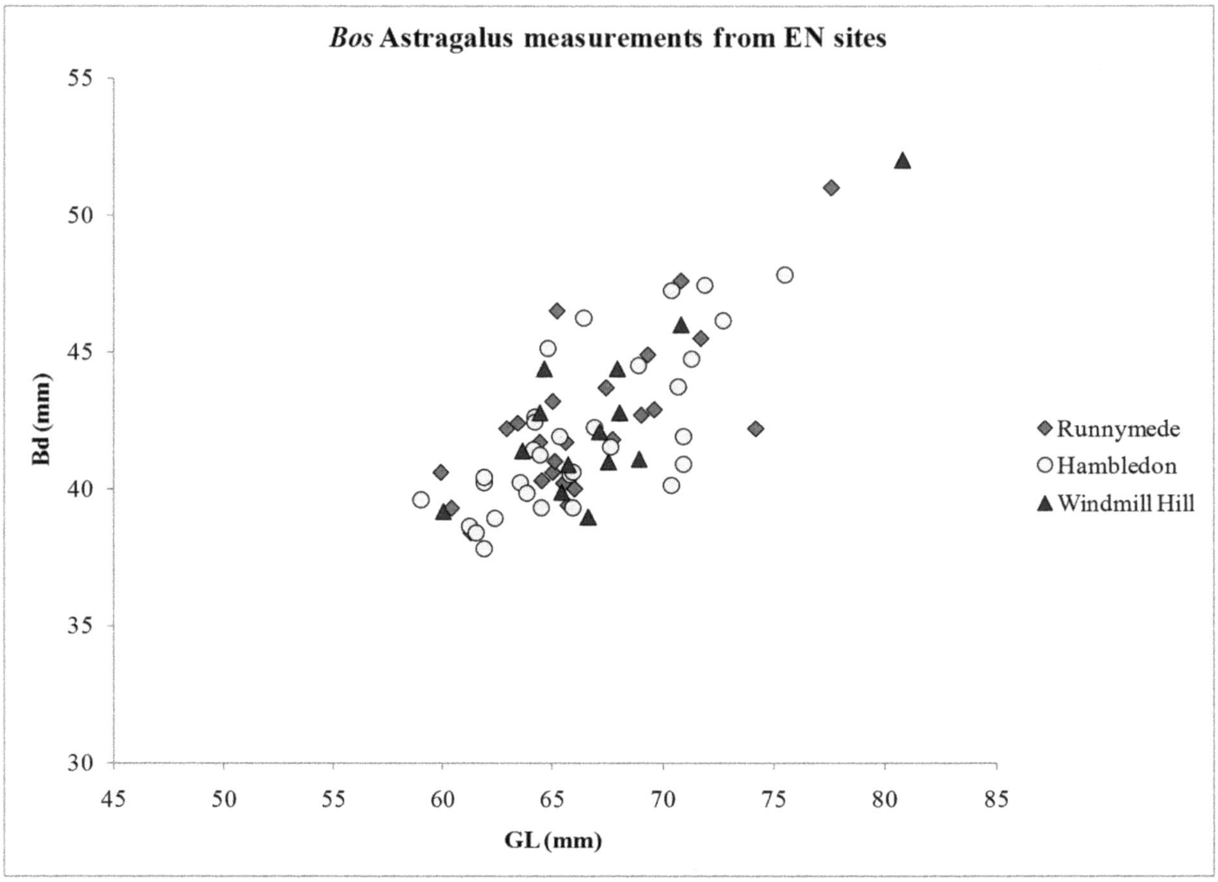

Figure 5.32 *Bos* astragalus measurements (GL = greatest length, Bd = distal breadth, von Den Driesch 1976) from British Early Neolithic sites.

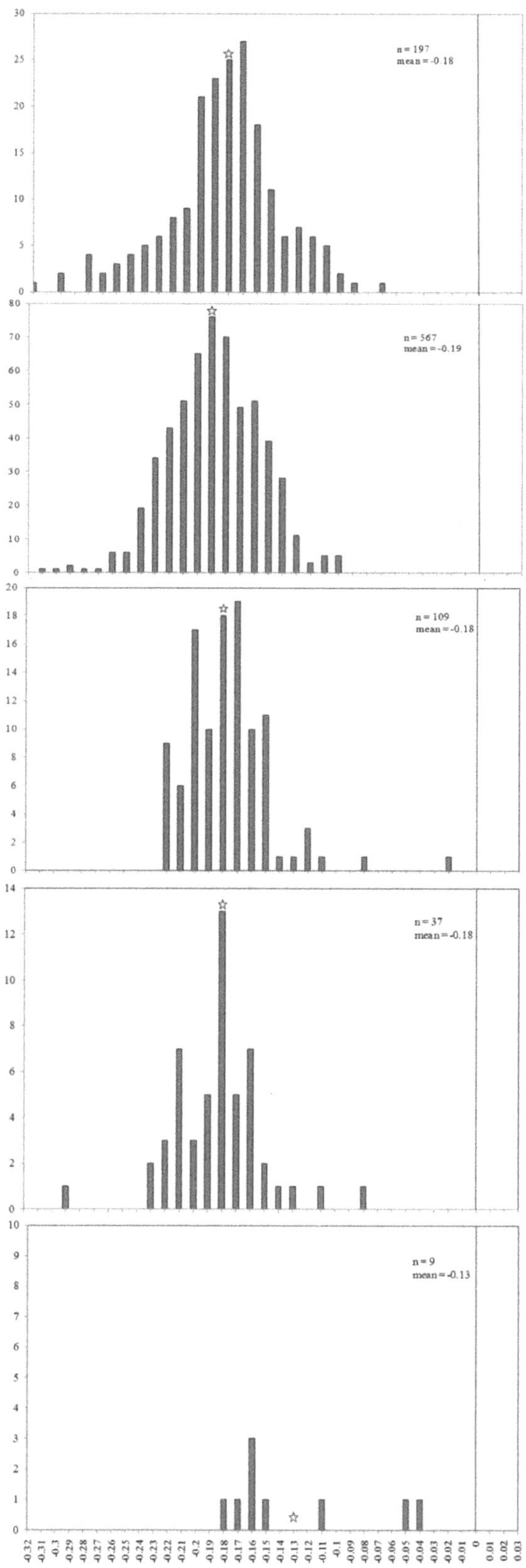

Runnymede Bridge

Hambledon Hill

Coneybury Anomaly

Windmill Hill, 1929-25 excavations.

Windmill Hill (Whittle *et al.* 1999).

Ascott-Under-Wychwood (Benson and Whittle 2007).

Figure 5.33 Log ratio comparison of cattle postcranial measurements from British Early Neolithic sites. Fully fused specimens (excluding the scapula and horn cores) have been included. The mean is marked by a star, and the standard measurement by the black vertical line at 0.

Due to the absence of more diverse measurements from the comparison sites, biometrical analysis of teeth will be confined to the mandibular third molar. Mandibular third molars provided a similar picture to the postcranial elements when compared to a standard (Figure 5.34). The length of mandibular third molars from Hambledon Hill, Runnymede Bridge and Windmill Hill have a distribution that is largely indistinguishable. The similarity between the plots suggests that the animals are from the same or closely related populations. The single specimen marked with a triangle on the Windmill Hill plot was identified as aurochs by the original investigator, an identification based on its large size compared to other specimens from the site (Grigson 1999). The aurochs specimen falls outside the range of cattle third molars recorded from both Hambledon Hill and Runnymede Bridge, but it is within the distribution of Mesolithic aurochsen presented in Chapter 4. The location of this specimen outside of the distribution of cattle from the two main Early Neolithic sites studied as part of this study, and that of Windmill Hill, supports the interpretation that most of the animals on Early Neolithic sites were from domestic cattle populations.

This research will now turn to a comparison of the biometrical data from Early Neolithic sites from Britain with roughly contemporary material from mainland Europe. Two main sites will be investigated: Cerny, France (Tresset 1997); and Schipluiden, Holland (after Zeiler 2006); along with some specimens from Danish Køkkenmøddinger (Degerbøl and Fredskild 1970).

In Figure 5.35 cattle astragalus measurements from the Early Neolithic in Britain have been compared to measurements from mainland Europe. The two largest specimens from Early Neolithic Britain (one each from Runnymede Bridge and Windmill Hill) are almost certainly aurochs bones. Of great interest are the bones from Schipluiden, Cerny and the Danish Køkkenmøddinger sites. Danish cattle overlap with the British specimens, but are towards the larger end of the British Early Neolithic cattle plot. Work by Tresset (2003) suggested that the similar size of cattle from the Early Neolithic in Britain when compared to late 5th millennium BC cattle from the Paris Basin is evidence in support of the hypothesis that British domesticated types originated there. This data, however, indicates that the Cerny cattle were large compared to British ones (they all plot in the upper half of the British distribution), and no small French specimens are similar in size to the smallest British specimens. Conversely, the animals from Schipluiden (which fall almost exclusively outside the range of French cattle and are significantly different from the Danish Køkkenmøddinger specimens) plot in the middle of the British distribution and are therefore probably similar in average size, though this interpretation is tentative due to the small Dutch sample.

In Figure 5.36 postcranial measurements from British Early Neolithic sites have been compared to remains from Schipluiden using the log ratio technique. Although the mean is slightly higher at Schipluiden suggesting that the animals might be slightly larger overall than British animals, this evidence reinforces that from the astragalus that cattle from the Neolithic of Holland compared well with the British material.

There are two main implications of these results. The first is that the homogeneity expressed in the British Early Neolithic is common to parts of Europe. Some small variation does exist, for instance the French and Danish cattle bones are all within the British distribution, but at the largest end. The second implication concerns Tresset's work on the origins of British cattle. The overlap between British and French cattle is undeniable, but the French appear to be on average clearly larger. This opens up three possible interpretations:

- The French and British animals are in fact similar in overall size but those from France appear to be larger because of a greater proportion of males in the sample (we have seen that females predominate in the British sample). If this is the case the French cattle may indeed represent a possible source of origin for the British ones.
- The original British domestic cattle do indeed derive from the French cattle by the time the accumulation of cattle bones occurred at Hambledon Hill and Runnymede Bridge cattle has been subject to a more rapid size decrease than it has occurred in France (we must remember that both British sites date to the latter part of the Early Neolithic).
- British cattle do not derive from French cattle, or at least not from those populations that also gave origin to the Bercy animals.

Danish animals appear to be even larger than French ones and therefore represent an even less likely source for the British cattle, but the sample is small and it therefore requires verification with additional data. The Dutch cattle have also provided a small sample but this is the most similar to the British material, and, on the basis of the current - admittedly limited -comparison, represent the most likely ancestors for the British animals.

5.8 British Cattle in the Late Neolithic

Introduction

Durrington Walls is the only Late Neolithic faunal assemblage that provided primary data for this investigation. By drawing comparisons between Durrington Walls and other sites of the same period it will be determined whether the animals at Durrington Walls can be regarded to be typical of the Late Neolithic as a whole or if large differences existed between sites. Faunal assemblages from the Late Neolithic are as limited in both number and extent as those from the Early Neolithic. Comparison can be made between the remains recorded from Durrington Walls and those from West Kennet (Edwards and Horne 1994) and Barrow Hills (Levitan and Serjeantson 1999). Both age and sex information was extremely limited at West Kennet and

Barrow Hills, therefore the following analysis will concentrate on biometrical information.

Biometry
Astragalus measurements from Durrington Walls and other Late Neolithic sites were compared to a standard and plotted as a log ratio diagram in Figure 5.37. The astragalus measurements suggest that bones from the Late Neolithic sites (Barrow Hills and the West Kennet palisade enclosure) were comparable. The greatest lateral length and greatest medial length have been plotted in Figure 5.38. Very few specimens provided both of the measurements necessary to be included in this plot, hence the small sample size, but it does serve to reinforce the similarity of specimens from the Late Neolithic in general. The only specimen from the Grooved Ware pit at Puddlehill is also within the range of Late Neolithic domestic cattle.

A single mandibular third molar from West Kennet Palisade Enclosure (Edwards and Horne 1994; Horne 1994) had a length of 37.8 mm, and fell within the range already established from Chapter 4.

Newgrange, Ireland, was chosen for comparison with Durrington Walls for a number of reasons. A number of similarities exist between the two sites, both in chronology and form. Newgrange and Durrington Walls provided evidence of mortuary/ceremonial practice and closely associated domestic structures (still somewhat rare in both Britain and Ireland for the period in question). In practical terms, there are no other large Late Neolithic assemblages from Ireland with which Durrington Walls could have been compared. Durrington Walls was dominated by *Sus* remains, whereas at Newgrange the emphasis appears to have been on *Bos* exploitation (van Wijngaarden-Bakker 1986). Finally, it is useful to compare Durrington Walls with an assemblage that with certainty derived wholly from domestic animals. Aurochsen were absent from Ireland after the end of the Pleistocene (van Wijngaarden-Bakker 1986, 42). Excluding the possibility that aurochs were deliberately brought to the Irish mainland from elsewhere during the Late Neolithic these domesticates could not have mixed with wild populations.

Late Neolithic remains presented in the previous chapter led to the conclusion that there was little mixing between domestic cattle and wild aurochs during the period. Some large outliers were ambiguous, and by comparison with earlier Mesolithic aurochsen remains, could have belonged to either small female aurochs that were hunted or the largest members of a domestic population. A log ratio comparison of cattle mandibular third molar lengths is provided in Figure 5.39, this time the horizontal bar indicates the range of measurements taken by van Wijngaarden-Bakker (1986) from Newgrange (based on 109 specimens). As was clear in Chapter 4 the tooth lengths of cattle from Durrington Walls fall well to the left of the standard and are much smaller than Pleistocene aurochs (as expected). The single outlier at -0.02 was interpreted as a possible aurochs specimen due to its large size and separation from the other specimens. However, when compared with the domestic animals from Newgrange, all of the third molar measurements fall well within the established range. The possibility arises, then, that the upper size limit of Late Neolithic cattle has been underestimated. This is perhaps because of the tendency for larger numbers of female cattle have been kept to maturity than males, resulting in the remains of fewer male animals on Late Neolithic sites.

The breadth of the humerus distal trochlea also highlights the overlap between Durrington Walls cattle and those domestic cattle from Newgrange. Distal humerus measurements from Durrington Walls range from 69mm to 88mm, and fall into two groups. These groups were interpreted in Chapter 4 as small domestic females and larger domestic males. The possibility that they were different populations (perhaps wild and domestic) was ruled out by comparison with Mesolithic aurochs remains. When these measurements were compared with the Newgrange results (Figure 5.40) it was apparent that the Durrington Walls measurements were limited compared to the range exhibited at Newgrange (based on 43 specimens). Unfortunately without greater detail (only the range, not specific measurements were provided in the publication) it is impossible to say whether a similar distinction between female and male cattle existed in the BT measurements.

Not all data from Durrington Walls fell clearly within the range of domesticated cattle from Late Neolithic/Beaker Newgrange. For instance, Figure 5.41 presents distal tibia breadth measurements from the two sites. Durrington Walls measurements lie between 55mm and 67mm, while Newgrange has a much larger range. The mean of the Durrington Walls measurements falls at the smaller end of the Newgrange distribution.

A number of Dutch sites can also provide a direct contemporaneous comparison with the Durrington Walls cattle material. Late Neolithic Durrington Walls will be compared to Late Neolithic and Eneolithic sites from Holland. In Figure 5.42 the comparison of Durrington Walls astragalus measurements with specimens from Vlaardingen (Clason 1967), Zandwerven (Clason 1967) and Hekelingen (Prummel 1987). The astragalus measurements clearly overlap, but a number of the Durrington Walls specimens are smaller than the smallest Dutch specimens and are outside the range of measurements from Hekelingen (shown by the horizontal bar). For the mandibular third molar lengths (Figure 5.43) there is no clear difference between animals at any of the sites. In general the cattle from Durrington Walls are more similar to Dutch specimens than they are to those from Newgrange.

5.9 Aurochsen in the British Neolithic

Aurochsen were present at Windmill Hill and Ascott-under-Wychwood according to the identification of the original investigators (Grigson 1999; Mulville and Grigson 2007, 249). One of these, a calcaneum from Windmill Hill was of sufficient length to be recorded as

Bos primigenius. However, when compared using the log ratio technique above it is clear that these specimen falls well within the range of domestic cattle from the site itself, and those from other contemporaneous sites. Three phalanges were identified as *Bos primigenius* by Jope (1965, 144).

At Ascott-under-Wychwood two postcranial measurements came from specimens identified as aurochs (Mulville and Grigson 2007, 249). These specimens are well removed from the other measurements in the sample. Interestingly, although they are separate from the probable domestic animals at Ascott-under-Wychwood, they are not as strongly differentiated from the main distributions at Runnymede Bridge (largest -0.07), Windmill Hill (largest -0.08) and Hambledon Hill (largest -0.10). The largest specimens at these sites could belong to the smaller end of the aurochs scale or towards the larger end of the assumed domestic distribution.

Maiden castle produced two measureable aurochs bones, a tibia and an axis, that were reported as early prehistoric in date. Pit 6 at Puddlehill contained a minimum of 4 *Bos primigenius* individuals (Grigson 1976, 15). This pit was dated to the Late Neolithic based on the presence of Grooved Ware pottery (Smith 1976, 9). An aurochs radius was recovered in the main enclosure ditch at Mount Pleasant (Harcourt 1979, 215), an identification based on its large size. According to the author, aurochs was more abundant in the subsequent Beaker phase of the site and the collection consisted of a distal humerus, mandibular third molar and two astragali. Later in the Bronze Age at Irthlingborough, Northamptonshire, six aurochs bones and teeth were found within a rich domestic cattle assemblage (that contained the skulls of at least 185 animals) (Davis and Payne 1993; Towers *et al.* 2010).

These examples illustrate the record of aurochs in the Neolithic in Britain. In general, the presence of aurochs on archaeological sites is extremely limited and seems to be confined to a few, isolated specimens. However, they do appear to be consistently present on sites of the period, suggesting that their exploitation was regular, though scanty, throughout the Neolithic. This pattern is apparent on the sites that I have studied for this research, and by comparison with the available publications of other sites.

5.10 Conclusion

This chapter has explored the relationship between the animals of Mesolithic and Neolithic Britain, and those found in continental Europe. It has found variation within the wild populations extant in Europe during the Mesolithic, and reinforces the assertion that comparisons between wild and domestic animals need to make use of local wild populations, not those found in other geographical areas. It has also highlighted the similarities and differences between early domesticates in the areas adjacent to Britain, and commented on the likely origins of domesticates based on their biometrical characteristics. Finally, the role of aurochs in the Neolithic of Britain was briefly discussed.

The next chapter will explore the mobility of cattle in the Late Neolithic, and following this the final chapter of the volume will discuss the results of this chapter along with the findings of Chapter 3 and 4.

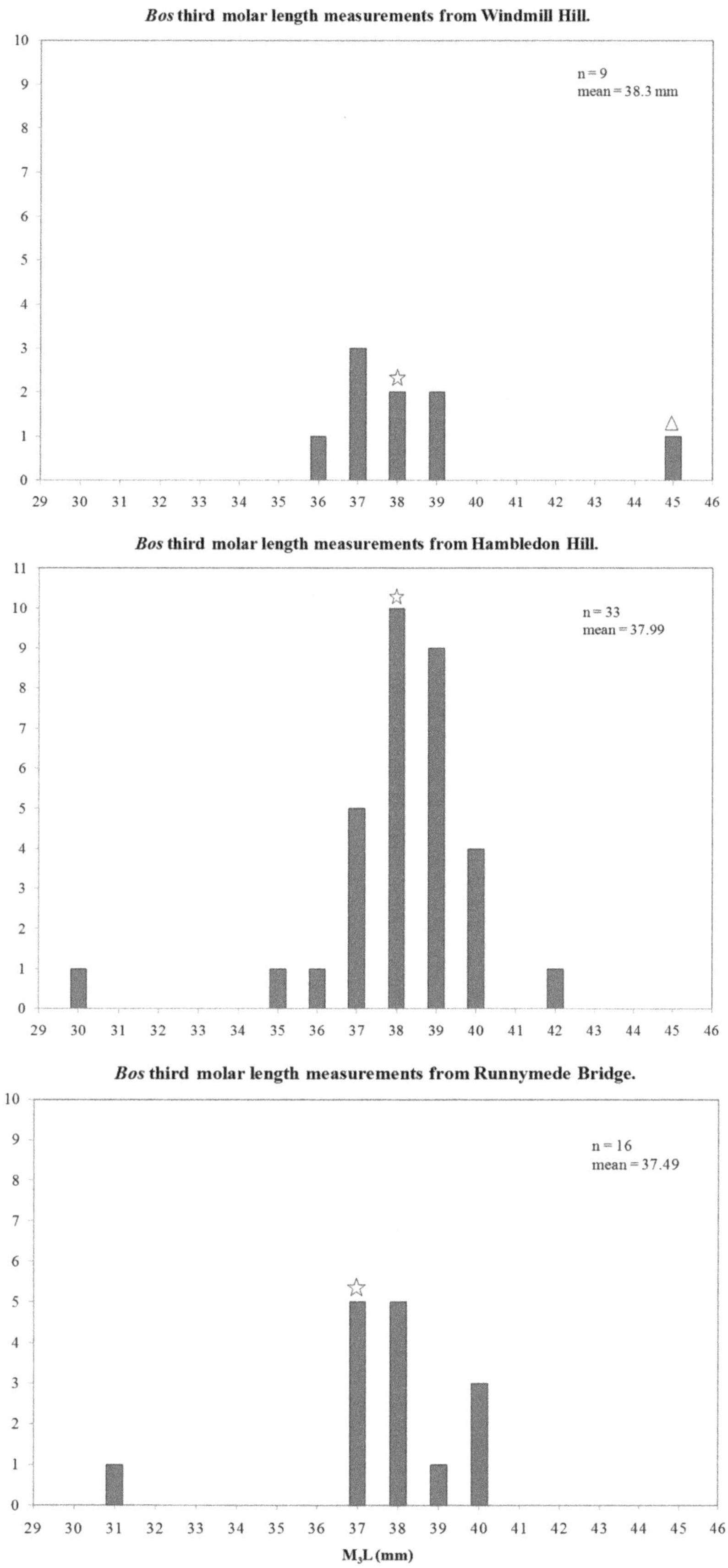

Figure 5.34 ***Bos*** **mandibular third molars from Windmill Hill (upper diagram, data from Grigson 1999), Hambledon Hill (middle diagram) and Runnymede Bridge (lower diagram). Heavily worn teeth were excluded from analysis. The specimen identified as aurochs by Grigson (1999) is indicated by a triangle.**

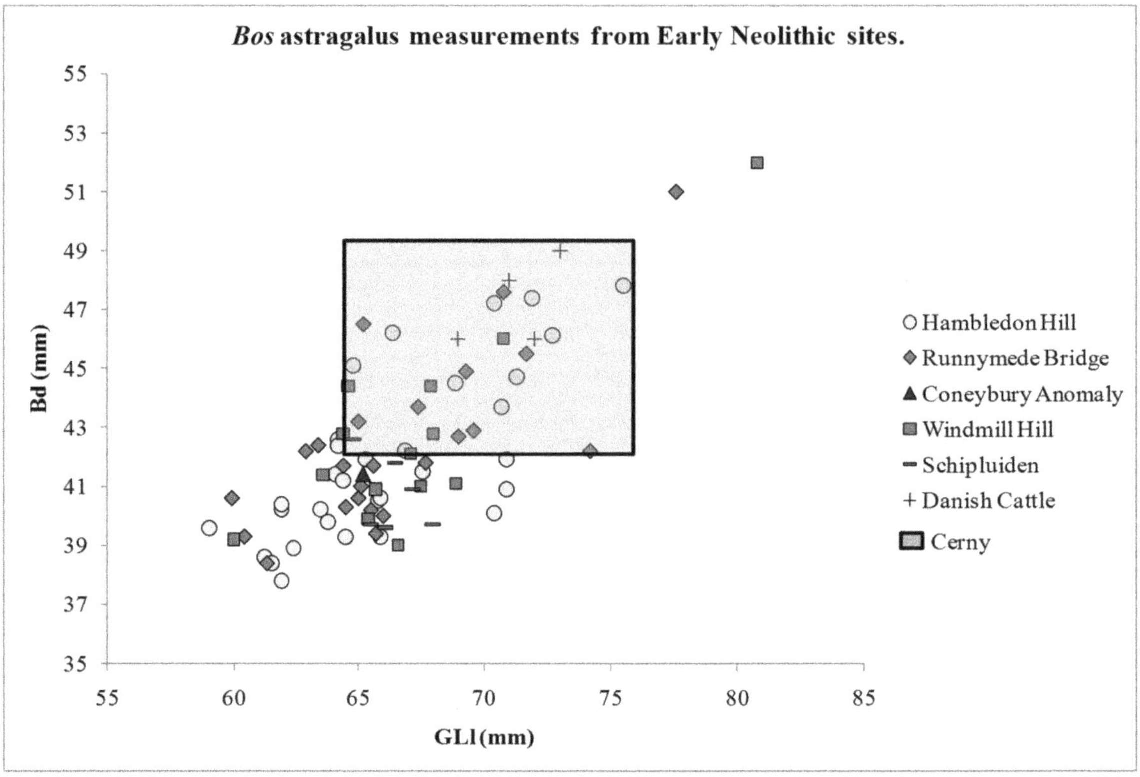

Figure 5.35 Cattle astragalus measurements (GLl = greatest lateral length, Bd = breadth of the distal end, von Den Driesch 1976) from British Early Neolithic (Windmill Hill after Grigson 1999), Dutch Schipluiden (Zeiler 2006), Danish Køkkenmøddinger, and the range of measurements from Cerny, France (Tresset 2005).

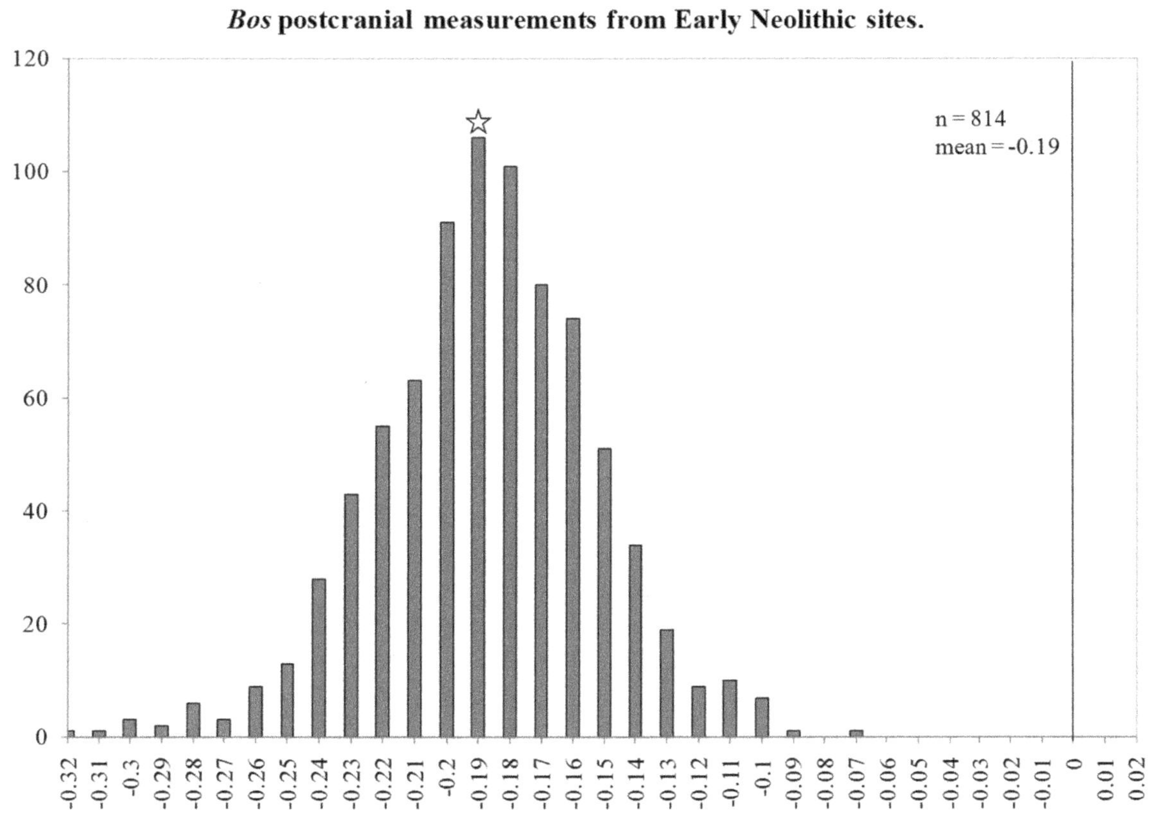

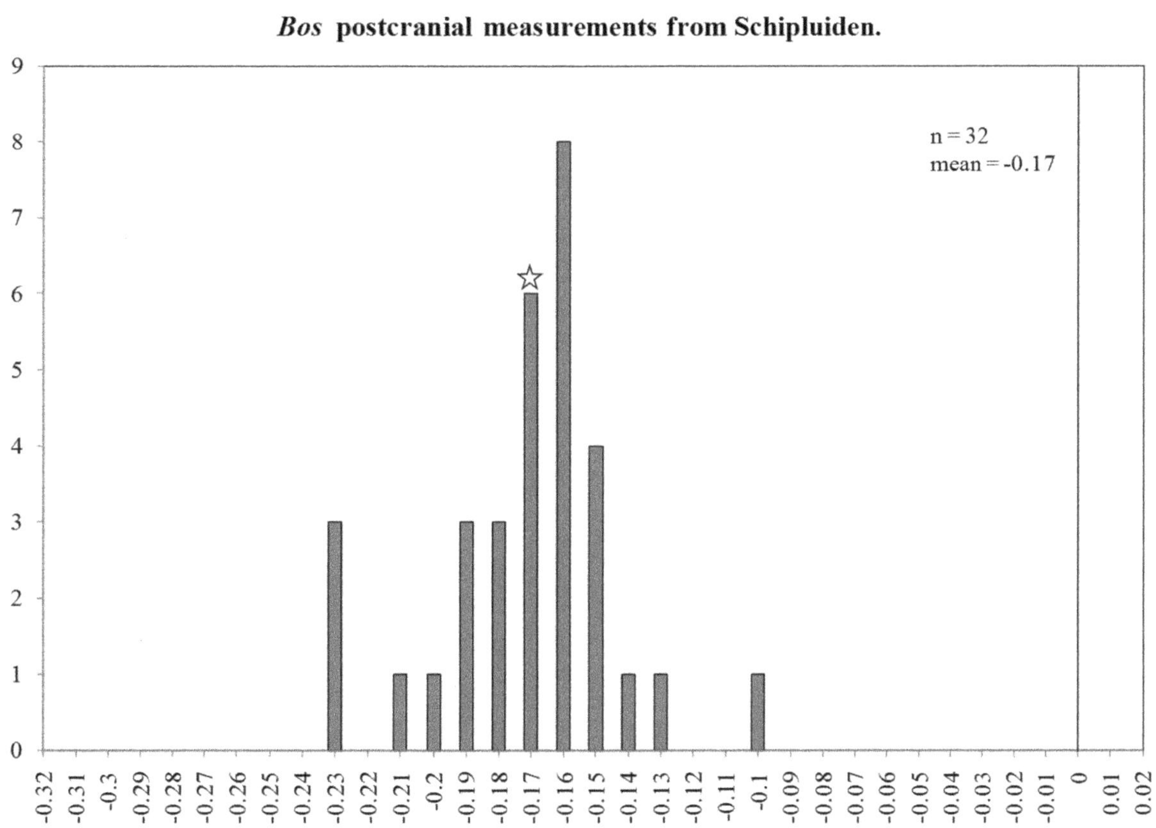

Figure 5.36 Bos postcranial remains from British Early Neolithic sites and Eneolithic Schipluiden (Zeiler 2006). Only fully fused bones have been included.

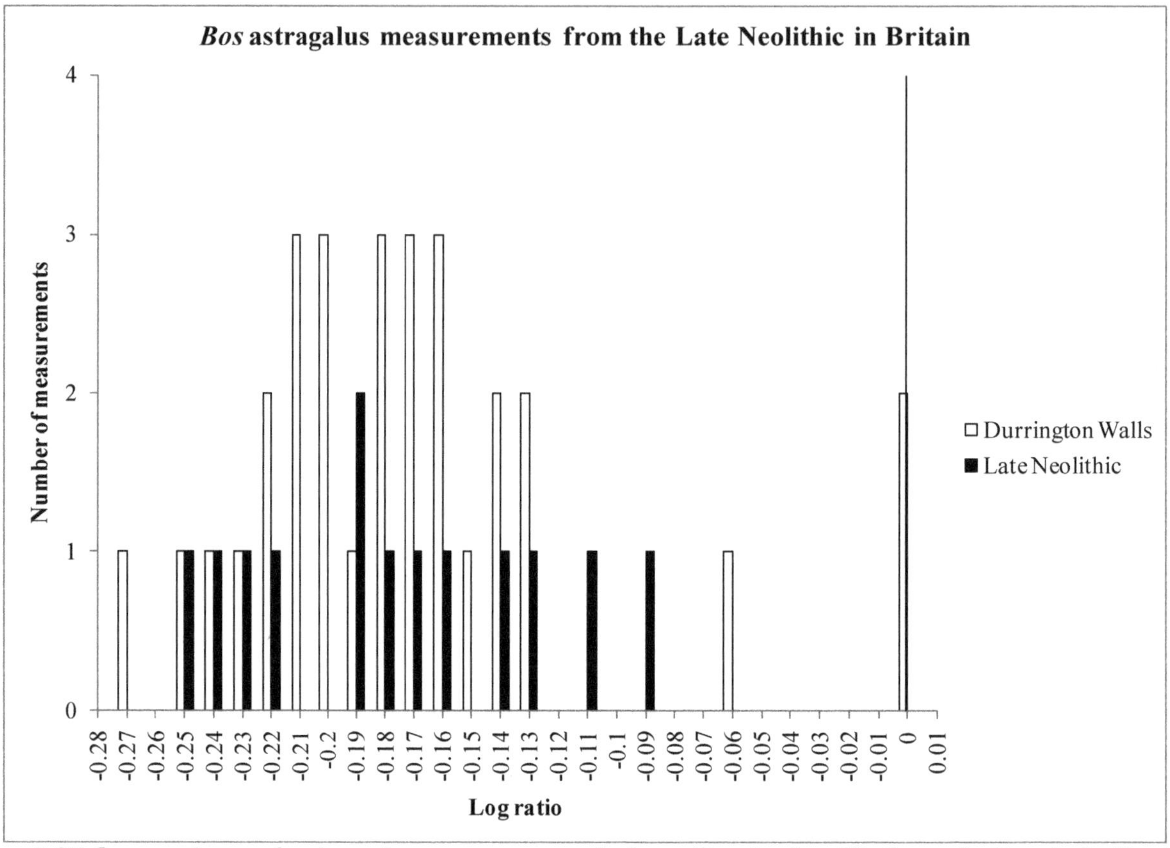

Figure 5.37 Comparison of astragalus measurements from Durrington Walls and other Late Neolithic sites (West Kennet Palisade Enclosure after Edwards and Horne (1994), and Barrow Hills after Levitan (1999).

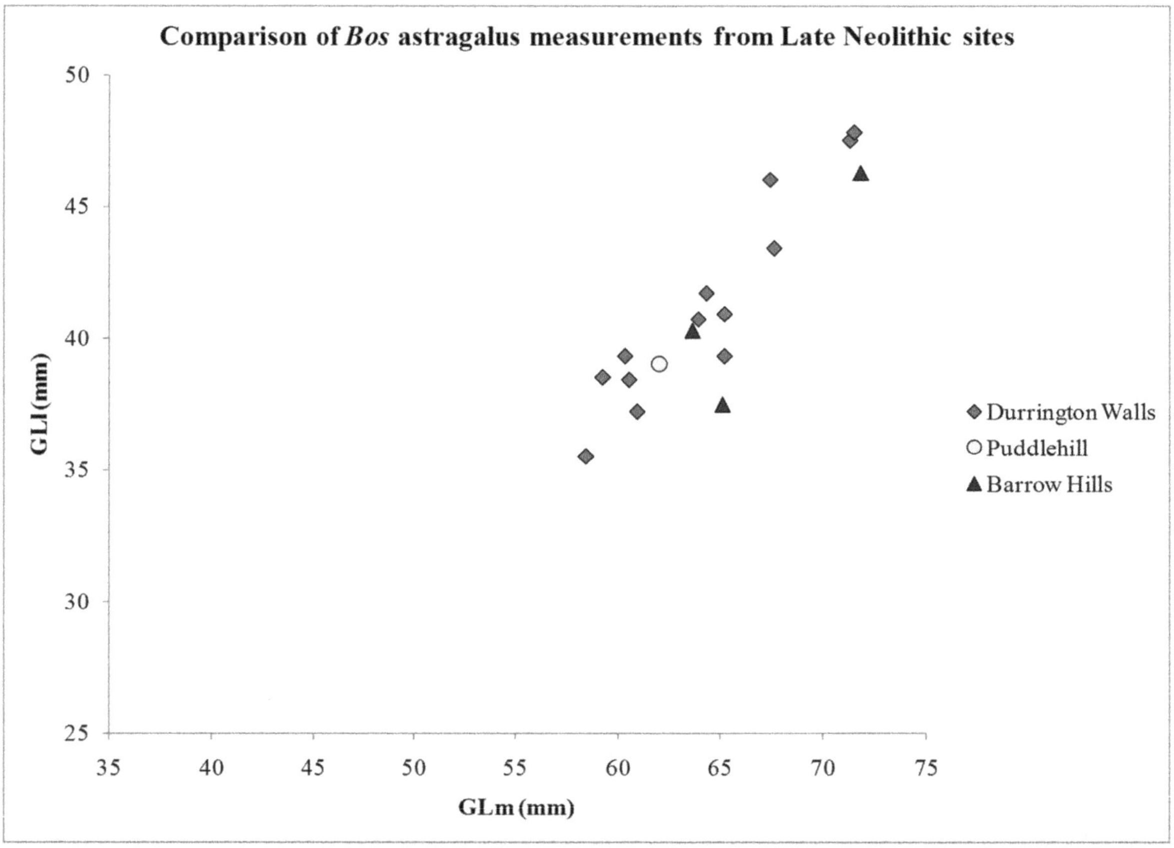

Figure 5.38 *Bos* astragalus measurements (GLm = greatest medial length, GLl = greatest lateral length, von Den Driesch 1976) from the Late Neolithic sites of Durrington Walls, Puddlehill and Barrow Hills.

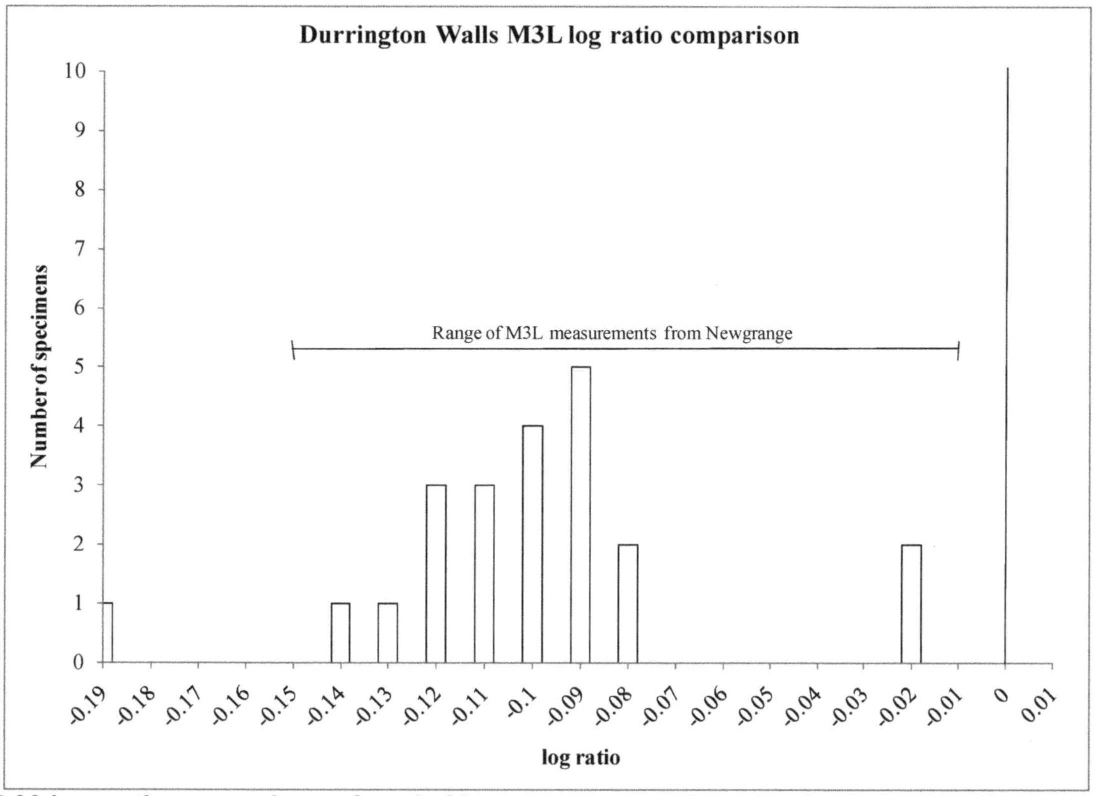

Figure 5.39 Log ratio comparison of cattle M_3 measurements (von den Driesch 1976) from Durrington Walls. The horizontal line indicates the range of measurements taken by van Wijngaarden-Bakker from Newgrange, Co. Meath, Ireland (1986, 42), based on 109 specimens.

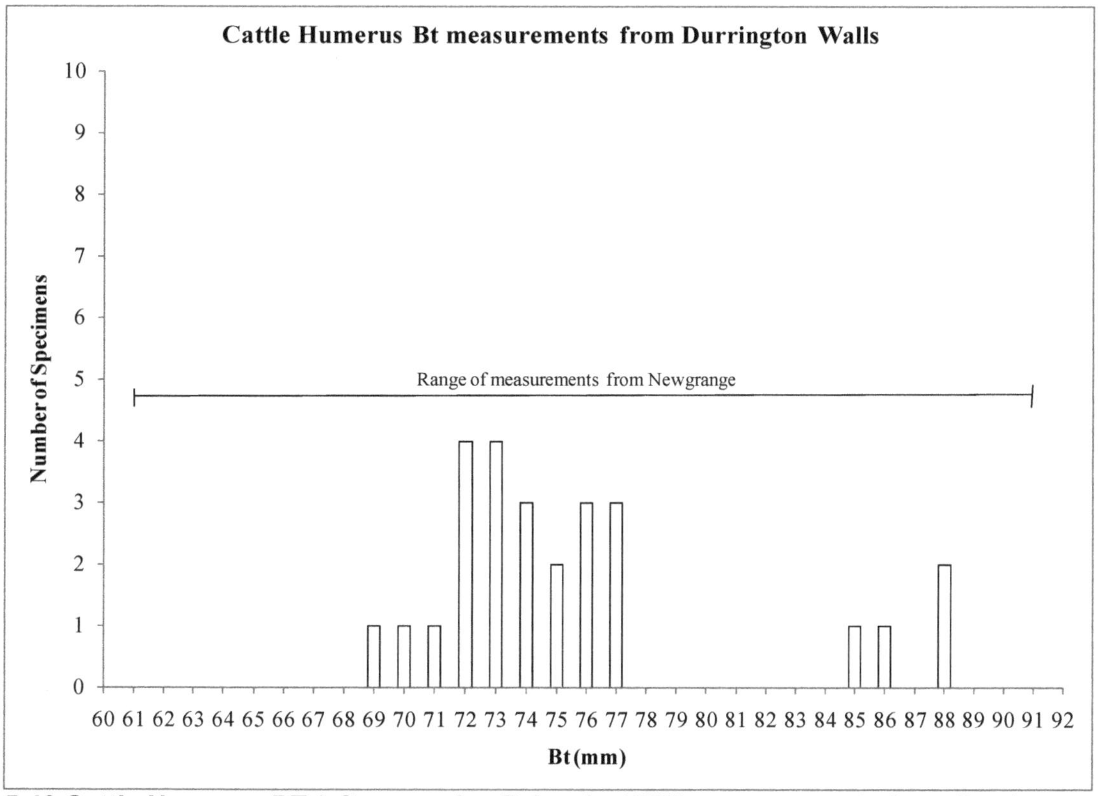

Figure 5.40 Cattle Humerus BT (after von den Driesch, 1976) measurements from Durrington Walls. The horizontal line indicates the range of measurements from Newgrange, Co. Meath based on 43 specimens (van Wijngaarden-Bakker 1986, 42).

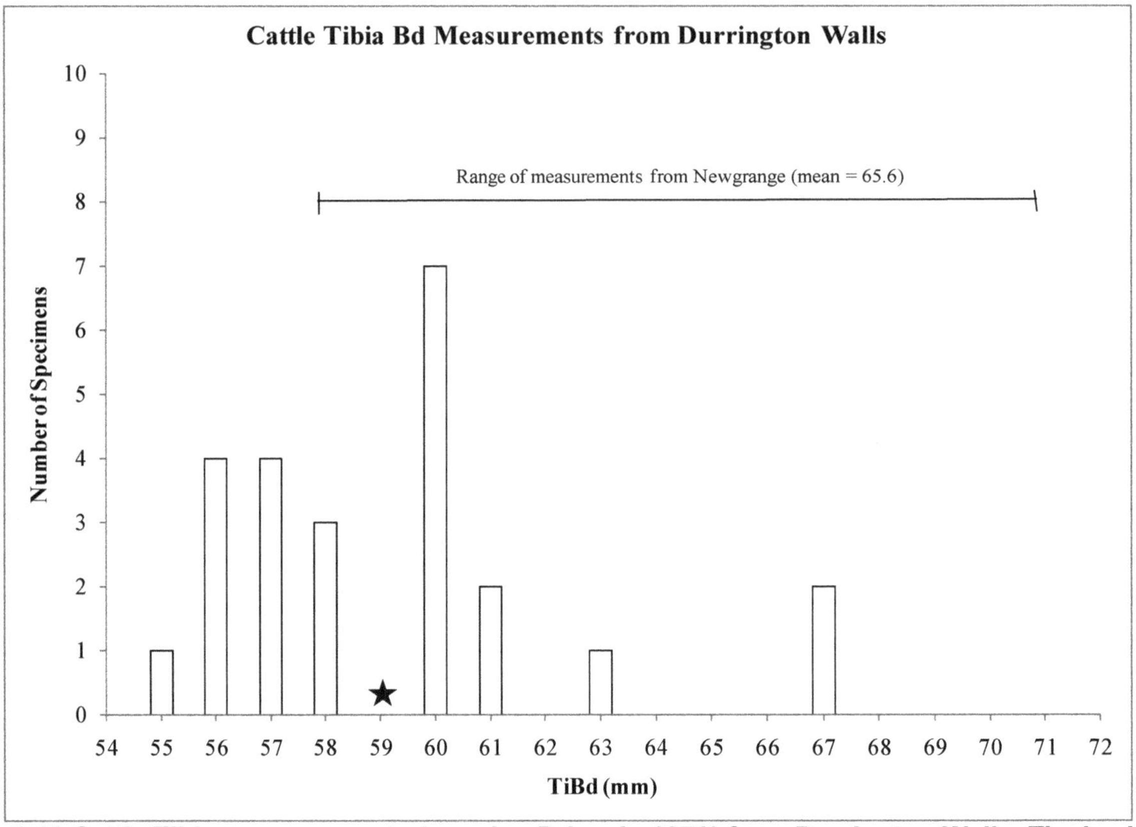

Figure 5.41 Cattle Tibia measurements (von den Driesch, 1976) from Durrington Walls. The horizontal line indicates the range of measurements from Newgrange, Co. Meath, Ireland (van Wijngaarden-Bakker, 1986, 42).

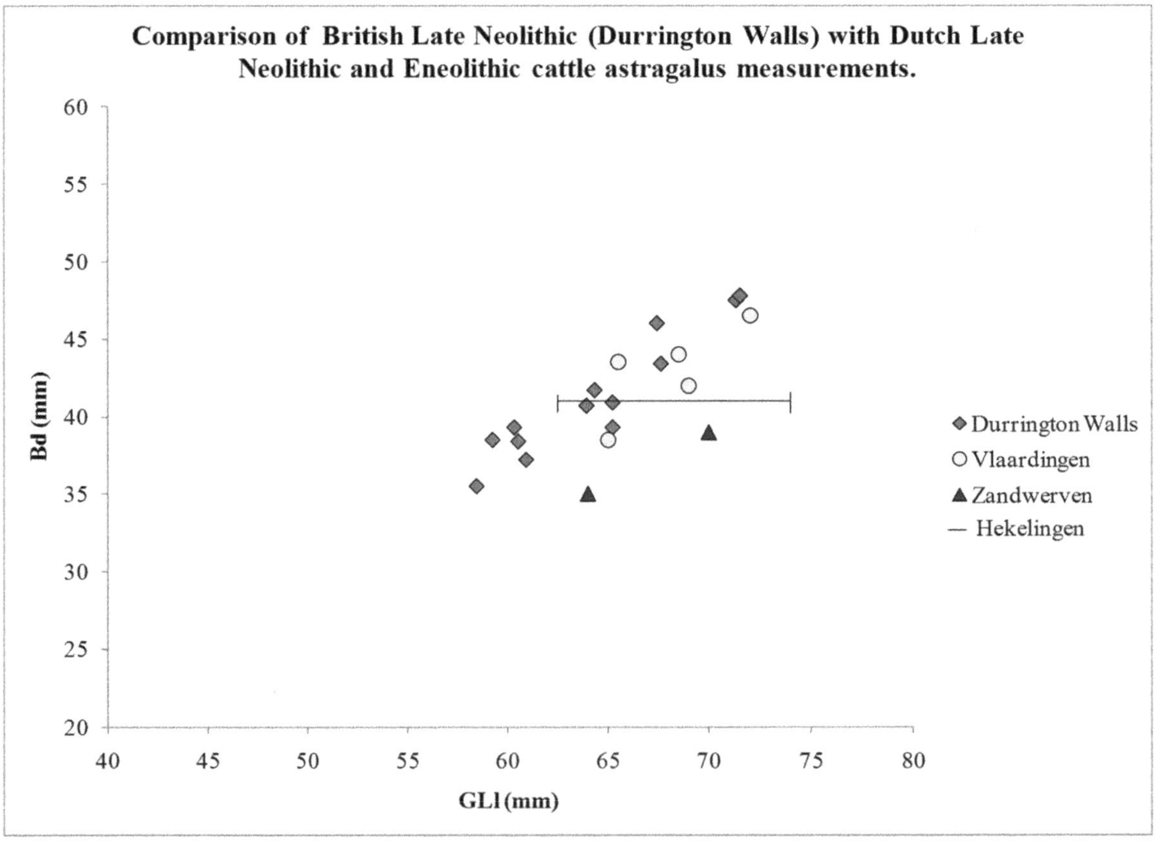

Figure 5.42 Comparison of astragalus measurements from Durrington Walls (Wiltshire), Vlaardingen (South Holland), Zandwerven (North Holland) (Clason, 1967) and Hekelingen (Prummel 1987).

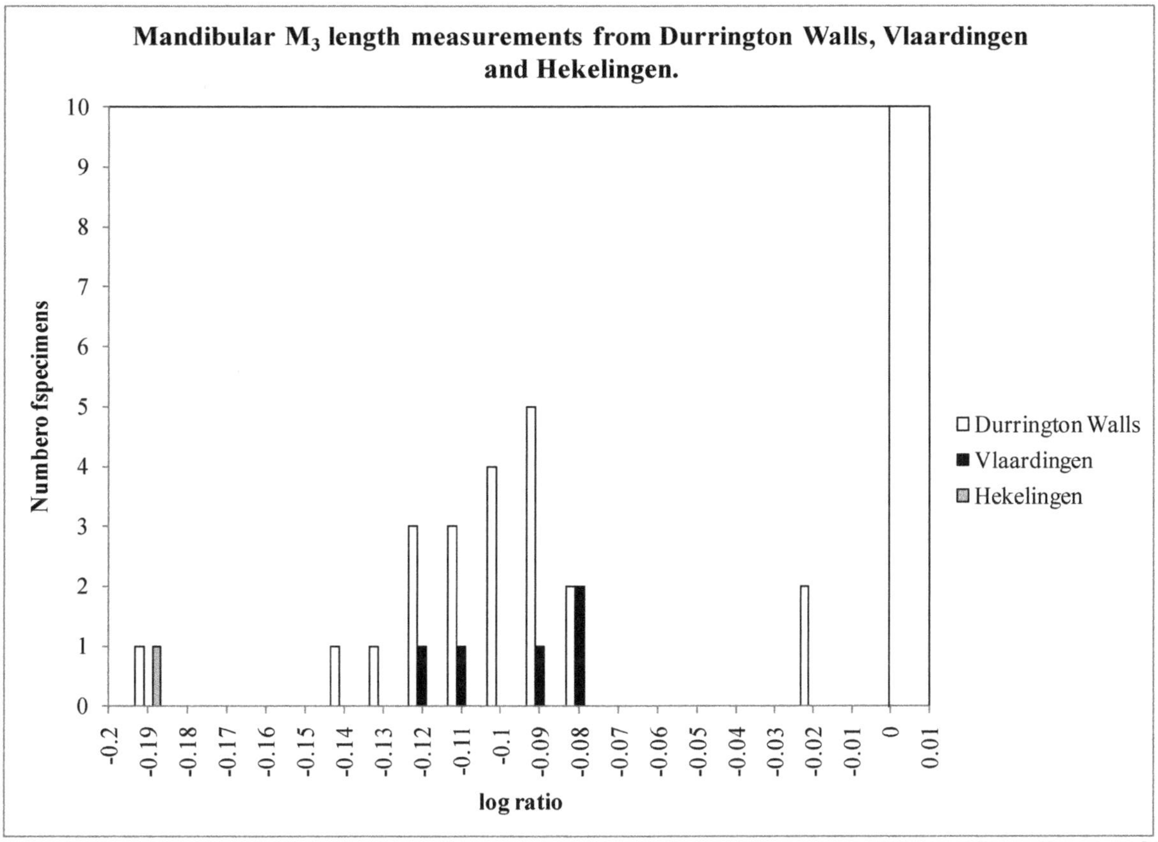

Figure 5.43 Mandibular third molar length measurements from Durrington Walls, Vlaardingen (Clason 1967) and Hekelingen (Prummel 1987).

Chapter 6
Strontium isotope analysis

6.1 Introduction

One of the central aspects of cattle husbandry that needs to be explored is mobility. Cattle have the propensity to the driven over long distances, and the custom of droving cattle in Britain, as elsewhere, is well known from historic times (*e.g.* Fox 1996). Its prehistoric undertaking is less well known, but evidence for the movement of animals would have significant implications for our understanding of the human groups that kept them. An important role has been envisaged by many researchers for cattle in Neolithic Britain (for instance Ray and Thomas 2003; Serjeantson 2006). Economic and non-economic reasons may both have provided the motivation to move cattle through the landscape and it is probable that cattle were moved between different areas of pasture during the Neolithic. Depending on specific circumstances the distances over which cattle could have been moved are substantial.

Despite the importance of cattle in the Neolithic and the potential for movement that they have, the application of isotope analysis for the investigation of their mobility has so far been limited. This chapter will present the results of two related studies into the movement of cattle during prehistory based on strontium (Sr) isotopic analysis. As part of this research a small pilot study that analysed teeth from three sites of different chronologies will be presented first followed by the results of a more detailed study of Late Neolithic remains from Durrington Walls in Wiltshire (see Chapter 2 for details of the site. Prior to the presentation of results, a brief discussion of the previous application of strontium to cattle remains will be included.

Finally, a small number of pig teeth from Late Neolithic Durrington Walls were also analysed. The results gathered from these teeth and their implications will be discussed in the final part of the chapter.

Some of the results included in this chapter were published in 2010 (Viner *et al.* 2010). Those figures that have been published before are identified in the relevant caption.

The application of Sr isotope analysis to animal mobility

Although a great deal of work has focused on investigating human mobility through strontium isotope analysis (for example Bentley and Knipper 2005; Bentley *et al.* 2003; Evans *et al.* 2006; Montgomery *et al.* 2000; Nehlich *et al.* 2009; Price *et al.* 1994; Price *et al.* 2001; Price and Gestsdóttir 2005; Richards *et al.* 2008), comparatively few studies have been undertaken that specifically aim to understand the movement of cattle in the past. The work by Evans *et al.* (2007) on cattle teeth found at two Anglo-Saxon sites in central England. Based on $^{87}Sr/^{86}Sr$ ratios, animals from Empingham (6-7th century) and Ketton (10-12th century) were interpreted as having been reared on pasture-land of a type not found in the local area. In addition to the identification of non-local animals, the use of $^{87}Sr/^{86}Sr$ in conjunction with Sr concentration in tooth enamel identified changes in diet during the period of tooth formation. Inclusion of grass in the animal's diet in the spring and summer, and dried food such as hay in winter could have produced the variations observed (Evans *et al.* 2007). This paper was one of the first investigations to concentrate efforts on cattle mobility, and its success provides support for the use of Sr analysis to shed light on husbandry practices.

Recently published work on cattle (and aurochs) from Gayhurst, Buckinghamshire, and Irthlingborough, Northamptonshire, found that most animals were of local origin (Towers *et al.* 2010). Only two (one from each site) were considered exotic, and an origin for these animals to the west and north of the sites studied, or further to the southeast in the Chiltern Hills was postulated by the investigator. The aurochs sampled were also found to have local $^{87}Sr/^{86}Sr$ values.

A further example of specifically cattle oriented research is that conducted by Schweissing and Grupe (2003). The main body of research investigated human migration during the Roman period, but an additional aim was to highlight the long distance trade of domestic cattle in early modern Austria. Horncore material was sampled with the aim of tracing changes in the $^{87}Sr/^{86}Sr$ signature through time (as the tip of the horncore is the youngest, and the base is the oldest part (Schweissing and Grupe (2003). We have, however, now become aware that horncore, like postcranial bone, is inappropriate for strontium isotope analysis due to its susceptibility to diagenetic alteration. Therefore, the results of this study cannot be considered reliable.

Other large mammal species have been the focus of strontium isotope investigations and these indicate the usefulness of the approach to understanding animal mobility. For instance, Britton *et al.* (2009) investigated the movement of modern caribou (*Rangifer tarandus granti*) in Alaska. Strontium variation exhibited in tooth enamel correlated well with the known movement of the caribou herd (Britton *et al.* 2009). Similarly, the migration patterns of late glacial horse (*Equus hydruntinus*) and red deer (*Cervus elaphus*) from central Italian archaeological sites were investigated through the analysis of strontium isotope ratios and concentration, and the results were used to test established models of human subsistence (Pellegrini *et al.* 2008). Studies have also been conducted to understand the movement of animals under human control. Tooth enamel from sheep at Kasteelberg, South Africa, provided evidence of different movement patterns in animals deposited at the same site (Balasse *et al.* 2002). A single cattle tooth included in the study also indicated that the animal had moved between inland and coastal areas during its life. A great deal of similarity in the development and character of sheep and cattle teeth exists, and as a result analysis of sheep/goat teeth can be a useful source of comparison. The study of Kasteelberg sheep also highlighted the need to test multiple individuals from archaeological sites, as there is no guarantee that animals deposited at the same site were treated in the same way during life, and numerous husbandry methods could have been applied that result in different $^{87}Sr/^{86}Sr$ results.

In Britain strontium isotope analysis was used to determine that fallow deer specimens found on first century AD archaeological sites (in particular Fishbourne, Sussex) had a variety of origins (Sykes *et al.* 2006). On the basis of $^{87}Sr/^{86}Sr$ results, along with other lines of evidence, it was concluded that at least one deer was unlikely to have originated in Britain and was probably imported as a juvenile. Another (later) specimen was considered local to the area, and that it had a limited home range compared to wild animals.

Studies that focus mainly on human migration have also contributed to the body of information available for cattle by using them (and other mammals) as a source of comparison. The analysis of tooth enamel strontium isotope composition in archaeological and modern animals is often undertaken in an attempt to define the local signature of a site (*e.g.* modern steenbok (*Raphicerus campestris*, Balasse *et al.* 2002), modern snails (Bentley *et al.* 2003), and archaeological specimens of domesticated animals (Bentley *et al.* 2003; Bentley *et al.* 2004; Sykes *et al.* 2006)). However, as the work by Bentley *et al.* (2003) highlights, a local origin for domesticated animals found on archaeological sites cannot be assumed. If animal remains are to be used as a proxy for the local signature, detailed investigation must ensure that the animals had not been introduced to the site from elsewhere/do not have an external origin. The cattle tooth used in Bentley *et al.*'s study was originally intended to be a proxy for the local $^{87}Sr/^{86}Sr$ signature. As the results obtained did not concur with the expected local signature the researchers were led to postulate a mobile pastoralist system for *Linearbandkeramik* (LBK) groups in southern Germany (Bentley *et al.* 2003, 483).

Rather than using modern or archaeological animal remains as a proxy for local available strontium, the present study will rely on modern vegetation samples (discussed below in Section 6.5) along with previously conducted strontium isotope analysis of human and animal remains in the research area (such as Evans *et al.* 2006a; Montgomery *et al.* 2000), and information from the map of strontium variation in Britain produced by Evans *et al.* (2010). The map was produced by measuring the strontium isotope composition of plants from different biosphere compositions.

The use of Sr to investigate mobility in British prehistory

Cattle movement has yet to be studied in detail within the study area and period with which this research is concerned, but research using strontium isotope analysis from the region has provided evidence of human movement between and within the study area, between southern Britain and other parts of the country, and between Britain and other parts of Europe. A number of studies of human skeletal material have found that human migration took place both before and subsequent to the Late Neolithic period that forms the focus of the present study. Sites such as Boscombe, Normanton Down and Stonehenge, all in Wiltshire and close to Durrington Walls, have been the subject of human strontium isotope studies (Evans *et al.* 2006a). Patterns of migration during childhood were identified in some Bronze Age humans, while others were found to have lived locally during the period of tooth development. Another important study from southern Britain focused on the mobility of humans from the Early Neolithic site of Monkton-up-Wimbourne in Dorset (Montgomery *et al.* 2000). Humans at the site had migrated over significant distances during their lifetime, and an area of the Mendips 80km from the site was identified as the most likely origin (Montgomery *et al.* 2000). As well as providing concrete evidence of migration between different parts of southern Britain, a further implication of the research is the establishment of a local strontium signatures for the immediate research area, and the region as a whole. Evidence from these studies will be used to corroborate the results of cattle studied at Late Neolithic Durrington Walls.

Migration, of humans at least, has thus been identified at prehistoric sites in the study area. These studies have shown the potential of strontium isotope analysis to identify movement between geologically diverse environments. Strontium isotope analysis has been successfully applied to cattle teeth, and in other large mammals it has been productive in investigating movement between different areas and pasture lands. This chapter aims to investigate potential movement of cattle between areas of differing geological conditions during the Late Neolithic period. It will also work to identify the minimum distances involved in the movement of cattle and suggest possible areas of origin for any individual animals that are found to be non-local.

6.2 The pilot study - Materials

Archaeological background information

Three sites were investigated during the pilot study, each is in a different part of Britain (locations shown in Figure 6.1), and they come from two main periods: the Late Neolithic and the Late Bronze Age/Early Iron Age. The main aim of the pilot study was to find evidence for cattle mobility at these the three sites by comparing the Sr composition of cattle tooth enamel with the known local geological signature. Work was carried out by S. Viner and J. Evans in 2007 at the NERC Isotope Geosciences Laboratory in Keyworth, Nottinghamshire. Although the pilot study was limited by the small sample size the presence of a non-local animal at any of the sites would have provided a starting point for further investigation and justification for more detailed analysis.

The earliest site investigated as part of the pilot study was late Neolithic Durrington Walls. As the site has also formed a significant part of the zooarchaeological aspect of this research, details of its excavation, date and interpretation have been presented in Chapter 2 and will not be revisited here. Durrington Walls is located in an area of chalk downlands in Wiltshire, Southern England. The importance of Durrington Walls to the investigation of strontium isotope ratios lies in its status as one of very few Neolithic habitation sites. In addition, Durrington Walls clearly also has a ceremonial function as evidenced by the monuments at the site. Cattle recovered here may not be typical of the Neolithic domestic economy. The tooth used in the pilot study came from the most recent period of archaeological investigation that was conducted as part of the Stonehenge Riverside Project.

Poulton Chapel House Farm is a multi-period site in Cheshire (Figure 6.1). The site has been under investigation since 1995, initially as a joint University of Liverpool and Chester Archaeology initiative, and more recently under the direction of Mike Emery as part of the Poulton Research Project. The tooth chosen for analysis was recovered from a ring-ditch that has been dated (based on associated ceramic material) to the Late Bronze Age/Early Iron Age (K. Cootes pers comm.). Poulton Chapel House Farm was selected for inclusion in this study due in part to the presence of numerous cattle remains at the site, including some well-preserved mandibular specimens that are appropriate for strontium analysis. In addition, the Late Bronze Age/ Early Iron Age period in Cheshire is poorly understood due to a lack of excavation of relevant sites in the county. Knowledge of cattle husbandry practices in the area was extremely limited and no information regarding the potential movement of cattle into or within the area during the period was available.

Welland Bank Quarry was a Late Bronze Age/Early Iron Age site in the Lower Welland Valley, Lincolnshire (NGR TF183079). Excavations were conducted from 1993-97, in advance of gravel quarrying that led to the complete destruction of the site. Direction of the excavations was undertaken by Francis Pryor and Tom Lane on behalf of Archaeological Project Services. Based on ceramics associated with them the faunal remains from Welland Bank Quarry date to the Late Bronze /Early Iron Age and are therefore broadly contemporary with the specimen from Poulton Chapel House Farm. Unlike Poulton Chapel House Farm, Welland Bank Quarry provided numerous features that were interpreted as 'domestic' in character. The investigation provided an important non-ritual specimen to the study.

Specimens chosen for analysis
Numerous studies have suggested that the composition of both tooth dentine and bone make these substances more susceptible to diagenetic alteration than tooth enamel (Budd *et al.* 2000; Evans *et al.* 2007, 1996; Sillen and Sealy 1995). Some treatment methods have been developed that allow the removal of strontium of diagenetic origin. These approaches usually involve the removal of diagenetic strontium using acetic acid, but the validity of such approaches is still unclear (see Nelson *et al.* 1986; Sillen and Sealy 1995 for a discussion). Therefore, samples of tooth enamel were used for strontium isotope analysis because of its known resistance to alteration in the burial environment, and mandibular bone was used as a comparison (see below).

To avoid any potential variation between maxillary and mandibular teeth and so that a comparative period in each animal's life was investigated, the same tooth from each mandible was selected. Mandibular second molars (M_2) were chosen for the pilot study. These permanent teeth develop between birth and ten months of age (Bentley and Knipper 2005). In addition to tooth enamel samples, a piece of bone from the associated mandible was taken for each tooth used in the pilot study. Bone (like dentine) is altered diagenetically and will provide an indication of the local strontium signature. The enamel signature can be compared to this, along with other sources of geological information, to highlight whether cattle exhibit local or exotic readings.

6.3 Pilot study- results and discussion

The methods used for both the pilot study and the main study, along with more specific information about the sampling strategy were discussed in Chapter 2 and will not be presented again here.

Cattle tooth enamel from Poulton Chapel House Farm corresponded very closely to the signature from associated bone material (Figure 6.2). $^{87}Sr/^{86}Sr$ values from the tooth enamel ranged from 0.7119 to 0.7124. The mandibular bone provided a result of 0.7122, a signature that is indistinguishable from the enamel samples. The similarity between the enamel strontium signature and that of the local environment (*i.e.* the bone sample) provides evidence that the animal is likely to have spent the period during which its tooth developed in an area with geological conditions very similar to those found at Poulton Chapel House Farm. It is most likely that the animal was born and reared locally, and then deposited at the site on its death. There is no evidence for long distance movement of cattle, but of course further analysis of animals from the site could provide different results. The tooth was sampled six times from the base to the cusp and variations in the $^{87}Sr/^{86}Sr$ value of different samples suggests that no significant variation in the conditions experienced by the cattle took place during the period of tooth development. The small variation in signatures seen is not linear (*i.e.* the $^{87}Sr/^{86}Sr$ ratio does not go from low to high or *vice versa*, but is more erratic - first decreasing down the tooth, and then increasing). Perhaps, then, the animal was moved frequently between areas that provided a very similar $^{87}Sr/^{86}Sr$ composition, without any clear directionality identifiable in the tooth enamel. General similarity between the enamel signatures and the associated bone suggest that the difference in conditions was slight, and perhaps the distances moved were negligible. All of the samples from the Poulton Chapel House Farm tooth are within the expected variation of the strontium domain that the site is located in.

Six samples were also taken from the Welland Bank Quarry specimen, and at first tooth enamel appears to be different from the $^{87}Sr/^{86}Sr$ gained from the associated mandibular bone. The least radiogenic signal from tooth enamel was 0.71032 compared with the much lower value of 0.70942 produced by the diagenetically altered bone sample. Welland Bank Quarry is located in a strontium domain in which the underlying geology is likely to provide $^{87}Sr/^{86}Sr$ values of between 0.709 and 0.710 (according to the map provided by Evans *et al.* 2010). Although some of the enamel samples are slightly elevated in terms of the underlying geology, they are appropriate for animals that were raised in the area. The results from Welland Bank Quarry are therefore somewhat ambiguous, it is possible that the tooth enamel is exhibiting a local signature, but the difference between the enamel and the bone could indicate some variation in the geological conditions prevalent during tooth formation. Therefore, it is possible that the animal from Welland Bank Quarry was reared for part of its life away from the site, although not under vastly different conditions. There is no evidence that this animal moved over long distances to be deposited at Welland Bank Quarry and evidence of movement within the animal's life is limited. There is a general trend towards less radiogenic readings through time, but the variation is slight and cannot be said to provide evidence of movement during tooth formation. It is less pronounced than that from Poulton Chapel House Farm, but does have greater coherence in the direction of the variation. Welland Bank Quarry is on the edge of the Fenland area, and the study of the animal bones has revealed that wetland and coastal resources, such as beaver and seal, were exploited, and that livestock accustomed to drier conditions, such as goats, were also used. The catchment area of the site was therefore environmentally diverse and its exploitation may have required a rather high level of local mobility. The high frequency of horse bones at the site also suggests that the need to move around in the countryside must have been a pressing concern (Albarella *et al.* in prep.). Unlike the other two sites, the enamel from Durrington Walls provided a strontium signature very different from the associated bone and also the expected chalk land signature (considered to be no higher than 0.709, Evans *et al.* 2006b). Enamel signatures were extremely closely confined, ranging from 0.71013 to 0.71019. When compared to known geological information it can be suggested that this animal was raised in one of a variety of geographic areas, including central England, small parts of southern Wales, Devon and parts of the east of England as far north as Norfolk. It certainly could not have been raised on chalkland like that surrounding

Durrington Walls and other parts of southern England. In contrast, the associated bone sample gave a significantly different strontium signature to the enamel, but one that was similar to a chalk value. The very narrow range of $^{87}Sr/^{86}Sr$ ratio signatures from the tooth enamel is an indication that the animal was subject to very little movement during the time - between birth and 10 months - that the second molar developed.

These three teeth allow some general conclusions to be drawn concerning the movement of cattle during prehistory, although the small sample size should caution against over interpretation. No evidence was uncovered to suggest that cattle were moved between areas of substantially different geological conditions during the Late Bronze Age/Early Iron Age. Poulton Chapel House Farm and Welland Bank Quarry are in different parts of England, and are a different type of site, yet they led to similar interpretations of the strontium data. More detailed investigation of cattle husbandry practices during the period is essential to determine whether this is a general phenomenon. It is also possible that animals were moved within different areas of similar geological conditions during the period of tooth development, a type of mobility that would be invisible in the results of strontium isotope analysis. In contrast, values from the Durrington Walls tooth indicate that the area inhabited during tooth development and the area of deposition were different geologically. All of the samples from the chosen tooth fall within the same non-local band, strongly suggesting an external origin for the cattle. Clearly this animal must have been brought over a substantial distance from the area in which it fed during growth. Whether transportation occurred during the animal's life or happened after its death is unknown, but the former is a much more likely hypothesis as cattle can relatively easily be moved on the hoof, while carrying the carcass – or even parts of it - over long distances would have proved rather difficult.

The pattern of movement exhibited by the comparison of cattle tooth enamel with local geological conditions and diagenetically altered mandibular bone from Durrington Walls led directly to the need for a more detailed study of Late Neolithic husbandry practices at this site. The remainder of this chapter will provide contextual information about the specimens used, present the results and discuss the implications of them for our understanding of the Late Neolithic.

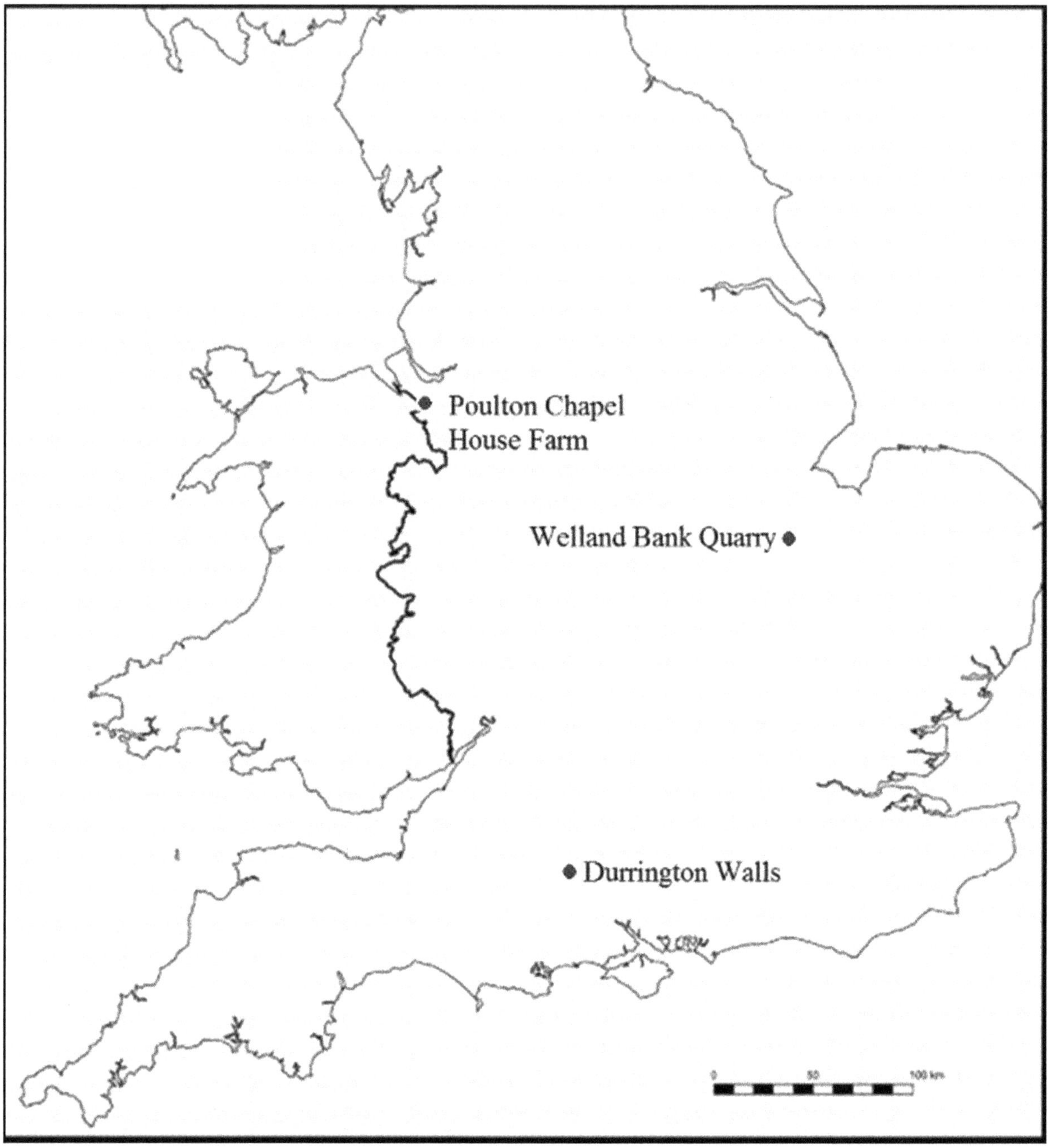

Figure 6.1 Location map showing the three sites used in the pilot study, Poulton Chapel House Farm (Cheshire), Welland Bank Quarry (Lincolnshire) and Durrington Walls (Wiltshire) (Viner 2010, Figure 1).

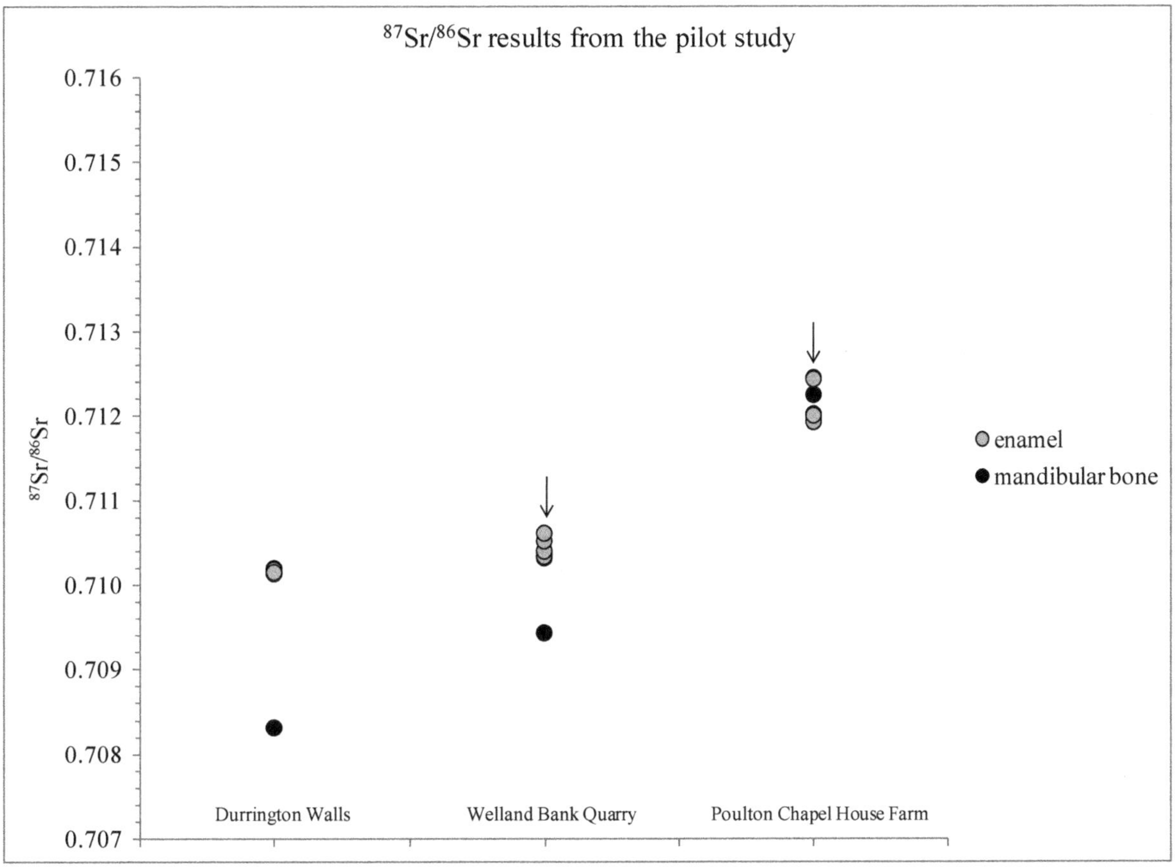

Figure 6.2 $^{87}Sr/^{86}Sr$ ratios of teeth used in the pilot study. Grey points are enamel samples and black dots the associated mandibular bone. Arrows indicate the direction from base to tip of the tooth

6.4 Detailed study of Durrington Walls cattle teeth – materials

Contextual information

Cattle teeth selected for Sr isotopic analysis came from a number of different archaeological contexts at Durrington Walls, though most of the specimens were chosen from context 593. This was the upper level of a large midden located to the north of the avenue, which provided a large amount of bone material. Other teeth derive from context 652, *i.e.* the upper layer of the midden located to the south of the avenue, and context 665 which, although initially thought to be the fill of a feature, was later re-classified as an old ground surface. The final context from which teeth were drawn was 770, a midden deposit located below the henge bank, which was excavated to the north of both the avenue and the house structures (Parker Pearson *et al.* 2005).

Context 593 was chosen because of its importance in the make-up of the site and also because it is the context that has provided the most abundant faunal material to date. Material from other contexts was chosen to provide a potential contrast to 593, and the choice of specific context relied on their importance, location and stratigraphic integrity and on the availability of suitable material.

Specimens chosen for the study

12 teeth from Durrington Walls (numbered DW1-DW12 in the tables and figures of this chapter) were chosen for strontium isotope analysis. All of the specimens came from excavations that were undertaken as part of the Stonehenge Riverside Project during the 2005, 2006 and 2007 field seasons.

In order to ensure that the results from different teeth were as comparable as possible consistency of both specimen selection and sampling was a central concern of the study. In line with the pilot study the second permanent molar of the mandible was initially considered for analysis. However, when loose this tooth can be difficult to distinguish from the first permanent molar, therefore only those found in jaws were considered for inclusion. It soon became apparent that too few appropriate specimens were available and the focus of the study was therefore shifted to the third permanent mandibular molar (M_3). This tooth can be easily identified based on morphological characteristics, even when it is loose. Mandibular teeth were used throughout the study, partly due to variation in the defining morphological characteristics that can hinder identification in upper permanent third molars (*i.e.* the posterior ridge that distinguishes the upper third molar from other molars is not always clearly developed) and partly because wear stage – consistent with standard zooarchaeological practices – was only recorded for the lower molars. All of the enamel slices were removed from the lingual part of the anterior cusp of the tooth, with the exception of tooth DW5 in which the posterior cusp was used because of abnormal wear on the anterior cusp. Instead of the bone sample that was taken for comparison during the pilot study, dentine was used in the main study. This was because most of the specimens in the main study were loose, and therefore no associated bone was available.

The teeth were in various states of wear, from un-erupted to moderately well worn. Wear stages (Table 6.1) were recorded before destruction of the teeth, along with measurements, following the method outlined in Chapter 2. Teeth used for strontium isotope analysis have also been included in the zooarchaeological analysis presented in Chapter 4.

Table 6.1 Teeth chosen for Sr isotope analysis from Durrington Walls, including the year of excavation, context, loose/jaw status and wear stage (following Grant, 1982). E=erupting (see Chapter 2 for a definition of 'jaw').

Tooth number	Year	Context	Loose/Jaw	Wear
DW1	06	593	l	E
DW2	05	652	l	g
DW3	05	652(d)	j	g
DW4	06	593	l	j
DW5	05	593	j	E
DW6	05	593	l	broken
DW7	06	593	l	k
DW8	07	593	l	d
DW9	07	593	l	k
DW10	05	665	j	l
DW11	05	770	j	h
DW12	07	593	l	g

6.5 Results of the main study

Introduction

The results of the pilot study identified that one cattle from Late Neolithic Durrington Walls had moved from an area with a different Sr composition to the area in which it was found, and while areas with appropriate Sr signatures can be found within 30km of Durrington Walls, the animal could have come from much further away. Following on from this discovery the more detailed study aimed to investigate whether the animal was an isolated example of movement or whether a more wide ranging pattern could be identified. In addition, if movement was a common feature of Late Neolithic husbandry at Durrington Walls, it would be necessary to investigate the variety of cattle origins. By comparing the $^{87}Sr/^{86}Sr$ composition of cattle teeth from the site with the known values of different geographic areas, the origin of individuals can, to some extent, be pinpointed. An important question is whether a single exotic origin can be identified for non-local individuals, or whether animals were derived from a variety of geographical areas.

Defining the 'local' signature

For the results from Durrington Walls to be understood it is necessary to gain an idea of the likely Sr signature of the site itself. As was mentioned in the discussion of the pilot results the expected upper limit for chalkland areas is 0.709. To gain a more accurate impression of the conditions local to Durrington Walls two main sources of information were used: modern vegetation and the signature of humans that had lived under similar conditions. Eight samples of surface vegetation were collected from the vicinity of the site in 2006; a description of the location of each sample can be found in Table 6.2. In addition, research by Evans *et al.* (2006b) focused on conducting Sr isotope analysis on human remains dating to the Roman period from the site of Lankhills, Hampshire, which, like Durrington Walls, is on chalkland. Tooth enamel from individuals that lived locally provided values of $^{87}Sr/^{86}Sr$ = 0.7085±0.0004 (Evans *et al.* 2006b: 270). A further study of Bronze Age individuals from Normanton Down ($^{87}Sr/^{86}Sr$ = 0.707837, $^{87}Sr/^{86}Sr$ = 0.708042) and Stonehenge ($^{87}Sr/^{86}Sr$ = 0.708226, $^{87}Sr/^{86}Sr$ = 0.70791) identified the signatures as consistent with time spent on chalkland (Evans *et al.* 2006a, 313). The map recently produced by Evans *et al.* (2010) would indicate values of between 0.708 and 0.709 for the chalkland areas of Wiltshire. Based on these sources of evidence it can be ascertained that animals living in the chalk areas of southern Britain will provide $^{87}Sr/^{86}Sr$ from around 0.7078 to no higher than 0.7090. The grey band on Figure 6.3 represents the chalk signature expected to be prevalent in Durrington Walls material.

Table 6.2 Details of vegetation samples from the vicinity of Durrington Walls with location description and their $^{87}Sr/^{86}Sr$ values.

Sample	Material	Sample	$^{87}Sr/^{86}Sr$
DWplant-	Plant	Stonehenge	0.707936
DWplant-	Plant	Durrington	0.707667
DWplant-	Plant	Woodhenge	0.708700
DWplant-	Plant	Stonehenge	0.707937
DWplant-	Plant	Countess	0.707863
DWplant-	Plant	Bulford	0.708385
DWplant-	Plant	South of	0.708745
Average			0.7082

$^{87}Sr/^{86}Sr$ results

To ease understanding of the results presented in this section each tooth will be discussed as part of a subgroup of teeth which all appear to provide similar isotopic results. These subgroups are not intended to be inflexible and are merely a way of organising the interpretation of the material, rather than a strict categorisation of the teeth into groups. The dentine sample for each tooth is shown as a black dot on Figure 6.3, and are included in Table 6.3. The dentine and bone samples from the study are consistent with the expectation that these tissues are susceptible to post-depositional alteration and that they will provide an indication of the $^{87}Sr/^{86}Sr$ signature of the burial environment.

In contrast with the dentine signatures the results from tooth enamel provide a picture of great variability in cattle teeth from Durrington Walls (Figure 6.3). Enamel $^{87}Sr/^{86}Sr$ values vary a great deal from highly radiogenic signatures that represent grazing on areas of old rocks (that contain high levels of rubidium), to much lower signatures that suggest areas of young rocks with low rubidium content (Evans *et al.* 2006b, 266). While some of the teeth with low $^{87}Sr/^{86}Sr$ values came from animals that were grazed on areas consistent with Durrington Walls during the period of third molar development; others came from areas that had $^{87}Sr/^{86}Sr$ signatures that were very different and clearly exotic. In some cases there were also substantial differences in the extent of within-tooth variation.

Table 6.3 $^{87}Sr/^{86}Sr$ values of samples from Durrington Walls cattle teeth (slice 1 is the uppermost part of the tooth, slice 3 the one closest to the root enamel junction)

Ctx	Tooth number	Subsample	$^{87}Sr/^{86}Sr$
593	DW1	1	0.70973
		2	0.70973
		3	0.70956
		dentine	0.70854
652	DW2	1	0.71257
		2	0.71252
		3	0.71136
		dentine	0.70837
652	DW3	1	0.71021
		2	0.70952
		3	0.71081
		dentine	0.70872
593	DW4	1	0.71356
		2	0.71203
		3	0.71305
		dentine	0.70833
593	DW5	2	0.7092
		3	0.7089
		dentine	0.70837
593	DW6	1	0.71014
		2	0.7104
		3	0.71066
		dentine	0.70867
593	DW7	1	0.71066
		2	0.71059
		3	0.71051
		Dentine	0.70872

Table 6.3 continued

593	DW8	2	0.710515
		3	
		4	0.709495
		Dentine	0.708018
593	DW9	1	0.708719
		2	0.70881
		3	0.708933
665	DW10	1	0.710501
		2	0.71133
		3	0.712199
		Dentine	0.708996
770	DW11	1	0.710935
		2	0.71066
		3	0.710815
		Dentine	0.708612
593	DW12	1	0.714826
		2	0.714805
		3	0.714826

Durrington Walls tooth 9 (DW9) came from an animal that was raised on chalkland during the period of its tooth development. No dentine sample was processed for this tooth as the large number of samples from other teeth was considered adequate to provide an indication of local burial conditions. It provided samples that ranged from 0.708684 to 0.708898 (Figure 6.3, Table 6.3), and which coincide with those from the analysis of modern vegetation and the dentine signatures recovered from the site. Although it is not possible to rule out a more distant chalk origin for this specimen, the evidence is consistent with local grazing, an interpretation which represents the most economical understanding of the results.

Enamel from tooth DW5 overlaps with a number of the dentine values from Durrington Walls, and is within the range established for chalkland. It also concurs broadly with the results from DW9 discussed above, though one of the two available subsamples provided a more radiogenic result, which falls on the upper limit of what can be considered a chalk signature. Despite the slight variation in the signature of the enamel samples, it remains likely that the individual was raised on chalkland during the period of tooth development.

In contrast to DW9 and DW5, the rest of the teeth from the site exhibit no consistency with the expected chalk signature. The three enamel subsamples from tooth DW1 are tightly clustered and exhibit more radiogenic values than those from the teeth that have already been identified as having chalk signatures. All of these signatures are too high (*i.e.* too radiogenic) to be of chalk origin, therefore it can be concluded that this animal spent a period of its life in an area of different underlying geology to both of the chalk grazing animals, DW9 and DW5. This area cannot have been in the immediate vicinity of Durrington Walls.

As has been discussed previously the pilot study subsamples all provided $^{87}Sr/^{86}Sr$ values outside the range of chalkland, and indicate that the animal grazed on land unlike that found in the vicinity of Durrington Walls. The cattle used for the pilot study is part of a subgroup of animals that all came from similarly radiogenic areas. DW3, DW6, DW7, DW8 and DW11, along with DW1 and the pilot specimen all have upper limits of between 0.710480 and 0.710900. DW7 provided three enamel samples, all of which are too radiogenic to be of chalk origin, an indication that the animal grazed in a non-chalk area for the whole period of third molar development (between 10 months and two years of age, Bentley 2006). DW6 concurs well with DW7, having an almost identical upper boundary. However, the samples are less tightly confined; the least radiogenic of the three provided a value of 0.71014, while the most radiogenic had a value of 0.710399, a difference of only 0.000259. This greater variation in enamel signature will be discussed in the following section. A higher level of variation exists within DW3. The most radiogenic section of the tooth makes DW3 highly comparable with the other teeth in the subgroup, but one of the subsamples is substantially less radiogenic. This subsample is, however, still too radiogenic to be considered as having chalk signature and we can therefore be confident that the animal was not raised on chalkland during the period of tooth development. DW11 also overlaps with samples from DW8, DW7, DW6 and DW3, although it has slightly higher, more radiogenic values. This very slight tendency is not significant enough to suggest a different origin for this specimen. In terms of both its values and the distance between samples DW11 most closely resembles DW7. Like DW7 it provided three very similar results from analysis of tooth enamel, the least radiogenic of which is clearly above the chalkland upper range.

None of the teeth from this subgroup provide evidence that cattle were grazed on chalkland during the development of the third molar. Further to this, they all provide very similar Sr values (although there is some disparity between the internal variations of specific specimens). By comparison with known geological distributions in Britain the results provide evidence that cattle were grazed in an $^{87}Sr/^{86}Sr$ domain with values above 0.709 and an upper limit of no more than 0.711. This range of values is common to areas of central England, and can also be found in small areas of the southern and eastern coastal regions, as well as Devon and Cornwall. To reach an area with the appropriate $^{87}Sr/^{86}Sr$ signature a distance of at least 30km must be travelled from Durrington Walls.

DW10 and DW2 form a further subgroup and are all significantly more radiogenic than those non-chalk specimens discussed above. DW10 provided three enamel subsamples, which have fairly diverse $^{87}Sr/^{86}Sr$ values, but are all much higher than would be expected from tooth development in a chalk area. DW2 results are comparable to DW10, although, on average, slightly more radiogenic (*i.e.* providing higher readings). This tooth too shows more significant internal variation than was seen in the other non-chalk teeth. By comparison with known $^{87}Sr/^{86}Sr$ domains, DW10 and DW2 were probably in development in an area with values of between 0.7100 and 0.7130. This kind of signature could be obtained by being grazed in areas of Devon and Cornwall, Wales or as far north as Cumbria. The closest area that would provide this kind of signature is almost 90km away (Gloucestershire/south Wales), followed by parts of Devon that are almost 100km distant.

The final two specimens that were included in the study (DW4 and DW12) have much higher readings and are therefore much more radiogenic than any of the other specimens. DW4 provided three subsamples, two of which overlap with the results from DW10 and DW2, and therefore provide evidence of grazing within the same domain as these two specimens. However, the most radiogenic sample has a much higher value than those encountered previously. This value allows a more specific identification of the likely origin of the animal. Development probably took place in an area similar to that of DW10 and DW2, however the very high signature of one of the samples indicates grazing on very old rocks in areas with values of above 0.713.

DW12 provided distinctively high $^{87}Sr/^{86}Sr$ values and is from the most radiogenic area encountered during the research. The three enamel results were very similar to one another (two of the specimens are so similar that they appear as a single point on Figure 6.3), all of which had very high $^{87}Sr/^{86}Sr$ values above 0.714. The possible origins of DW4 and DW12 are extremely limited and according to the map provided by Evans *et al.* (2010) the only origin for such values is some distance to the north in Cumbria, Northumberland and Scotland. However, there is some evidence to suggest a small parcel of very old (*i.e.* very radiogenic) rocks exist in the area of the Malvern Hills (Worcestershire) (based on modern vegetation samples Chenery *et al.* in press), and this would seem a more likely origin for the cattle at Durrington Walls.

Within tooth variation

The purpose of taking numerous samples from each tooth was so that variation during the period of tooth development could be understood (as discussed in Chapter 2). Cattle tooth enamel develops sequentially from the apex to the root (Balasse *et al.* 2002; Bentley 2006; Fricke and O'Neill 1996) and therefore chronological variation in the Sr signature of enamel can be investigated. However, because of the way that cattle tooth enamel develops it is not possible to isolate discrete periods of tooth development and it is impossible to get an exact picture of the variation in Sr through time. A further complicating factor is the degree of wear on the occlusal surface of the teeth. The teeth selected were at different stages of wear that ranged from entirely unworn to in an advanced stage of wear (wear stages are provided in Table 6.1 and follow the system of Grant 1982). As a result all teeth were sampled at the root crown junction and further samples were taken further up the tooth. For some specimens the unworn tip of the tooth was available whereas others had to be sampled further down the tooth. The samples were not taken at the same point on each tooth and are therefore not directly comparable.

Two main issues arise from the internal variation of cattle teeth at Durrington Walls. The first is that from apex to root some specimens decrease $^{87}Sr/^{86}Sr$, while others increase $^{87}Sr/^{86}Sr$. Therefore while some specimens became gradually less radiogenic (*i.e.* closer to a chalk value) others became more radiogenic (*i.e.* further away from the chalk signature). There is no pattern governing the direction of the vector for the teeth, and teeth from the different areas discussed above have vectors in both directions.

The second issue is the extent of variation within each tooth. Although as mentioned caution must be exercised due to sampling difficulties, broad differences can be seen between the level of difference within teeth of different subgroups. DW2, DW10 and DW4 come from areas that are more distant from Durrington Walls and they also see the greatest internal variation in strontium value.

Concentration
Sr concentrations vary from tooth to tooth at Durrington Walls, and range from extremely high (for example DW2 that has Sr concentration values in the tooth enamel of 211-305 ppm) to extremely low (for instance DW4 that provided concentrations of 90-94 ppm) (Figure 6.4). Strontium concentration is affected by a number of factors including trophic level of the individual (Price *et al.* 1994), weaning and lactation (Blakely 1989), the Sr concentration of food consumed (Evans *et al.* 2007) and in the case of dentine the local burial conditions (Evans *et al.* 2007). It can therefore be difficult to identify the reason for a change in concentration. When the strontium concentration (Table 6.4) of each tooth is compared with the likely origin of each tooth as identified above there is no pattern that links strontium concentration and likely geographical origin and there is also no obvious correlation between the amount of within tooth variation in $^{87}Sr/^{86}Sr$ values and strontium concentration.

Pig teeth and strontium isotope analysis
In addition to the twelve cattle teeth that were subjected to strontium isotope analysis, a further two pig teeth were included. Pigs were not a major focus of the study, not because of disinterest in the information that they would provide, but because of more complex taphonomic problems that affect *Sus* teeth compared to that of other mammals. There is some evidence to suggest that because *Sus* teeth do not mineralise to the same extent as the teeth of other mammals that are used for strontium isotope analysis they are more susceptible to diagenetic alteration than for instance cattle teeth are. Therefore, pig teeth are more likely than other mammal teeth to provide an $^{87}Sr/^{86}Sr$ signature that includes strontium of diagenetic origin.
Tooth enamel and bone from the associated mandible was processed from two pig teeth (both from context 593). A single sample of enamel was included from each tooth as pig teeth do not develop incrementally and therefore there is no potential to investigate change through time by sampling along the tooth crown (as was the case for cattle).

The results of pig tooth analysis are shown in Table 6.5 and Figure 6.5. For tooth DW14 the enamel and bone gave similar $^{87}Sr/^{86}Sr$ values, and are both consistent with the chalkland signature. It is impossible to know whether these values are the result of diagenetic alteration of both the tooth enamel and the mandibular bone, or whether the $^{87}Sr/^{86}Sr$ of the enamel sample reflects the strontium consumed during tooth development of the pig. Tooth DW13 in contrast provided a substantially different enamel value compared to the mandibular bone, a value that is much higher than would be expected for a site located on chalkland such as Durrington Walls. If the results are accurate this pig could have been raised in the area immediately adjacent to the chalkland (in the domain that includes values of 0.709-0.711 Evans *et al.* 2010) or further afield in Cornwall and the southern coastal area of Devon (domain values from 0.711- 0.712).

6.6 Discussion
The primary aim of the strontium isotope analysis of cattle teeth from Durrington Walls was to investigate patterns of mobility of the animals that were eventually disposed of on site. Such investigation was undertaken with the awareness that the evidence so gathered could provide insights not only on the nature of the use of the site but also of strategies of cattle husbandry in Neolithic Britain. Of the 12 teeth that were analysed, only two had Sr signatures that were consistent with a period spent in an area of underlying chalk geology. One of the teeth (DW5) was un-erupted at the time of the animal's death, and it can be concluded that the entire period of third molar development (10 months-2 years) was spent in chalkland areas. DW9 also provided chalkland values, but the situation is complicated by fairly heavy wear on the tooth (Table 6.1). The top part of the tooth – which is the first to develop - is therefore missing due to wear. This may have provided a different signature, therefore the only safe interpretation is that the animal spent at least a substantial part of its second year in chalk conditions,

but there is the possibility that movement occurred. The most likely scenario is that both of these animals were grazed in chalk areas.

These animals might have grazed in the vicinity of Durrington Walls. However, the domain that would provide chalk signatures has a broad geographical distribution, and, given the significant distances over which other cattle at the site moved, it would be flawed to assume that no movement of chalk signature animals had taken place. The domain that Durrington is located in stretches from North Yorkshire, almost continually along the east coast of Britain to the north of Norfolk. Chalkland is also prevalent in many counties of the midlands and the southwest including Cambridgeshire, Oxfordshire, Greater London, as far east as Kent and as far south as Sussex. The Isle of Wight would produce similar signatures in cattle teeth as those seen in the chalkland teeth, as would large areas of northern France. So although the animals provide a signature that could place them within the vicinity of Durrington Walls during tooth development, they could equally have travelled over substantial distances to reach the site.

Most of the cattle from the study originated in a geological domain that is adjacent to the chalkland, and there are areas with appropriate geology relatively close to Durrington Walls. Despite this proximity, the transport of cattle over even these relatively short distances would have required substantial effort and organisation, even if they were moved as live animals, as is likely. A smaller number of individuals probably came from further away (parts of Wales, Gloucestershire or Devon and Cornwall). Two of the teeth (DW4 and DW12) suggest the distances moved could have been of a much greater scale. The cattle that have moved further represent a greater investment in time and energy than those that grazed in the vicinity of the site. Movement of cattle in the Late Neolithic may have been more difficult, due to the higher forest coverage, and certainly time consuming. In some cases the minimum distances that must have been traversed by cattle is almost 100km in a straight line (and in fact much further if the results from DW4 and DW12 are accepted). If natural features and barriers are also considered (*e.g.* hills, valleys, stretches of water *etc.*) the actual distance travelled would have been much greater. The movement of cattle would have been an arduous task and it must have occurred because there were very good reasons to cover such distances. As mentioned before the animals were likely to have been driven, as the transportation of whole carcasses would have represented a difficult and unnecessary effort. The possibility that only parts of the animals were introduced to the site is also unlikely because most parts of the cattle skeleton are represented in the zooarchaeological sample, including teeth and cranial elements. This is significant as the heads are very heavy and not as productive as other parts of the body as meat bearers. It seems thus most likely that the animals were brought to the site alive, and were killed and butchered in the vicinity. In the absence of methods to preserve meat, this system would also make most sense as the meat from animals killed before a 100km journey would not be fresh by its arrival in Wiltshire.

Table 6.4 Durrington Walls cattle teeth with strontium isotope concentration for each sample.

Tooth number	Subsample	Sr concentration (ppm)
DW1	1 2 3	110 136 146
DW2	1 2 3	305 254 211
DW3	1 2 3	98 109 69
DW4	1 2 3	93 94 90
DW5	2 3	217 226
DW6	1 2	140 142
DW7	1 2 3	185 179 161
DW8	2 3 4	166 153 151
DW9	1 2 3	108 121 126
DW10	1 2 3	183 126 96
DW11	1 2 3	257 219 210
DW12	1 2 3	218 179 197

Table 6.5 Pig teeth from Durrington Walls with concentration and $^{87}Sr/^{86}Sr$ values for both tooth enamel and associated mandibular bone.

Tooth number	Subsample	Concentration	$^{87}Sr/^{86}Sr$
DW13	enamel bone	164 358	0.710984 0.708406
DW14	enamel bone	117 288	0.708908 0.707995

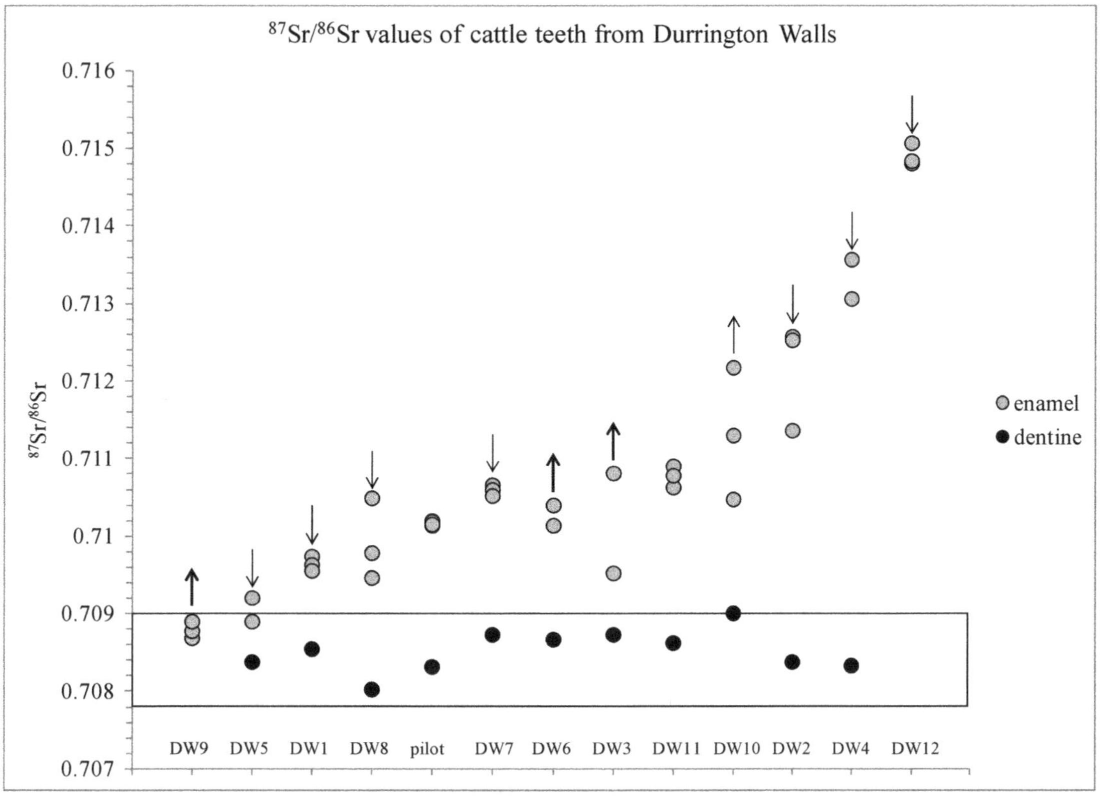

Figure 6.3 $^{87}Sr/^{86}Sr$ ratios of cattle teeth from Durrington Walls. DW4, DW12 and the pilot specimen have points that are closely spaced and obscure one another. Grey points are enamel samples and black dots the associated dentine. Arrows indicate the direction from tip to base of the tooth. No dentine sample was processed for DW9 or DW14.

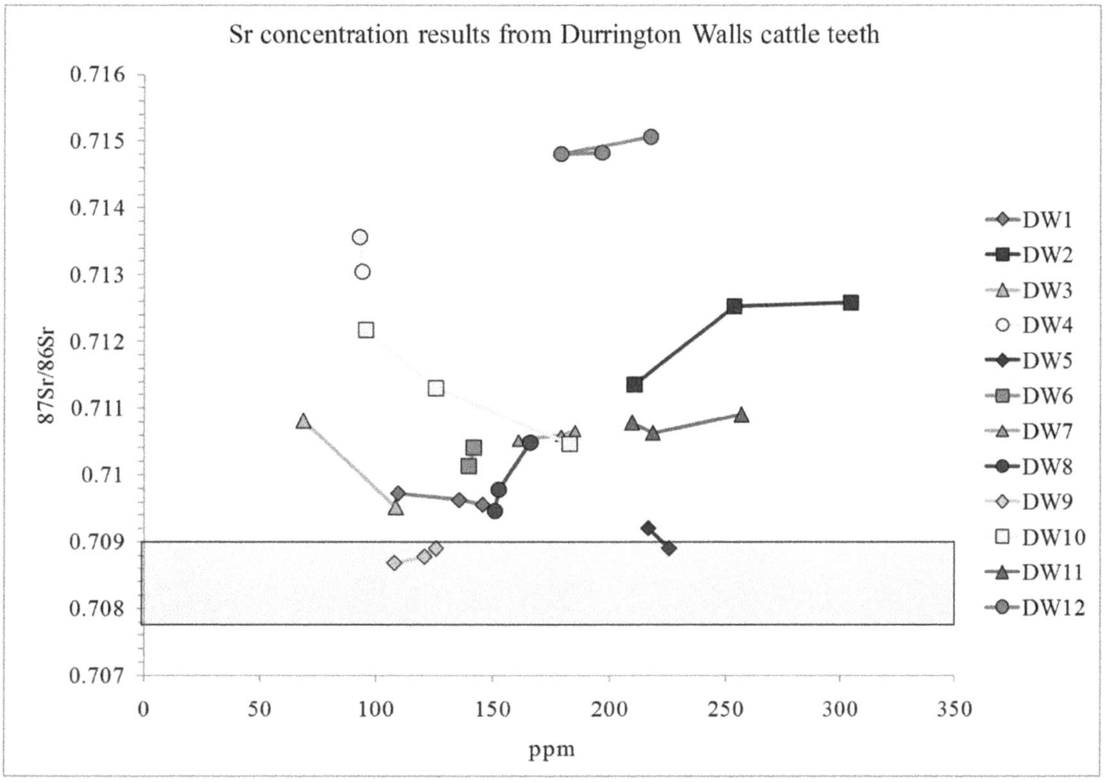

Figure 6.4 Strontium concentration plotted against $^{87}Sr/^{86}Sr$ results from Durrington Walls cattle teeth.

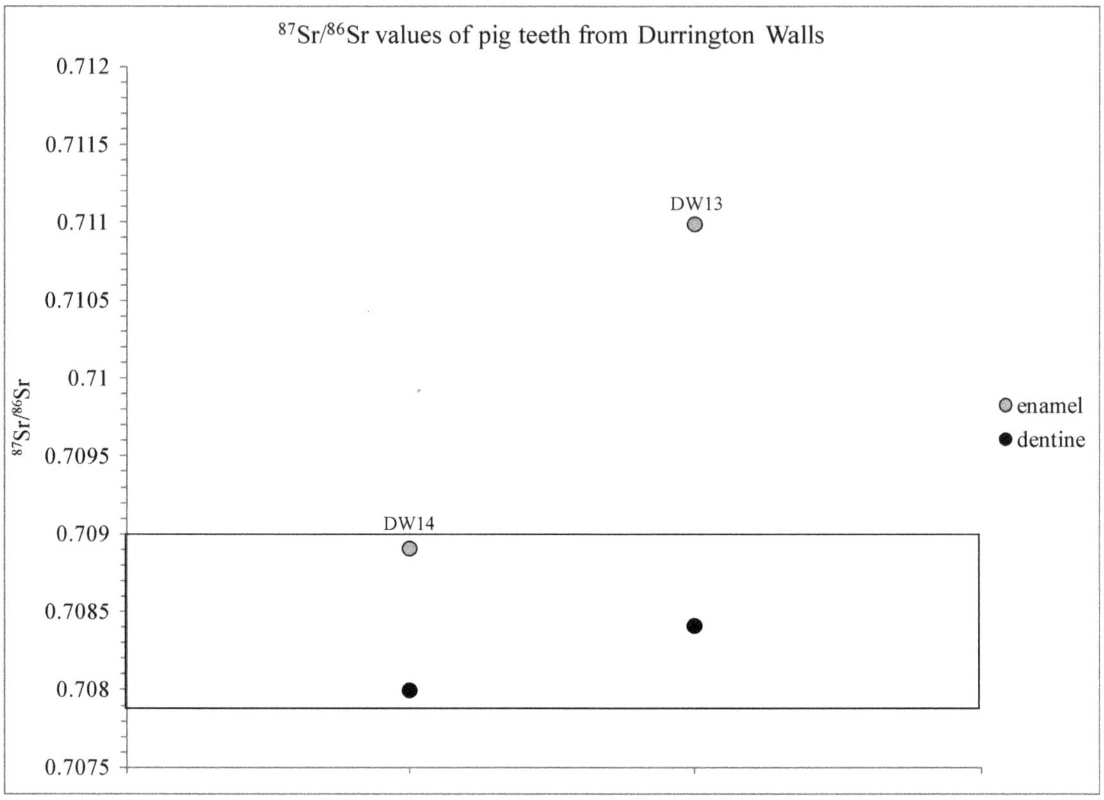

Figure 6.5 $^{87}Sr/^{86}Sr$ values of pig tooth enamel and dentine from Durrington Walls.

In the Late Neolithic, Durrington Walls was probably the site of large scale feasting activities. Zooarchaeological analysis has identified large numbers of pig bones, many of which had been butchered and bore signs of roasting (Albarella and Serjeantson 2002). Fewer cattle were recovered, but even these smaller quantities would have provided extremely large amounts of meat. As Serjeantson has argued, the slaughter of a single head of cattle could have provided feast like quantities of meat (Serjeantson 2006). Given the other evidence for feasting at the site it seems likely that the cattle used in the Sr study were deposited as a result of the same process. The evidence here provides another facet to the story of feasting at Durrington Walls in the Late Neolithic. Not only were huge numbers of animals slaughtered for the feasts, but the origin of the animals may have been important too. Cattle, and therefore probably people, were converging on Durrington Walls. Contributions were made from numerous geographical regions, beginning with the area immediately surrounding the site, and from a number of more distant parts of Britain. No evidence is available yet specifically about the Late Neolithic movement of humans in the area that this study is concerned with, but it seems likely that if cattle were being driven, some humans at least would have had to drive them. In general the movement of humans and cattle is similar, some were found to be local, while others came from some distance away.

Other motivations for the movement of cattle in the Late Neolithic need to be considered. One possible scenario is that rather than being moved for the explicit purpose of feasting at the site, cattle were moved through different geological areas in search of pastureland. Examples in the ethnographic and historical record show that movement between summer and winter pasture is a feature common to many stock herding societies. But in the case of Durrington Walls the specific areas from which cattle were moved make it unlikely that pasture was the primary motivation for mobility. Ample suitable pasture areas would have been available in the original areas. In some cases there was no gradual movement towards less radiogenic (more chalk like) values. The implication of this result for cattle husbandry in the Late Neolithic is that there was generally no gradual movement towards Durrington Walls as teeth developed. For at least half of the animals movement towards Durrington Walls began at some point after the third molar was completely developed.

The results from pig tooth analysis were surprising. The single tooth that provided a non-local signature is potentially very important, and provides one of very few examples of non-local pig teeth in the strontium isotope analysis literature. The single sample could mean that the result is anomalous, but the result provides evidence that a detailed study of the susceptibility of pig tooth enamel to diagenetic alteration is urgently needed.

6.7 Conclusion

Strontium isotope analysis has provided evidence that cattle were being moved over significant distances during the Late Neolithic period in Britain. No single origin can be identified for the cattle deposited at Durrington Walls. Some provided evidence for grazing in conditions very similar to those found in the local area of the site, while others must have spent the time between 10 months and 2 years much further away, in some cases at least 100km from the site. These results emphasise the importance of cattle during the period and suggest the presence of a compelling motivational factor for cattle herders. The animals are likely to have been moved to the site while still alive, and contributed to the feasting activity that has been identified through zooarchaeological studies. The variation identified within tooth enamel suggests that different animals were exposed to different patterns of movement during the period of tooth development. In some cases the animals clearly began a general movement towards more chalk like areas while the tooth was in growth, but for others tooth development was complete before movement in the direction of Durrington Walls began.

Chapter 7
Discussion and conclusions – *Sus* and *Bos* exploitation in Britain from the Early Mesolithic to the Late Neolithic.

7.1 Introduction

Cattle and pig husbandry in Neolithic Britain has been revealed as a highly complex phenomenon that varied in its nature between time periods, sites and species. This work began by outlining the reasons for its undertaking, and the aims of the research as a whole. This final chapter will bring together the various strands of evidence presented for the beginnings of husbandry practices in Britain, and discuss the implications of the research for our understanding of animal/human relationships in the prehistoric period. Beginning with the evidence for pre-Neolithic exploitation of the two ancestral wild species (wild boar and aurochs), it will go on to consider the transition from the Mesolithic to the Neolithic period, including the evidence collected in opposition to the idea of a local domestication event, discuss the evolution of animal husbandry in the Late Neolithic and finally move on to a detailed discussion of cattle secondary product use and mobility. As part of the discussion the limitations of the research will be identified, and areas of future work that are essential to expanding the understanding of changing animal exploitation during the prehistoric period will be highlighted.

The separation of pigs and cattle for most of this inquiry was necessary in part due to the methodology that was developed to analyse faunal assemblages, and also for ease of presentation in Chapter 3 and Chapter 4. Differences between cattle and pig husbandry, and their possible causes will be explored in this chapter. The relationship between humans and these two animals was different throughout the period investigated. Cattle produce meat and can be a source of milk, traction and manure. Pigs, providing no dairy products, are used mainly for meat, but like cattle might also be used exploited for blood, manure and perhaps labour for clearing land. But it is in their dietary needs and habitat preferences that the greatest distinction between the two species can be found. Omnivorous pigs are extremely versatile in their adaptation to different food sources (Schley and Roper 2003, and see Chapter 1), while cattle are far more specialised and tend to require greater human care. Their varying behaviours and social structures also impact on husbandry strategies (Noddle 1989, 185). From ethnographic analogy it is apparent that pigs can live close to or within human habitation areas (Rappaport 1968) or can be left to find their own food further from human control, and can easily be trained to return to human settlements. Pig herding is not unknown, but in general the species is not easily moved in large groups or over long distances (Malcolmson and Mastoris 2001). In contrast, cattle are often kept in large groups, and they do not usually graze within human settlements. Their more specialised diet makes cattle more susceptible to periods of resource stress than pigs, and they may have required additional fodder especially during winter. Evidence for additional foddering has been found in the European Neolithic (Akeret *et al.* 1999; Akeret and Rentzel 2001) and movement of cattle through the landscape in search of sufficient grazing could have been a necessary practice. To an extent then, differences in management between the two species are to be expected.

7.2 Wild animals in Britain – constancy and change

Introduction

Two main aims led to the investigation of Mesolithic aurochs and wild boar exploitation in Britain: the first aim was to provide a secure baseline in terms of biometrical information and population structure with which to compare the Neolithic zooarchaeological record; and the second was to identify the place of British aurochs and wild boar within the known record of Europe.

Size change from the Pleistocene to the Holocene

There is ample evidence from both zooarchaeological and palaeontological sources that European aurochs diminished in size from the Pleistocene to the Holocene (*e.g.* Grigson 1989, 78; van Vuure 2005). British aurochsen were not immune to the environmentally induced size reduction that affected their cousins from the rest of Europe, and they diminished in size from the Pleistocene to the Mesolithic (Chapter 4). The observed size reduction was not uniform, and the bones of wild cattle tended to be more severely affected than their teeth. There is consensus in the zooarchaeological literature that the observed size reduction in aurochs from the Pleistocene to the Holocene was a result of climatic amelioration, with increasing temperature across northern Europe at the onset of the Holocene the most likely cause of diminishing body size in the wild cattle and other mammals (Kýsely 2008; Uerpmann 1978).

In contrast, there is no evidence of a reduction in size of wild British cattle during the Holocene itself. Aurochs remains from post-Mesolithic sites are comparatively rare, but those that are available, specifically the Uskmouth specimen (Alice) and the few available Neolithic archaeological specimens, are not demonstrably different from Mesolithic aurochs in body size.

Size change in wild boar is more difficult to assess because of the lack of pre-Mesolithic osteological material. No Pleistocene specimens were included in this study and it is therefore impossible to directly assess long-term size change of this species in Britain. It is clear though that a reduction in size did affect wild boar in other parts of Europe (*cf.* Albarella *et al.* 2009; Davis 1981; Ducos and Kolska Horwitz 1998) at the beginning of the Holocene. Post-Mesolithic wild boar measurements are rarely available for comparison, a problem that is compounded by the similarity between domesticated types and wild boar identified in the course of this research. An assessment of wild boar size variation during the Holocene in Britain is difficult to make, but again, the variability of wild boar in other parts of Europe can be used to infer change in British animals too. Variation in the size of wild boar during the Holocene has been observed for instance in Italy (Albarella *et al.* 2006a, 213) and Portugal (Albarella *et al.* 2006b).

The place of British animals in the European record

Chapter 5 (sections 5.2 and 5.6) compared British Mesolithic animals to the aurochsen and wild boar from other parts of the European continent. Biometrical analysis of wild boar remains indicated that British specimens tended to be intermediate between the small animals from Iberia and the large animals from Scandinavia, overlapping with the latter to a lesser extent. The few Dutch specimens were towards the lower end of the Scandinavian and the upper end of the British distribution – but the addition of more specimens might alter the situation considerably. Differences in the extent of variation of teeth and bones were also identified.

Aurochs were found, as expected, to be smaller than the animals from Denmark. The small size of British aurochsen compared to Danish ones was known prior to this investigation (*e.g.* Grigson 1989), and the small size of British wild boar compared to other European animals was also suspected (Albarella *et al.* 2009). The data presented here serve to emphasise and clarify the specific ways in which animals from Europe differed to those from Britain. The implication of these findings is that to understand British prehistoric husbandry it is important to use local wild boar as a comparison with later animals, especially if domestication is inferred from reduced body size, as there is considerable variation of body size over geographical areas.

Geographical variation of wild boar and aurochs *within* Britain must be considered, but has been difficult to assess from the available faunal material. More is known about the variation of wild boar in Holocene Europe than aurochs, as various strands of research have led to a detailed understanding of wild boar history in some areas. Denmark is one such area and it has been shown that environmental changes and insularism during the Mesolithic (Ertebølle) period affected the wild boar over relatively small areas in different ways (Rowley-Conwy and Dobney 2005). Variation over the distances between the important sites discussed in this research should be expected. Temperature (and rainfall) is extremely variable in Britain, with current mean winter temperatures of 7°C in the west and 4°C in the east of the country, and mean summer temperatures of 16-17°C in the south and 13-15°C in the north (Barker 1985, 193). In particular, then, there could be a significant difference between animals in southern Britain (*i.e.* the Mesolithic sites in the Kennet Valley), southern Wales (*i.e.* Goldcliff) and northern England (*i.e.* the Vale of Pickering sites). The problem in terms of understanding wild mammal variation, however, is that most of the zooarchaeological aurochs remains come from Yorkshire, in the north east of England, while most of the pig remains come from a restricted area of the Kennet Valley, in Berkshire. With such a biased record there is no way at present to identify whether significant variability existed between the different areas of Britain in the past. This should be an area of future research.

Conclusion

- Size change from the Pleistocene to the Holocene seen in aurochs in Europe also occurred in the British record, and is most likely linked to changing temperature.
- No size reduction between the Early Mesolithic and later periods was found in aurochs, but more biometrical information about later animals is needed to adequately define their size through time.
- Comparisons between British wild boar and aurochs and their counterparts from mainland Europe are very useful in order to clarify the characteristics of these animals and the eventual transition to husbandry. It is, however, also true that regional variation should make us wary of using European animals as a proxy for British ones.
- Finally, a great deal of further work is required to understand the potential differences between animals in various parts of Britain itself. Homogeneity between animals from different parts of the country cannot be assumed.

7.3 The nature of the transition to animal husbandry

Introduction

The transition to animal husbandry is one of the conundrums of European prehistory. The introduction of animals from an external geographic source has traditionally been seen as the most likely explanation, and, as will be discussed below, this is probably the case for Britain. In other areas of Europe it is, however, likely that local wild animals also contributed to domestic pig, and possibly even cattle, stock during the Neolithic. The domestication of European wild animals without any contribution from introduced types (for instance as suggested in the Carpathian basin by Kyselý 2008) is a theoretical possibility, as is the local domestication of some and the introduction of others. Introgression is not precluded even if the majority of animals were introduced in domesticated form, and small scale mixing of wild and domestic animals is likely to have taken place. That Neolithic animal keepers would have welcomed or found it necessary to improve their domestic stock with the addition of wild animals is widely acknowledged in the case of cattle (*e.g.* Glass 1991). This section will discuss changes in practice through time, and consider the likelihood of local domestication, introduction of animals from continental Europe and the contribution from wild animals to domestic stock in Britain.

The Mesolithic exploitation pattern

In order to argue that in the Early Neolithic animals began to be husbanded it is necessary to determine what were the prevalent practices before the start of the period. Most evidence from the Mesolithic in Britain suggests that hunting was the principal – if not the only - system of mammal exploitation. Claims of incipient husbandry or herding of red deer before the Neolithic (*e.g.* Jarman 1976, 87) based on the exploitation of specific age groups have been argued against (Simmons *et al.* 1981, 123), and there is no unambiguous evidence that dispels the traditional view of wild exploitation by Mesolithic people in Britain. Zvelebil's argument (1995) for the development of pig taming in pre-Neolithic Scandinavia is based on intensification of pig exploitation, morphological changes in pig skeletal elements (the lower third molar) prior to the beginning of the Neolithic and on the presence of large numbers of juvenile animals within assemblages. However, changes in the skeleton of wild boar could have a number of causes, one of which is the influence of human control, but more detailed biometrical analysis is necessary before husbandry can be conclusively determined (especially given the tendency of the third molar to be influenced by environmental change). Intensification of pig exploitation, and the increase in juvenile animals at archaeological sites can both be explained as part of a hunting system, and there is no need to postulate taming in order to explain these changes (*cf.* Rowley-Conwy *et al.* 2010). In Britain it can be assumed that hunting of wild pigs was the norm. Within the assemblages studied wild boar were of mixed age at death, and archaeological material indicates that males were exploited preferentially, a pattern that is not inconsistent with hunting practice. Aurochs exploitation also took the form of hunting. The inclusion of projectile points in Upper Palaeolithic and Mesolithic lithic assemblages (Friis-Hansen 1990), the occurrence of animal remains in Northern Europe with evidence of having been killed with arrows (Noe-Nygaard 1974), and numerous depictions of humans killing large mammals with arrows (*e.g.* Zaliznyak 1998) are among a number of different lines of evidence that indicate that hunting was the dominant system prior to the Neolithic.

Evidence of local domestication versus outside introduction

Evidence from different parts of Europe suggests that the contribution made by local animals to domesticated populations varied greatly between different areas. A consensus view of the local contribution of aurochs and wild boar is not possible because of the great variability that existed. Therefore, we can make analogies between the British situation and that in other parts of the continent, but ultimately each area must be viewed as a unique case.

This research aimed to investigate the initial stages of animal husbandry in Britain, and the evidence gathered supports, to an extent, the traditional views of the appearance of animal husbandry as a largely exogenous phenomena. Nevertheless the findings also demand greater acknowledgment of complexity than has been previously envisaged. The presence of sheep and goats in Neolithic Britain is incontrovertible evidence that at least one element of Neolithic animal husbandry was introduced to Britain from outside. As concerns domestic cattle and pigs, their entrance was also apparently abrupt, as the animals from the Mesolithic and Early Neolithic are significantly different from one another in statistical terms. No evidence of intermediate forms of either pigs or cattle occurs in the British zooarchaeological record, and cattle in particular exhibited clearly differentiated groups in the Mesolithic and Early Neolithic. Changes in both postcranial bones and teeth identified through biometrical analysis are consistent with an exogenous origin for domestic stock. In both species bones from Neolithic sites were smaller than those from the Mesolithic and in pigs the reduction in tooth size is even more substantial. . This is in contrast with what would be expected from a domestication-induced gradual size reduction from local wild boar stock, which would be likely to have acted more intensively on the more environmentally sensitive postcranial bones. A more rapid decrease in the size of the more conservative teeth is therefore likely to be the result of the appearance of introduced stock. This trend also contrasts with the one identified in other European areas where Neolithic size reduction in bones is greater than in teeth and as a consequence it has been interpreted as evidence of a local contribution to the domestication process (*e.g.* Hadjikoumis 2010).

A difficulty with all interpretations is, however, caused by the significant gap in our zooarchaeological knowledge between the latest known Mesolithic and earliest known Neolithic. This gap could mask a gradual change, which remains undetected. The local domestication hypothesis remains, however, less likely, especially considering the substantial amount of time that, in other parts of the world, seems to have lapsed before the domestication process produced detectable morphological changes, particularly in the more conservative teeth (Zeder 2006, 172).

Early Neolithic *Sus* remains from archaeological sites in Britain can less incontrovertibly be attributed to the domestic form than their counterparts of other species (*i.e.* sheep, goat and cattle). However, Early Neolithic and Late Neolithic pigs were different to their predecessors and it has become apparent that the Mesolithic system was replaced in the Early Neolithic with a system that included a greater use of young animals. This pattern could fit with Hesse's idea (1982) about the shift to a greater control of the animals as a consequence of husbandry practices. It is, however, the case that, in wild boar, variation in hunting practice can also lead to an increase in the number of juvenile animals exploited (*e.g.*, Rowley-Conwy *et al.* 2010). This can potentially be misinterpreted as the result of pig husbandry. Although the wild or domestic status of pigs cannot easily be determined an element of change in the Early Neolithic is apparent in terms both of kill-off patterns and size reduction, in particular the teeth.

Extensive use of wild boar continued into the Neolithic at some European sites (for instance Rocadour, France and Cova Fosca, Spain (Hadjikoumis 2010). At these sites there is no evidence that domestic pigs – and in fact any other domesticates - were in use. Biometrical information is consistent with the view that these were wild animals. Conversely, initial suggestions that there were no domesticated pigs during the Early and Middle Neolithic at Arene Candide, Italy (Rowley-Conwy 2000) have now been reconsidered. There was a general pattern of wild boar size increase after the Mesolithic period in Italy (Albarella *et al.* 2006, 217), which indicates that middle Neolithic pigs from Arene Candide are too small to belong entirely to the wild form, though they may include, and probably do, some wild boar. The absence of domestic pigs and the presence of wild boar at Neolithic sites would seem to be confined to those at which hunting continued as the major method of meat procurement. It remains essential, however, to determine whether the pigs from British Neolithic sites were domesticated as it cannot be assumed that pig husbandry was ubiquitous in Early Neolithic Europe. Chapter 3 established that, despite the similarity of Mesolithic and Neolithic pigs, there is sufficient evidence to conclude that animals in the Neolithic period were in fact mainly domestic.

Pigs overlapped to a much greater extent with the wild animals of previous periods than did cattle, a fact that makes assertion of exogenous origins more difficult to uphold. In addition, variation between different Early Neolithic sites suggests that the situation was not uniform, and perhaps the trajectory of animal husbandry was different in different parts of Britain. In particular, the differences at Early Neolithic sites between the teeth and bones (animals with relatively smaller teeth at Runnymede Bridge than Hambledon Hill) indicates that pig husbandry might have varied substantially on the site level. As has been argued previously, the pattern could suggest that mixing of domestic pigs and wild boar occurred at a different rate at these two sites, presumably with more inclusions from wild boar at Hambledon Hill where tooth size is greater. A possible scenario sees both sites supplied with domestic pigs of allochthonous origins, but with the Hambledon Hill pigs being less tightly controlled by humans, and therefore more prone to mixing with local wild boar. Mixing of wild boar and domestic pigs would have provided some benefits and could have been a deliberate or incidental. The classic ethnographic study with which to draw a comparison is New Guinea. In New Guinea, pig exploitation was practiced in many forms with varying levels of control over pig populations. Hunting of feral pigs (wild boar were not indigenous to New Guinea, and populations resulted from released domesticated animals) over which no control was exercised formed the mainstay of some human groups. In some cases all or some domestic males were prevented from reproducing through castration, which led to the frequent mixing of domestic females with feral males (Brown 1978; Rappaport 1968, 70-1; Rosman and Rubel 1989, 25-6). A further system saw the use of only domestic animals with no contribution from feral individuals.

A substantial contribution from wild boar is supported by the evidence from ancient DNA analysis. Work on modern *Sus* DNA by Larson *et al.* (2005) found evidence for the European domestication of pigs, but a later study, also based on ancient DNA, found that both local and introduced domestication had contributed to the early husbandry of pigs in Europe. British pig bones from Durrington Walls that were tested as part of the same project had European rather than Near Eastern haplotypes, suggesting domestication within the continent. According to the results of analysis presented in this volume, it seems unlikely that the domestication of pigs took place in Britain, therefore pigs must have been domesticated in continental Europe and then introduced into the study area. In the absence of a larger sample of analysed pig specimens we cannot, however, rule out that some pigs of Near Eastern origins may have made their way into Britain. It is also worth pointing out that the above mentioned study also indicated that, from the Late Neolithic onwards (therefore including the time when Durrington Walls was settled), pigs with a Near Eastern ancestry had become very rare in Europe (among hundreds of analysed samples only a medieval and a modern specimen both from Corsica were found).

In contrast, there seems to be little reason to assume significant mixing between wild aurochs and domestic cattle in the Early Neolithic of Britain. The separation of cattle and aurochs is clear, and the very different age and sex profiles indicate differing exploitation practices. Several aurochsen have been found within Early and Late Neolithic contexts, and it is clear from this research and the work of others that it continued to be exploited to a limited extent until its extinction. DNA analysis focused on the Y chromosome (*i.e.* paternal inheritance) indicates that mixing between domestic cattle and local aurochs occurred in northern Europe (Götherström *et al.* 2005). Investigations of mtDNA suggested a uniquely Near Eastern origin for the domestic cattle of Europe (Bollongino and Burger 2007; Bruford *et al.* 2003; Edwards 2007; Troy *et al.* 2001), but work specifically on British animals has identified a wild contribution in early domestic cattle (Bailey *et al.* 1996). In addition, animals from northern Europe and those from the Mediterranean exhibit different microsatellite markers supporting the proposition that there were two directions of spread into Europe, a northern Danubian expansion and a southern Mediterranean route (Cymbron *et al.* 2005; Götherström 2005; Tresset and Vigne 2007). However, evidence for mixing has been identified from central Europe (Glass 1991), and in historical records too.

Problems of dating the introduction of animal husbandry have meant that usually animal husbandry in Britain is presented as having occurred in a single event. It was, however, not necessarily spread in unison and in some areas the main domesticates may have been first exploited at different times. Although plagued by stratigraphic problems (Rowley-Conwy 1995), Geddes argument that transhumant sheep/goat use in the Pyrenees predated the appearance of cattle and eventually pigs (Geddes 1983; 1985) has yet to be conclusively disproven. In the west Mediterranean there is evidence that goats and emmer wheat formed the first part of Neolithic exploitation (Rowley-Conwy 2000), while further to the west in Ireland, the remains of domesticated cattle and sheep/goats have been found on coastal Mesolithic sites. In the earliest contexts the main domesticates do not always occur together and it is possible that they were introduced under different systems of exploitation. All of the assemblages from the Early Neolithic included in this study contained all of the main food animals (cattle, sheep and pigs), but they were not from the earliest part of the Neolithic. If the herding of cattle and keeping of pigs involved separate systems of exploitation, as has been suggested here, there is little reason to assume that the timing of their introduction was identical. Assemblages from the very beginning of the period – currently missing in Britain - are necessary to investigate the synchronicity of husbandry introduction.

If cattle and aurochs seem to be quite clearly morphologically distinct, this is not the case for wild boar and domestic pig, and it is therefore likely that in this latter species more mixing occurred. Although, for both species, a largely introduced domestication has been suggested for Britain, it is not impossible that in pigs an element of local domestication also occurred, thus explaining the similarity with wild boar morphology. This could be in line with the genetic evidence available for Europe that sees a scenario in which the domestication of pigs is represented by a combination of autochthonous and allochthonous elements while in cattle the latter seems to be clearly predominant.

The transition to husbandry in other parts of Europe

In continental Europe variable trajectories of Neolithisation are apparent in the archaeological record (Zvelebil 1996). These differing trajectories can be seen in particular in western Europe where it is expected that a significant contribution to Neolithisation was made by local Mesolithic groups (Tresset and Vigne 2007). Western Europe was also at the confluence of the Danubian (northern) and Mediterranean (southern) Neolithisation routes. Animal husbandry could also follow various trajectories and the transition does not appear to have been uniform in different parts of the continent. The evidence presented in this volume indicates that in Britain the process of Neolithisation differed from that of some of its western European neighbours. Chapter 5 highlighted the differences between British and Iberian pigs. In Britain, there is evidence of a mainly exogenous origin for domesticates, while in contrast, the work of Hadjikoumis (2010) in Iberia has led to the identification of a substantial element of local as well as exogenous domestication events. Ancient DNA analysis found Near Eastern pigs in southeastern parts of Europe (*e.g.* Romania and Croatia) as might be expected, but interestingly also in early Neolithic France and Germany (Larson *et al.* 2007). Consequently it does not seem that the same mechanism can explain the origins of husbandry in all European areas, at least as far as pigs are concerned. Genetic, biometric and ageing evidence point out that in pigs both local and introduced factors contributed to domestication, but the nature and extent of this contribution varies according to area and site. In cattle, despite the high likelihood of aurochs introgression, the introduced element is clearly predominant and clearly homogenous across regions.

Tresset (2003) has argued that, on the basis of biometrical similarity between Neolithic animals, the origin of British livestock can be traced to the Paris Basin. Chapter 5 compared British cattle and pigs to animals from other parts of the continent, specifically its closest neighbours (Iberia, France and the Netherlands) and more distant areas such as Scandinavia. As observed by Tresset, there was considerable similarity between the cattle of the Paris Basin and those from southern British Neolithic sites, and the distribution of the two overlapped. However, the Paris basin specimens tended towards the larger extreme of the British distribution, and in fact British specimens were more similar, in terms of their size, to animals from Holland. It must therefore be acknowledged that there is more than one possible origin for British domestic animals. Wider morphometric comparison between British and European animals will be necessary to answer the question of livestock origins.

Conclusion

- The main exploitation of both wild boar and aurochs in the Mesolithic was through hunting.
- Local domestication of either species in Britain is unlikely. The gap in archaeological evidence between the Late Mesolithic and the Early Neolithic encourages caution, but the balance of the evidence indicates that domestic cattle and pigs were mainly introduced into Britain.
- Pigs were certainly domesticated during the Early Neolithic as evidenced by both changes in their skeleton and variation in age through time.
- The complexity of the pig record in Britain can be explained through different contributions from wild boar at different archaeological sites, as well as diversified husbandry regimes.
- In conjunction with mtDNA evidence, the results of pig analysis demonstrate that the systems prevalent in the Neolithic of Britain are not directly comparable to ethnographic comparisons from Papua New Guinea.
- The genetic evidence from the Late Neolithic suggests that the domestic pigs introduced into Britain derive from animals that had originally been domesticated in mainland Europe; but only further genetic work on the Early Neolithic sites will clarify whether pigs of near Eastern ancestry were also introduced. In terms of the mechanisms of pig domestication in Britain this will make little difference as in both scenarios the animals were imported from the main European continent.
- The current evidence indicates that British animals are biometrically similar to animals from different parts of Europe. Wider morphometric comparison between Britain and the rest of Europe is necessary to provide solid foundations for this hypothesis.

7.4 Exploitation change from Early to Late Neolithic

Introduction

The Neolithic is often viewed from the perspective of its monumental architecture, and the move from the Early to the Late Neolithic is commonly seen in these terms. But the period also saw clear changes in other aspects of the archaeological record. Archaeobotanical and palynological evidence has led to speculation that plant cultivation became less important, and that a greater reliance on livestock was common in the Late Neolithic (Barker 1985, 204; Hey and Barclay 2007, 408), but this stance has been challenged by Bogaard and Jones 2007. In terms of animal husbandry, one of the drastic changes during the Late Neolithic is the emphasis placed on pigs at the expense of cattle (the reverse of that found in the preceding Early Neolithic) (Grigson 1982b). A similar change has been identified in the Late Neolithic of western continental Europe (Tresset and Vigne 2007). Many archaeological sites during the Late Neolithic are dominated by pig remains, but there are exceptions to the dominance of pigs during the period (Serjeantson 1996). The question to be addressed in the following section is how the changing emphasis of cattle and pig use corresponds to different husbandry practices in the Early and Late Neolithic periods.

Pigs

As mentioned, in a zooarchaeological sense one of the major changes from the Early to the Late part of the Neolithic is the shift in prominence from cattle to pigs. Grigson (1982b) compared the frequency with which the main domesticates occurred at Early Neolithic and Late Neolithic sites in southern England and Yorkshire, and found that pigs became more prevalent during the later part of the period. Durrington Walls was one of the sites used by Grigson to illustrate the change in emphasis through time, and further excavations after the publication of Grigson's work led to an even more pronounced dominance of pig bones at the site (Albarella and Serjeantson 2002). Work in progress on material from more recent excavations at Durrington Walls indicates that pig is more than ten times as abundant as cattle on the basis of the number of identified specimens (Albarella and Viner in prep.) The same pattern has been found in other Late Neolithic contexts. Mount Pleasant, Dorset (Harcourt 1979) was dominated by pig remains (17 pig and 8 cattle by MNI), and the West Kennet palisade enclosures contained the remains of many pigs (Edwards 1994; Edwards and Horne 1994; Horne 1994). Pig dominance is not ubiquitous in the Late Neolithic, and some sites that are roughly contemporary with Durrington Walls do not conform to the expected pattern (Grigson 1982b, 307-8; Serjeantson 1996). For instance the henge at Marden, Wiltshire had a similar number of pigs and cattle (in terms of MNI as the number of specimens of each species was not provided; Wainwright *et al.* 1971). Neolithic faunal assemblage composition is more complex than merely a change from cattle emphasis to pig emphasis. Figure XIII.1 illustrates the relative abundance of cattle, pigs and sheep at Late Neolithic sites in southern England (after Albarella, unpublished). At some sites cattle are clearly dominant, while at others pigs are predominant and at a number there is no tendency towards greater numbers of either species.

Some researchers (*e.g.* Parker Pearson 2003) have seen a difference between the deposition of cattle and pigs in the Late Neolithic, but the problem of trying to identify patterns of deposition is compounded by the difficulty of establishing criteria by which so-called 'domestic', 'funerary' and 'ceremonial' sites can be defined (*cf.* Bradley 2005). Parker Pearson (2003) argues for a link between cattle and funerary activity, and pigs with ceremonial activity. Unpublished work by Albarella (FigureXIII.2 andXIII.3) provided evidence of differential deposition of cattle and pigs at ceremonial/funerary and domestic sites in the Late Neolithic. The difference is rather slight (41% cattle, 53% pigs at domestic sites, 53% cattle, 40% pigs at funerary/ceremonial sites), but such a distinction would suggest the use of cattle more frequently in funerary/ceremonial situations, and the more frequent use of pigs in daily life. A dominance of pig remains at Grooved Ware sites has also been identified by some researchers. At the sites investigated by Grigson, pigs made up 60% of the faunal remains at Grooved Ware sites compared with only 33% at non-Grooved Ware sites. A more recent survey based on a larger number of sites confirmed this trend (Figure XIII.4 and XIII.5 from Albarella, unpublished). At Yarnton, Oxfordshire, pigs made up a significant proportion of the animal bone assemblage, and were noted as particularly prevalent in pits with Grooved ware (Hey *et al.* 2003, 84), suggesting a link between Grooved ware ceramics and pigs even within single sites. This association of pottery and pig remains is an important factor in considering the reason for an increase in pig remains at some Late Neolithic sites. Durrington Walls was the site of large scale feasting during the Late Neolithic (Albarella and Serjeantson 2002), and the large number of pig remains might reflect feasting events rather than a real increase in pigs during the Late Neolithic. The association with Grooved Ware has been seen as support for the idea that pigs were used in feasting events during the Late Neolithic.

Changing environmental conditions could have led to the increased reliance on pigs, along with wild plants and animals, during the Late Neolithic (Grigson 1982b, 308). An increased reliance on pigs could have been linked to their propensity to thrive on bracken, which is poisonous to other ungulates such as cattle and sheep, and for which there is evidence of an increase during the Late Neolithic (Grigson 1982b, 309). The fact that pigs could make use of bracken growing in regenerating woodland clearings at a time when wild woodland resources were of increasing importance to human populations might have encouraged their significance as a food species.
Increasing emphasis on pigs during the Late Neolithic corresponded to some, limited, changes in the husbandry practices observed at Durrington Walls compared to Early Neolithic sites. More young animals were found within the assemblage at Durrington Walls than at earlier sites. In conjunction with this tendency to use younger animals, Late Neolithic pig bones were generally smaller than Early Neolithic ones, but no significant difference could be found between the teeth of the two periods. This pattern might be explained by increased importance of meat (that would have led to the slaughter of more animals at younger ages) and feasting during the Late Neolithic. Tighter control of pigs during the Late Neolithic might also have caused a reduction in body size during the period. It is unlikely that the younger age of the animals at Durrington Walls could explain the smaller size of postcranial bones, with no associated reduction in tooth size, as it is also observed in postcranial measurements that are known to increase little after fusion.

At the time of this study there was some debate around whether pig tooth enamel was suitable for strontium isotope analysis due to its porosity. The issue has been investigated, and the reliability of porcine enamel for strontium analysis has been shown experimentally (Madgwick *et al.* 2012). The movement of pigs during the Neolithic is a possibility and historical sources, too, indicate that pig movement was possible, but the scale, in historic times at least, was not comparable to that of cattle movement. Two pig teeth from Durrington Walls were sampled for strontium isotope analysis, and, despite the small sample size, the results hint suggest that at least one animal was reared away from the chalklands that surround the site. Potential mobility of pigs in prehistoric Britain will require significant future research.

Cattle
Unlike pigs, the changes from the Early Neolithic to the Late Neolithic in Britain were not accompanied by physical changes in the skeleton of the cattle. Both teeth and postcranial bones remained similar throughout the Neolithic. Age and sex of cattle also remained constant from the earliest to the latest part of the Neolithic in Britain. The methods and also the desired output of cattle husbandry were the same in the two periods, and will be discussed in more detail in the following section.

Finally, cattle exploitation remained largely unchanged from its introduction in the Early Neolithic until the Late Neolithic, and was homogenous between most of the sites studied here. This is in contrast to pig husbandry which, as we have seen, underwent some variation in the Late Neolithic, reflecting the greater importance of the animal as a source of meat. It also varied significantly between the two main Early Neolithic sites studied as part of this research. Therefore, one basic system of cattle husbandry seem to have survived throughout Neolithic Britain, while a greater variety of pig husbandry strategies may have been employed.

Conclusion

- The increase in pig numbers at some Late Neolithic sites was accompanied by certain changes in husbandry practices, *i.e.* the greater use of young animals and generally smaller boned animals compared with the previous period.
- The frequency of pig remains increased in the Late Neolithic period, probably suggesting an enhanced importance of pig husbandry, explained by environmental, economic or social factors, or more likely by a combination of them.
- Cattle husbandry did not change through time indicating that the motivations and methods employed in Neolithic husbandry remained largely the same.

7.5 Milk, meat and mobility – aspects of cattle husbandry in the Neolithic

The purpose of cattle production

In addition to meat, cattle provide a series of secondary products (*sensu* Sherratt 1981; 1983) that were potentially important components of the Neolithic economy in Britain. In the Neolithic as a whole, pigs were generally killed at an earlier point in their lives than cattle, a pattern which reflects the different products that the two species can potentially provide. The significant difference in the representation of the two sexes seen in cattle, but not in pigs, is best explained by the potential difference in the use of these two animals and the related mechanisms of supply to human settlements. Despite the extensive debate surrounding the practicability of dairying in prehistory (for instance Amoroso and Jewell 1963; Clutton-Brock 1981; Legge 1981; 2005; McCormick 1992; Ryan 2005; Tani 2005), there is now clear evidence of dairying activity in the Neolithic of Europe (Craig *et al.* 2005; Evershed *et al.* 2008; Vigne and Helmer 2007) including Britain (Copley *et al.* 2003; 2005; Dudd *et al.* 1999; Mukherjee *et al.* 2005). The importance of milk products in the Neolithic is difficult to assess, as much of the evidence comes from 'ritual' sites such as Hambledon Hill and Windmill Hill in the Early Neolithic (Copley *et al.* 2003) and could represent a ritual rather than daily activity. An important exception is Eton Rowing Lake, an Early Neolithic site with evidence of domestic activity, at which a large proportion of the ceramic shards analysed contained residue from dairy products (Copley *et al.* 2003).

Although the evidence is not incontrovertible the work carried out as part of this research supports the view of previous authors regarding the suggested exploitation of cattle milk in Neolithic Britain. The death of significant numbers of animals around 2-3 years at all sites studied is an indication that meat was an important resource, as at this age animals are at their optimum meat weight (Legge 2005). The high number of female animals compared to males on the sites investigated during the study, and highlighted by other zooarchaeologists, is suggestive, however, of an interest in milk production too. It has been argued here that the cattle from Neolithic sites (apart from Runnymede Bridge) were brought to the sites as adults. Along similar lines Legge (1981) has argued that the cattle found at some Neolithic sites represent the surplus from a milk producing economy. An emphasis on meat such as that found at the sites studied for this research does not preclude the use of animals for other purposes. It seems likely that cattle were exploited as a source of more than one product. The majority of animals that were deposited on sites were adult when they died, and it must be considered that other secondary products were exploited to varying extent.

Cattle are an important source of labour in many societies (Harris 1989). Cattle labour is thought to have contributed significantly to some prehistoric economies (*e.g.* Isaakidou 2006), but evidence for the use of traction in Neolithic Britain is lacking. Castration, which is commonly employed in systems that use cattle for traction, has so far not been securely identified in the archaeological record of Neolithic Britain. The occurrence of castrated animals has been suggested for Late Neolithic Skara Brae, Scotland (Watson 1931 cited in Jewell 1963, 85), but without original measurements it is impossible to assess this claim. Castration of cattle at Durrington Walls (Harcourt 1971, 344) has also been suggested, but no such evidence was uncovered during this study. However, the difficulty of distinguishing between the bones of female and male cattle at Neolithic sites in southern Britain (Chapter 4) could have been increased by the presence of castrated animals.

Blood must also be considered as a potential food product taken from live animals. Blood drawing is impossible to identify zooarchaeologically but is still carried out by the Masai of East Africa (Zeuner 1963a, 51-2), and was reported as recently as the 1830's in Ireland (Lucas 1989, 200-22). Finally, manure is likely to have been important. The conditions prevalent in prehistoric Britain make it unlikely that manure as fuel source was important, as ample wood and tinder could have been collected in most parts of the country, and the often wet conditions might have made drying of manure for burning difficult. But it was probably used as fertiliser, as is evidenced from the Early Neolithic in Europe (Bakels 1997).

Mobility

Historically, mobility has been an essential part of cattle husbandry in Britain. Transhumance was common in the mediaeval period and still practiced until very recently in Britain and Ireland (Dyer 1996; Fox 1996; Herring 1996). Arguably the large scale movement of livestock seen in southern and Alpine Europe (in which some systems involved the movement of animals up to 300km; Walker 1983, 41) would not have been possible or necessary given the different climate and topography of Britain (Bradley 1978, Walker 1983). Historical records and archaeological information belie this assertion. Very large numbers of live cattle were driven from Wales to rich pastureland in the midlands and southern England (and almost as far as the east coast) by the 17th century, and historical sources hint that the practice was already established *ca.*926 AD (Colyer 1976). Another common route brought cattle from Scotland and northern England to the Midlands where cattle were fattened before slaughter (Thirsk 1985), and typically the highland zone has provided animals to the lowland zone for fattening (Barker 1985). Although in later times this long distance movement may have been driven by the needs of the market, good pastureland for fattening animals was also a driving force behind their long distance movement. Long distance movement of cattle is potentially much older than written sources suggest.

Both pastoralism (the raising of livestock on unimproved pasture land; Salzman 2004, 1) and transhumance (seasonal, repetitive movement that includes all or part of a human community and their herd animals) are difficult to identify archaeologically (Bradley 1972, Walker 1983). The nature of the archaeological record of Neolithic Britain (*i.e.* dominated by monuments rather than domestic activity) has led to a view of the Early Neolithic in which mobility was important (Bradley 1978; Pollard 1999; Thomas 1991; Whittle 1997). Transhumance has been suggested for parts of northern Britain in the Early Neolithic (Waddington 1996) and Neolithic Scotland (Barclay 2005) as well as, in general an upland/lowland exploitation of the landscape (Pollard 1999). In Late Neolithic Wales 'flimsy' structures at high altitudes have been interpreted as the seasonal shelters of people who spent other parts of the year in valley bottoms where more substantial earthworks and timber structures were located (Pollard 1999, 80). Whittle (1997, 21-2) offers a scale of possible scenarios, from 'circulating mobility' to 'embedded sedentism' any or all of which could have been practiced in Britain during the Neolithic. Direct evidence of transhumance (of the kind gathered from some European Neolithic sites, see below) has yet to be found from the British Neolithic. Most arguments for a mobile Neolithic rely heavily on the absence of substantial domestic structures (*e.g.* Thomas 1999), the continued use of wild resources and a purely symbolic role for domestic animals and plants in the Neolithic. Although not common, Early Neolithic domestic structures have, however, been excavated throughout Britain (Rowley-Conwy 2004). The perceived lack of cereal remains used to argue against the habitual use of cultivated plant food in Neolithic Britain was challenged by Bogaard and Jones (2007), who highlighted the taphonomic and recovery bias that influence archaeobotanical samples. Results from this study, and many others, indicate that some hunting did take place during the Neolithic period, but it tended to be very limited in nature and assemblages are usually dominated by domestic animals. No sites have yet been discovered that contain exclusively wild animals akin to Cova Fosca in Spain and Rocadour in France, where hunting was clearly the predominant (if not exclusive) means of acquiring meat. Coneybury Anomaly is perhaps the only Early Neolithic site with a large faunal assemblage that is dominated by wild animals (roe deer in particular) (Maltby 1990), but these are mixed with the remains of many domestic animals. There is no evidence to suggest that the exploitation of wild resources was the most important subsistence practice.

More direct evidence of transhumance has been gathered from Europe. Geddes' argument that a transhumant system was established in the Mesolithic of the Aude Valley, that included movement of animals between lowland and upland pastures (Geddes 1983; 1985), is problematic due to the insecure context of sheep bones at the sites he used. More concrete evidence comes from the Black Forest region of Germany, where there was evidence for seasonal use of upland and lowland areas in the Late Neolithic (Kienlin and Valde-Nowak 2002). Strontium isotope analysis of human remains in Germany also identified transhumance (Bentley and Knipper 2005). In addition Akeret and Jacomet (1997) investigated macrofossils in sheep/goat faeces at the Swiss site of Horgen Scheller. The presence of macrofossils from plants that would have been consumed in the winter and spring led to the conclusion that the animals were kept away from the settlement during the summer and autumn. Perhaps the most interesting suggestion has come from Bentley (2007) who linked transhumance to different family groups in the LBK and suggested that there might be economic specialisation along family lines (Bentley 2007).

Cattle were moved over long and short distances in the Neolithic of Britain. In Chapter 4 there was some indication that very young animals were absent from sites that had been interpreted as having a ritual function (Runnymede Bridge, the only site interpreted as domestic in nature, provided greater evidence of young cattle than the ritual sites investigated). It is impossible, however, to determine whether the absence of very young animals was due to their absence from a site as a whole, or their survival into later life. Legge (1981) argued that the whole cattle population was not represented at many Early Neolithic sites, and that only specific members (adult females) were transported to sites from other areas. In the Late Neolithic, Durrington Walls has been interpreted as a *consumer* site rather than a *producer* site (Parker Pearson *et al.* 2006). Though still in progress, analysis of both cattle and pig bones from recent excavations at this site has indeed highlighted the almost complete absence of neonatal specimens despite careful recovery and the recording of thousands of specimens. This makes it probable that on-site breeding was either absent or rare. Under this interpretation animals would be born and raised away from the site, and brought to it at a more advanced age. This would concur with the pattern from the Early Neolithic, and suggests that cattle raising did not take place in the vicinity of the monuments that make up the bulk of the Early Neolithic archaeological record.

In the Late Neolithic analysis of the $^{87}Sr/^{86}Sr$ of cattle teeth highlighted the movement of cattle over potentially very long distances and in several different directions. This is not evidence of transhumance (as no repetition of the movement can be suggested by these results), but at least single movement. Within tooth variation in both $^{87}Sr/^{86}Sr$ and Sr concentration differed between animals. Thus it can be suggested that the regimes employed to keep different animals varied. Such dissimilarity between practices is supportive of ideas that propose multiple economic, mobility and possibly even transhumance patterns within the Neolithic period. More detailed work using the strontium isotope analysis of cattle is necessary to clearly understand the mobility involved in keeping this animal during the Neolithic. Two important questions will need to be addressed in the future: given the similarity identified between practices in the Early and Late Neolithic, did cattle mobility also play an important role in the Early Neolithic? Secondly, was mobility of cattle unique to Durrington Walls, or is it in evidence at other Late Neolithic sites?

Conclusion

- It is likely that cattle were used for secondary products (*e.g.* milk and perhaps manure and blood), but there is no zooarchaeological evidence on the ritual sites investigated here to support a specialised milk economy. Therefore, cattle likely had a number of uses.
- Evidence suggests some cattle were moved over long distances in prehistoric Britain, but the specific details of the system are unclear and, although a high rate of mobility is implied, this cannot be taken as a demonstration of a nomadic or even transhumant style of life.
- There are many differences in the way cattle and pigs were kept in prehistoric Britain and these can partly be explained as an unavoidable result of their inherent behaviour and needs.
- Cattle husbandry remained largely unchanged through time, and the use of cattle was generally similar at the Neolithic sites investigated. Runnymede Bridge provided the only variation.
- The asynchronous appearance of husbandry of different species in the Near East and parts of Europe is logical given the variation in husbandry strategies identified in this research, but at present there is no way to investigate the possibility of differential timing in Britain.
- Numerous pig husbandry systems were in place, and it is possible that there were as many pig keeping systems as there were archaeological sites in the Neolithic period. The importance of pigs as meat intensified during the Late Neolithic.

7.7 Final Summary

The traditional view of animal husbandry in Britain as a largely exogenous phenomenon is upheld by the results of this work. Cattle and pigs, along with sheep and goats, were brought to Britain from mainland Europe in domesticated form during a process that had much in common with other parts of Europe at the Mesolithic/Neolithic transition. But the British situation was unique, not only because of its geographical position on the western fringe of the continent, but also in the development and practice of husbandry once established within the island. Cattle and pigs were maintained in very different ways which led to perceptible differences in their contribution to the zooarchaeological record. While cattle were tightly controlled to the extent that very little mixing took place between wild and domestic types, pigs were allowed to range more freely and interbreeding with extant wild boar populations led to animals that were physically alike. Cattle were part of a system that was common to many Neolithic economies and that was maintained over long periods of time. Movement over long distances in the Late Neolithic is suggestive of cooperation between cattle herders in different parts of the country. In general, cattle husbandry was a homogenous pursuit with little geographical or chronological variation. The use of *Bos* contrasts sharply with *Sus* exploitation. Instead of a system that was common to most Neolithic sites, pigs were treated differently in different places and techniques were apparently variable. A major difference was in the level to which wild boar were allowed to mix with domesticated animals – an occurrence that was common throughout the Early Neolithic, but that varied in the intensity of the contact. As a result, the control exerted over pigs in the Early Neolithic was not constant, and pig keeping strategies may have been unique to individual sites.

The animals involved, due to their dissimilar characters, were treated differently and as such the process of Neolithisation for each was distinct. Within Britain the different approaches to animal husbandry on the site level need to be explored, as well as the investigation of larger scale systems that linked wide geographical areas.

Animal husbandry in Britain needs to be seen as a collection of related but distinct phenomena.

Appendix I
Protocol for recording *Bos* bones.

The protocol used for recording information from cattle bones and teeth will follow a system based on that outlined by Davis (1992) and Albarella and Davis (1996) with some modifications relevant to this research. This system is based on the identification and recording of only specific zones of a number of skeletal elements. These elements and zones are listed below.

All letters entered into the database will be in lower case, with the exception of tooth wear stages that are case specific and follow Grant (1982).

Postcranial bones
Atlas (at least half)
Scapula (when neck is measurable)
Humerus (proximal and distal) (at least half of an epiphysis)
Radius (proximal and distal) (at least half of an epiphysis)
Metacarpal (at least half of the epiphysis)
Phalanges 1 and 2 (at least half of the epiphysis)
Pelvis (ischial part of the acetabulum)
Femur (proximal and distal) (least half of an epiphysis)
Tibia (proximal and distal) (least half of an epiphysis)
Metatarsal (at least half of the epiphysis)
Astragalus (at least half)
Calcaneum (sustentaculum present)
Teeth
Upper and lower molars, fourth deciduous premolars and fourth permanent premolars (with at least half of the occlusal surface present). These teeth will be recorded as either loose (recorded as 'l') or jaws (recorded as 'j'). Jaws are defined by the occurrence of at least one tooth plus at least half adjacent tooth/alveolus or equivalent amount of bone. No attempt will be made to separate M1 from M2 in isolated teeth, but biometrical analysis in pigs may eventually lead to their identification.

Fusion
The state of fusion will be recorded for both distal and proximal ends of the bone. Pelvis, Metapodials, atlas, and scapula will be recorded as having only a distal end, while phalanges and calcanea will be recorded as having only a proximal end. Since no fusion data will be available for astragali, all records will include a characterisation as normal, light or porous, entered in the comments box.
The following fusion codes will be used:
f - fused
g - fusing
h - fused or fusing
ud - unfused diaphysis
ue - unfused epiphysis
ux - unfused, both diaphysis and epiphysis present

Eruption and wear
Eruption and wear stages are recorded using the system established by Grant (1982).
The presence of linear enamel hypoplasia (LEH) will be recorded on lower permanent molars only, as:
p- present, **pp**- more than 1 line present, **'blank'**- absent.

Measurements: Postcranial bones
The following measurements will be taken when possible on fused, fusing and unfused bones. Very porous astragali obviously belonging to very juvenile animals will not be measured. Measurements follow those defined by von den Driesch (1976), Davis (1992), and Albarella and Payne (2005). All measurements are taken to tenths of millimeters (mm).

Horn core:		min. and max. diameter of base, L
Atlas:		H, BFcr
Scapula:		SLC
Humerus:		BT, HTC
Radius:	GL	
MC:		GL, SD, BFd, BatF, a, b, 3, 6
Pelvis:		LA
Femur:	DC	
Tibia:		Bd, Dd, GL
Astragalus:		GLl, GLm, Bd (including those characterised as 'light', but excluding 'porous' specimens)
Calcaneum:		GL, GD
MT:		GL, SD, BFd, BatF, a, b, 3, 6

Measurements: teeth/mandible
The following measurements will be taken for upper and lower teeth. The length (L) and width (W) measurements of lower teeth are taken at the widest part of the tooth. All measurements are in tenths of millimeters (mm). No measurements will be taken if the enamel has been entirely lost.

$\mathbf{dP^4}$	W
$\mathbf{M^1}$	W
$\mathbf{M^2}$	W
$\mathbf{M^3}$	W
$\mathbf{dP_4}$	W
$\mathbf{M_1}$	W
$\mathbf{M_2}$	W
$\mathbf{M_3}$	L, W

The height of the mandible in front of M_1 on the buccal side.

Database fields – postcranial bones
The measurement GLl will be recorded under the heading GL. DC and GD will both be recorded in the DC column.

Rec Num
Site Name
Location
Box Number
Ctx Num
Bone Num
Phase
Dating
El
Taxon
Fus prox
Fus dist
Min diameter
Max diameter
BFcr
SLC
BT
HTC
GL
LA
DC
Dd
Glm
Bd
BatF
SD
a
b
3
6
Comments

Database fields - teeth
Rec Num
Site Name
Location
Box Number
Ctx Num
Bone Num
Phase
Dating
El
Taxon
J/L
dP4
dP4W
P4
M1
M1C
M1W
M1hyp
M2
M2C
M2W
M2hyp
M3
M3L
M3C
M3W
M3hyp
M12
M12W
M12hyp
Mand H
Comments

Codes
Postcranial Bones
at - atlas
sc- scapula
hu - humerus
othu - proximal humerus only
ra - radius
othra - proximal radius only
mc1- metacarpal (both condyles)
mc2- metacarpal (one condyle)
pe - pelvis
fe - femur
othfe - proximal femur only
ti - tibia
othti - proximal tibia only
as - astragalus
ca - calcaneum
mt1 – metatarsal (both condyles)
mt2 – metatarsal (one condyle)
p1 - 1st phalanx
p2 - 2nd phalanx
oth - specify element in comment
Teeth
l - loose tooth
j - jaw
x - maxilla
n - mandible
u - unknown

Taxon (as classified by the previous investigator of the material and not by us)
w – wild
d - domestic
'blank' – unknown

Appendix II
Protocol for recording *Sus* bones

The protocol used for recording information from pig bones and teeth will follow a system based on that outlined by Davis (1992) and Albarella and Davis (1996) with some modifications relevant to is research. This system is based on the identification and recording of only specific zones of a number of skeletal elements. These elements and zones are listed below.

All letters entered into the database will be in lower case, with the exception of tooth wear stages that are case specific and follow Grant (1982).

Postcranial Bones
Atlas (at least half)
Scapula (when neck is measurable)
Humerus (proximal and distal) (at least half of an epiphysis)
Radius (proximal and distal) (at least half of an epiphysis)
Metacarpal III and **IV** (at least half of the epiphysis) (the proximal end must also be present for the purpose of identification)
Phalanges 1 and 2 (proximal end) (at least half of the epiphysis)
Pelvis (ischial part of the acetabulum)
Femur (proximal and distal) (least half of an epiphysis)
Tibia (proximal and distal) (least half of an epiphysis)
Metatarsal III and **IV** (at least half of the epiphysis) (the proximal end must also be present for the purpose of identification)
Astragalus (at least half)
Calcaneum (sustentaculum present)

Teeth
Upper and lower molars, fourth deciduous premolars, fourth permanent premolars and canines (with at least half of the occlusal surface present). These teeth will be recorded as either loose (recorded as 'l') or jaws (recorded as 'j'). Jaws are defined by the occurrence of at least one tooth plus at least half adjacent tooth/alveolus or equivalent amount of bone. No attempt will be made to separate M1 from M2 in isolated teeth. Canines will be recorded as female (f), male (m), female alveolus (af) or male alveolus (am).

Fusion
The state of fusion will be recorded for both distal and proximal ends of the bone. Pelvis, Metapodials, atlas, and scapula will be recorded as having only a distal end, while phalanges and calcanei will be recorded as having only a proximal end. Since no fusion data will be available for astragali, all records will include a characterisation as normal, light or porous, entered in the comments box.
The following fusion codes will be used:
f - fused
g - fusing
h - fused or fusing
ud - unfused diaphysis
ue - unfused epiphysis
ux - unfused, both diaphysis and epiphysis present

Eruption and wear
Eruption and wear stages are recorded using the system established by Grant (1982).
The presence of linear enamel hypoplasia (LEH) will be recorded on lower permanent molars only, as:
p- present, **pp**- more than 1 line present, **'blank'**- absent.

Measurements: Postcranial bones
The following measurements will be taken when possible on fused, fusing and unfused bones. Very porous astragali obviously belonging to very juvenile animals will not be measured. Most measurements follow the definitions provided by Payne and Bull (1988) but for others see von den Driesch (1976).

Atlas:	H, BFcr
Scapula:	SLC
Humerus:	BT, HTC
Radius:	GL
MTC III and IV:	GL
Pelvis:	LAR
Femur:	DC
Tibia:	BdP, Dd, GL
Astragalus:	GLl, GLm (including those characterised as 'light', but excluding 'porous' specimens)
Calcaneum:	GL, GD
MTT III and IV:	GL

Measurements: teeth/mandible
The following measurements will be taken for upper and lower teeth as defined in Payne and Bull (1988). M3WC and M3WP are the greatest widths of the central and posterior cusps of the third molar and are taken following the same criteria as the other width measurements defined by Payne and Bull (1988). No measurements will be taken if the enamel has been entirely lost. No length measurements will be taken on upper and lower teeth when their wear is more advanced than stages "Wb" (as defined in Bull and Payne 1982) on both cusps (see Albarella an Payne 2005) and "g" (as defined in Grant 1982) respectively.

dP^4	L, WP
M^1	L, WA, WP
M^2	L, WA, WP
M^3	L, WA
dP_4	L, WP
M_1	L, WA, WP
M_2	L, WA, WP
M_3	L, WA,WC, WP

The height of the mandible in front of M_1 on the buccal side.

Database fields

The measurement GLl will be recorded under the heading GL. DC, BdP and GD will all be recorded in the DC column.

Rec Num
Site Name
Location
Box Number
Ctx Num
Bone Num
Phase
Dating
El
Taxon
Fus prox
Fus dist
BFcr
SLC
BT
HTC
GL
LAR
DC
Dd
Glm
X/N
J/L
C
dP4
dP4L
dP4WP
P4
M1
M1L
M1WA
M1WP
M1hyp
M2
M2L
M2WA
M2WP
M2hyp
M3
M3L
M3WA
M3WC
M3WP
M3hyp
M12
M12L
M12WA
M12WP
M12hyp
Mand H
Comments

Codes

Postcranial Bone

at - atlas
sc- scapula
hu - humerus
othu - proximal humerus only
ra - radius
othra - proximal radius only
mciii - 3rd metacarpal
mciv - 4th metacarpal
pe - pelvis
fe - femur
othfe - proximal femur only
ti - tibia
othti - proximal tibia only
as - astragalus
ca - calcaneum
mtiii - 3rd metatarsal
mtiv - 4th metatarsal
p1 - 1st phalanx
p2 - 2d phalanx
oth - specify element in comments

Teeth

l - loose tooth
j - jaw
x - maxilla
n - mandible
u - unknown

For the dP4, M1, M2, M3, M12 (M1 or M2) wear stages follow Grant (1982) and presence/absence using the codes:
p - present
'blank' - absent

For canines
f - female
m - male
af - female alveolus
am - male alveolus
'blank' - absent

Taxon (as classified by the previous investigator of the material and not by us)
w - wild
d - domestic
'blank' - unknown

Appendix III
Aurochs standard measurements

The following tables provide raw data collected from aurochs bones from the site of Ilford (dated to MIS7, Andy Currant pers. comm.). The mean values given below each measurement column are the standard measurements used as a source of comparison for log ratio diagrams presented in Chapter 4. All measurements are in mm, and follow von den Driesch (1976).

Table III.1 Astragalus measurements from Ilford aurochs.

Site Name	El	GL	Bd	GLm
Ilford	as	99.50	71.10	91.40
Ilford	as	102.30		95.70
Ilford	as	99.90		91.20
Ilford	as	92.80	66.70	86.40
Ilford	as	95.30	66.50	86.90
Ilford	as	97.60	67.20	85.30
Ilford	as	92.20	62.20	84.10
mean		**97.09**	**66.74**	**88.71**

Table III.2 Atlas measurements from Ilford aurochs.

Site Name	El	Fus prox	BFcr	H
Ilford	at	f	143.50	123.40
Ilford	at	f	143.20	119.40
Ilford	at	f	140.40	123.10
Ilford	at	f	133.60	118.00
Mean			**140.18**	**120.98**

Table III.3 Calcaneum measurements from Ilford aurochs.

Site Name	El	Fus prox	GL	Bd
Ilford	ca	f	193.00	77.90
Ilford	ca	f	188.00	72.10
Ilford	ca	f	184.00	71.30
Ilford	ca	f	200.00	75.30
Ilford	ca	f	200.00	78.70
Ilford	ca	f	193.00	75.60
Ilford	ca	f	201.00	77.80
Ilford	ca	f	196.00	77.60
Ilford	ca	f	197.00	76.20
Mean			**194.67**	**75.83**

Table III.4 Horn core measurements from Ilford aurochs.

Site Name	El	min diameter	max diameter
Ilford	hc	130.00	137.00
Ilford	hc	124.00	156.00
Ilford	hc	118.00	152.00
Ilford	hc	111.00	153.00
Ilford	hc	118.00	148.00
mean		**120.20**	**149.20**

Table III.5 Femur measurements from Ilford aurochs.

Site Name	El	Fus prox	Fus dist	GL	DC
Ilford	Fe	f	f		68.90
Ilford	Fe	f	f		67.80
Ilford	Fe	f	f		67.00
Ilford	Fe	f	f		63.90
Ilford	Fe	f	f	529.00	67.20
Ilford	othfe	f			69.80
Ilford	othfe	f			67.00
Ilford	othfe	f			70.20
Ilford	othfe	f			70.20
Ilford	othfe	f			63.70
mean				**529.00**	**66.96**

Table III.6 Humerus measurements from Ilford aurochs.

Site Name	El	Fus prox	Fus dist	GL	SD	BT	HTC
Ilford	Hu	f	f	416.00	72.00	117.30	49.90
Ilford	Hu	f	f		63.90		
Ilford	Hu	f	f		69.80	117.10	
Ilford	hu	f	f		66.40	114.50	47.40
Ilford	hu		f		63.40	114.60	55.40
Ilford	hu	f	f	451.00	68.70	118.40	52.90
Ilford	hu	f	f			117.40	48.10
Ilford	hu		f			114.40	49.50
Ilford	hu		f			113.30	48.60
Ilford	hu		f			109.60	50.50
Ilford	hu		f				49.00
Mean				**433.50**	**67.37**	**115.18**	**50.14**

Table III.7 Metacarpal measurements from Ilford aurochs.

El	Fus prox	Fus dist	GL	SD	BFd	BatF	a	b	3	6
mc1	f	f	267	56.30	95.00	81.10	44.20	46.00	47.60	48.60
mc1	f	f	256					46.90		43.80
mc1	f	f	251	56.00		81.90				
mc1	f	f	269	57.50	94.30	88.60	44.50	44.10	45.60	45.50
mc1	f	f	257	55.70	95.30	85.70	45.10	44.00	44.80	44.90
mc1	f	f		51.70		80.20				
mc1	f	f		54.50		86.10				
mc1	f	f	248	55.50		85.60	42.80		44.10	
mc1	f	f	253	53.40	85.50	80.20	40.00	39.30	42.50	41.80
mc1	f	f	268	56.60	92.80	85.80	43.60	43.10	43.00	42.50
Mean			**258.63**	**55.24**	**92.58**	**83.91**	**43.37**	**43.90**	**44.60**	**44.52**

Table III.8 Metatarsal measurements from Ilford aurochs.

El	Fus prox	Fus dist	GL	SD	BFd	BatF	a	b	3	6
mt1	f	F	300	47.10	82.90	78.10	38.40	38.80		42.40
mt1	f	F		47.20	78.40					
mt1	f	F	299	46.00	82.60	76.90	39.10	39.00		42.40
mt1	f	f	308	49.50	85.90	83.10	39.90	39.40	44.30	44.40
mt1	f	f	296	46.40	78.30	77.50	37.20	35.30	40.00	39.70
mt1	f	f	302	46.30	83.30	79.10	38.80	38.90	42.30	42.20
mt1	f	f	292	46.60	87.30	79.70	41.80		42.00	42.40
mean			**299.50**	**47.01**	**82.67**	**79.07**	**39.20**	**38.28**	**42.15**	**42.25**

Table III.9 Pelvis measurements from Ilford aurochs.

Site Name	El	Fus prox	LA
Ilford	pe	f	109.4
Ilford	pe	f	110.9
Ilford	pe	f	109
Ilford	pe	f	109.6
Ilford	pe	f	108.5
Mean			**109.48**

Table III.10 Radius measurements from Ilford aurochs.

Site Name	El	Fus prox	Fus dist	GL	SLC	SD
Ilford	ra	f	f	456		44.9
Ilford	ra	f	f	401	77.3	
Ilford	ra	f	f	406		72.2
Ilford	ra	f	f	379		70.1
Ilford	ra	f	f	367		70.6
Ilford	ra	f	f			73.4
Ilford	ra	f	f	416		73.1
Ilford	ra	f	f	339	50.9	
mean				**394.86**	**64.10**	**67.38**

Table III.11 Scapula measurements from Ilford aurochs.

Site Name	El	Fus dist	SLC
Ilford	sc	f	94
Ilford	sc	f	86.3
Ilford	sc	f	93.1
mean			**91.13**

Table III.12 Tibia measurements from Ilford aurochs.

Site Name	El	Fus prox	Fus dist	GL	Bd	Dd
Ilford	ti	f	f	470		
Ilford	ti	f	f	481		
Ilford	ti	f	f	473		
Ilford	ti	f	f		95.7	
Ilford	ti	f	f	501	92.2	67.4
Ilford	ti	f	f	481	94.5	68.8
Ilford	ti	f	f	460		49.5
Ilford	ti		f		89.8	
Ilford	ti		f		88.6	69.9
Ilford	ti		f		89.6	71.8
Ilford	ti		f		88.6	
mean				**477.67**	**91.29**	**65.48**

Table III.13 Maxillary measurements from Ilford aurochs.

Site Name	X/N	J/L	M2W	M3L	M3W
Ilford	x	l		39.8	31
Ilford	x	l		37.1	31.6
Ilford	x	j	31.5		
mean			**31.5**	**38.45**	**31.3**

Table III.14 Mandibular measurements from Ilford aurochs.

Site Name	X/N	J/L	M1W	M3L	M3W
Ilford	n	j	20		
Ilford	n	l		45.2	21.6
Ilford	n	l		46.7	18.2
Ilford	n	l		45.5	21.7
Ilford	n	l		49.2	20.1
mean			**20.00**	**46.65**	**20.40**

Appendix IV
Pig standard measurements

The standard measurements used for log ratio comparison of pig remains are those established by Albarella and Payne (2005, 598) from Late Neolithic Durrington Walls. All measurements are in mm and follow von den Driesch (1976) and Payne and Bull (1988).

Table VII.1 Standard measurements from Durrington Walls pig bones (after Albarella and Payne 2005). Those marked with * were not provided by Albarella and Payne (2005), but were calculated from Durrington Walls measurements collected for this research.

Element	Measurement	Standard
Astragalus	GLl	40.8
Astragalus	GLm	37.9
Atlas	H	46.9
Atlas	BFcr	56.8
Calcaneum	GL	79.3
Calcaneum	GD	31.1
Humerus	BT	31.3
Humerus	HTC	19.7
Pelvis	LAR	33.3
Scapula	SLC	23.8
Tibia	BdP	30.7
Tibia	Dd	26.7*
M^1	L	17.6
M^1	WA	13.7
M^1	WP	13.5
M^2	L	32.9
M^2	WA	17.0
M^2	WP	16.5
M^3	L	32.9
M^3	WA	18.5
dP_4	L	18.9
dP_4	WP	8.6
M_1	L	17.5
M_1	WA	10.3
M_1	WP	10.9
M_2	L	21.8
M_2	WA	13.7
M_2	WP	14.2
M_3	L	34.5
M_3	WA	15.7
M_3	WC	15.1*
M_3	WP	15.1*

Appendix V
Sus mandibles used in wear stage analysis

Table V.1 *Sus* mandibles from Hambledon Hill (DB = database ID, wear stage after Grant 1982).

DB ID	dP4	P4	M1	M2	M3
227			f	c	a
416				d	a
272	d		a	C	
329	d		b		
339	e		b	U	
417	d		a		
420	e		a		
440	d		a		
225				c	u
246				b	u
274			d	b	C
301		A	f	d	U
342			d	a	
407				b	U
439			d	a	
318		B	f		
376		B	e	d	
377				k	p

Table V.2 *Sus* mandibles from Runnymede Bridge (DB ID = database ID number, age stages after Grant 1982).

DB ID	dP4	P4	M1	M2	M3
184				e	c
189			d	a	
199		a	e	c	a
216		d	l	e	a
217			h	g	d
513			m	k	d
521		b	k	f	a
547					d
564		e	l	f	c
590					a
592			h	f	a
628					a
631		b	g	d	a
648				d	a
539	f		c	V	
193	e	C	b		
194					E
520			e	c	U
523			c	a	
532		b	e	b	V
536		b	d	b	
549	e		c	a	
647			c	a	
651			d	a	
180		c	j	c	
192		e	m		
209		d	h		
558			m	f	
567		b	e		
572			g	d	
574	d		b		
578	d		a		
579			d	b	
594			k	f	
608			d	21.9	
625		b			
633					

Table V.3 ***Sus*** **mandibles from Durrington Walls (DB ID = database ID, wear stage after Grant 1982).**

DB ID	dP4	P4	M1	M2	M3
813		c	g	e	a
2295		c	j	f	a
167		C	f	d	a
97			e	c	a
534		c	g	d	a
107			e	e	a
526			e	c	a
2351				d	a
31			d	c	a
1083				d	a
2686		E	k	e	a
1404		g	m	j	c
458				f	b
527				g	c
2150		g	l		d
1878			j	f	c
2149		a	l	f	b
138				e	d
1130			k	f	b
277				d	c
165		C	e	d	c
228			e	e	b
2580				f	b
1015					e
1489	e		b	V	
87	e		b	V	
264	f		C	V	
2475	f		c	V	
146			b	V	
2426	f	C	c	V	
2077	e		b	V	
1412			c	E	
1490	g		c	E	
1488	l		d	E	
1365	f		c	V	
964			d	U	
323	P		c	E	
611			c	E	
1014			d	U	

Table.3 continued

DB ID	dP4	P4	M1	M2	M3
319	f		b	E	
143	l	C	b	E	
683	g		d	E	
1814	h		e	U	
893	m		d	E	
2156			e	U	
2476	j		d	U	
1060	j		d	U	
1387			d	U	
2037			c	E	
1012		a	f	b	V
313		a	e	a	C
522		b	c	a	V
525	g		e	a	V
386				b	V
136		a	d	a	V
273			d	a	V
100			c	a	V
2441			e	a	V
2418			c	a	V
963				c	U
1592		b	j	d	U
1890		b	d	c	E
310					E
1813			j	e	U
1219			f	c	E
340		b	g	C	E
1774		e	j	c	E
374	k	C			
61	g		C		
1413			g	c	H
350			g	d	H

Appendix VI
Bos mandibles used in wear stage analysis

Table VI.1 *Bos* mandibles from Hambledon Hill (DB ID = database ID, wear stage after Grant 1982).

DB ID	dP4	P4	M1	M2	M3
333			l	k	
342			f	c	b
360		c	k	h	d
384		p	k	k	p
393					c
442			j	f	b
454		g	l	k	g
643		g	k	j	g
697	j	p	g	f	b
720			g	p	b
724		f	m	l	g
358		d	l	k	j
359		f	l	k	
441		g	m	k	k
648				k	k
719		h	l	k	p
220		g	l	k	
254	j		c		
260	j		g	c	
730		f	k		

Table VI.2 *Bos* mandibles from Runnymede Bridge (DB ID = database ID, wear stage after Grant 1982).

DB ID	dP4	P4	M1	M2	M3
119			k	g	c
841				g	C
538		e	k		
571			l	k	
854	j		b	V	
115	j	C			
848	a		V		
852	j		f	a	C
856	n		g	d	
542					K

Table VI.3 *Bos* mandibles from Coneybury Anomaly (DB ID = database ID, wear stage after Grant 1982).

DB ID	dP4	P4	M1	M2	M3
282			k	f	
310		d	k	g	g
284		g	l	k	
286			f	E	
300	e		a		
287	U				

Table VI.4 *Bos* mandibles from Durrington Walls (DB ID = database ID, wear stage after Grant 1982).

DB ID	dP4	P4	M1	M2	M3
877					d
895		h	l	k	
912		j	m		l
913				k	g
937				j	b
972		f		g	f
1081				k	h
1085			k	k	
1119		c	j	g	
1125		e	k	j	
1023			l	k	
1116					k
915	j		b		
983	k		j		
973	h		E		
986			g	b	
876		f	k	j	

Bibliography

Aaris-Sørenson, K. 1999. The Holocene history of the Scandinavian aurochs (*Bos primigenius* BOJANUS, 1827). In G-C. Weniger (ed.) *Archäologie und biologie des Auerochsen*, 49-57. Mettmann, Neanderthal Museum, Wissenschafliche Schriften 1.

Ahmad, E. Brooks, J.E. Hussain, I. Khan, M.H. 1995. Reproduction of Eurasian wild boar in the central Punjab, Pakistan. *Acta Theriologica* 40(2), 163-73.

Akeret, Ö. and Jacomet, S. 1997. Analysis of plant macrofossils in goat/sheep faeces from the Neolithic lake shore settlement of Horgen Scheller – an indication of prehistoric transhumance? *Vegetation history and archaeobotany* 6(4), 235-9.

Akeret, Ö. and Rentzal, P.2001. Micromorphology and plant macrofossil analysis of cattle dung from the Neolithic lake shore settlement of Arbon Bleiche 3. *Geoarchaeology* 16(6), 687-700.

Akeret, Ö. 1999. Plant macrofossils and pollen in goat/sheep faeces from the Neolithic lake-shore settlement Arbon Bleiche 3, Switzerland. *The Holocene* 9(2), 175-82.

Albarella, A. 1997. Shape variation of cattle metapodials: age, sex or breed? Some examples from Medieval and Postmedieval Sites. *Anthropozoologica* 25-26, 37-47.

Albarella, U. 2002. 'Size matters': how and why biometry is still important in Zooarchaeology. In K. Dobney and T. O'Connor (eds.) *Bones and the Man, Studies in honour of Don Brothwell*. Oxford, Oxbow.

Albarella U. 2010. Wild boar. In T.O'Connor & N.Sykes (eds.). Extinctions and Invasions: A Social History of British Fauna, pp.59-67. Oxford: Oxbow Books.

Albarella, U. and Davis S. 1996. *Mammals and Birds from Launceston Castle, Cornwall: Decline in Status and the Rise of Agriculture*. *Circaea* 12(1), 1-156.

Albarella, U. Davis, S.J. Detry, C. and Rowley-Conwy, P. 2005. Pigs of the "Far West": the biometry of *Sus* from archaeological sites in Portugal. *Anthropozoologica* 40(2), 27-54.

Albarella, U. Dobney, K. and Rowley-Conwy, P. 2009. Size and shape of the Eurasian wild boar (*Sus scrofa*), with a view to the reconstruction of its Holocene history. *Environmental Archaeology* 14(2), 103-36.

Albarella, U. Manconi, F. Rowley-Conwy, P. and Vigne, J. 2006. Pigs of Corsica and Sardinia: a biometrical re-evaluation of their status and history. In U.Tecchiati and B.Sala (eds.) *Archaeozoological studies in honour of Alfredo Riedel*, 285-302. Bolzano, Province of Bolzano.

Albarella U, Marrazzo D, Spinetti A and Viner S. In prep. *The animal bones from Welland Bank Quarry (Lincolnshire).*

Albarella U. and Payne S. 2005. Neolithic pigs from Durrington Walls, Wiltshire, England: a biometrical database. *Journal of Archaeological Science* 32(4), 589-99.

Albarella, U. and Serjeantson, D. 2002. A passion for pork: Meat consumption at the British Late Neolithic site of Durrington Walls. In P. Miracle and N. Milner (eds.) *Consuming passions and patterns of consumption*, 33-49. Oxford, Oxbow.

Albarella, U. Tagliacozzo, A. Dobney, K. and Rowley-Conwy, P. 2006a. Pig hunting and husbandry in prehistoric Italy: a contribution to the domestication debate. *Proceedings of the Prehistoric Society* 72, 193-227.

Ambers, J. and Housley, R. 1999. Radiocarbon dating. In A. Whittle, J. Pollard and C. Grigson (eds.) *The harmony of symbols: The Windmill Hill causewayed enclosure*, 116-20. Oxford, Oxbow.

Amoroso, E.C. and Jewell, P.A. 1963. The exploitation of the milk-ejection reflex by primitive peoples. In A.E. Mourant and F.E. Zeuner (eds.) *Man and Cattle*, 126-37. Royal Anthropological Institute Occasional Paper 18. London, Royal Anthropological Institute.

Armour-Chelu, M. 1991. The animal bones. In N. Sharples (ed.) *Maiden Castle excavations and field survey*, 131-51. London, English Heritage archaeological reports 19.

Arrington, D.A. Toth, L.A. and Koebel, J.R. 1999. Effects of rooting by feral hogs *Sus scrofa* L. on the structure of a floodplain vegetation assemblage. *Wetlands* 19(3), 535-44.

Bailey, J.F. Richards, M.B. Macaulay, V.A. Colson, I.B. James, I.T. Bradley, D.G. Hedges, R.E.M. and Sykes, B.C. 1996. Ancient DNA suggests a recent expansion of European cattle from a diverse wild progenitor species. *Proceedings of the Royal Society of London B* 263, 1467-73.

Bakels, C.C. 1997. The beginnings of manuring in western Europe. *Antiquity* 71, 442-5.

Balasse, M. Ambrose, S.H. Smith, A.B. and Price, T.D. 2002. The seasonal mobility model for prehistoric herders in the south-western Cape of South Africa assessed by isotopic analysis of sheep tooth enamel. *Journal of Archaeological Science* 29, 917-32.

Balasse, M. and Tresset, A. 2002. Early weaning of Neolithic domestic cattle (Bercy, France) revealed by intra-tooth variation in nitrogen isotope ratios. *Journal of Archaeological Science* 29, 853-9.

Balasse, M. Smith, A.B. Ambrose, S.H. and Leigh, S.R. 2003. Determining sheep birth seasonality by analysis of tooth enamel oxygen isotope ratios: the Late Stone Age site of Kasteelberg (South Africa). *Journal of Archaeological Science* 30, 205-15.

Barclay, G.J. 2005. The Neolithic. In K.J. Edwards and I.B.M. Ralston (eds.) *Scotland after the ice*, 127-49. Edinburgh, Edinburgh University Press.

Barclay, A. and Halpin, C. (eds.) 1999. *Excavations at Barrow Hills, Radley, Oxfordshire. Volume 1: The Neolithic and Bronze Age monument complex*. Oxford, Oxford archaeological unit.

Barker, G. 1985. *Prehistoric farming in Europe*. London, Cambridge University Press.

Barton, N. and Roberts, A. 2004. The Mesolithic period in England: Current perspectives and new research. In A. Saville (ed.) *Mesolithic Scotland and its neighbours*, 339-58. Edinburgh, Society of Antiquaries of Scotland.

Bartosiewicz, L. 1987. Cattle metapodials revisited: a brief review. *Archaeozoologia* I/1, 47-51.

Bartosiewicz, L. Van Neer, W. and Lentacker, A. 1997. *Draught cattle: their osteological identification and history*. Annalen zoölogische wetenschappen vol.281.

Bartosiewicz, L. 1999. Aurochs (*Bos primigenius* BOJANUS, 1827) in the Holocene of Hungary. In G-C. Weniger (ed.) *Archäologie und biologie des Auerochsen*, 103-50. Mettmann, Neanderthal Museum, Wissenschafliche Schriften 1.

Bar-Yosef, O. 1995. Earliest food producers – pre-pottery Neolithic (8000-5500). In T.E. Levy (ed.) *The archaeology of society in the Holy Land*, 190-204. London and Washington, Leicester University Press.

Bar-Yosef, O. and Meadow, R. H. 1995.The origins of agriculture in the Near East. In T.D. Price and G. Gebauer (eds.) *Last hunters, first farmers*, 39-94. Santa Fe, New Mexico, School of American Research Press.

Baubet, E. Bonenfant, C. and Brandt, S. 2004. Diet of the wild boar in the French Alps. *Galemys* 16, 101-13.

Bayliss A. Healy, F. Bronk Ramsey, C. McCormac, F.G. and Mercer, R. 2009. Interpreting chronology. In R. Mercer and F.Healy (eds.) *Hambledon Hill, Dorset, England. Excavation of a Neolithic monument complex and its surrounding landscape volume 1*, 378-411. Swindon, English Heritage.

Bell, M. Caseldine, A. and Neuman, H. 2000. *Prehistoric intertidal archaeology in the Welsh Severn estuary*. York, Council for British Archaeology.

Bellwood, P. 2005. *First farmers, the origins of agricultural societies*. Oxford, Blackwell Publishing.

Beja-Pereira, A. Caramelli, D. Lalueze-Fox, C. Vernesi, C. Ferrand, N. Casoli, A. Goyache, F. Royo, L.J. Conti, S. Lari, M. Martini, A. Ouragh, L. Magid, A. Atash, A. Zsolnai, A. Boscato, P. Triantaphylidis, C.Ploumi, K. Sineo, L. Mallegni, F. Taberlet, P. Erhardt, G. Sampietro, L. Bertranpetit, J. Barbujani, G. Luikart, G. and Bertorelle, G. 2005. The origin of European cattle: Evidence from modern and ancient DNA. *Proceedings of the National Academy of Science USA* 103(21), 8113-8.

Bender, B. 1978. Gatherer-hunter to farmer; a social perspective. *World Archaeology* 10(2), 204-22.

Benson, D. and Whittle, A. 2007. *Building memories: The Neolithic Cotswold long barrow at Ascott-Under-Wychwood, Oxfordshire.* Oxford, Oxbow.

Bentley, R.A. 2006. Strontium isotopes from the earth to the archaeological skeleton: a review. *Journal of Archaeological Method and Theory* 13(3),135-87.

Bentley, A. 2007. Mobility, specialization and community diversity in the Linearbandkeramik: isotopic evidence from the skeletons. In A.Whittle and V.Cummings (eds.) *Going over: The Mesolithic-Neolithic transition in north-west Europe*, 117-40. Oxford, Oxford University Press.

Bentley, R.A. and Knipper, C. 2005. Transhumance at the Early Neolithic settlement at Vaihingen (Germany). *Antiquity* 79 (306).

Bentley, R.A. Krause, R. Price, T.D. and Kaufmann, B. 2003. Human mobility at the Early Neolithic settlement of Vaihingen, Germany: evidence from strontium isotope analysis. *Archaeometry* 45(3), 471-86.

Bentley, R.A. Price, T.D. and Stephan, E. 2004. Determining the 'Local' 87Sr/86Sr range for archaeological skeletons: a case study from Neolithic Europe. *Journal of Archaeological Science* 31(4), 365-75.

Bertaux, D. and Micol, T. 1992. Population studies and reproduction of the feral cattle (*Bos taurus*) of Amsterdam Island, Indian Ocean. *Journal of Zoologists of London* 228, 265-76.

Blakely, R.L. 1989. Bone Strontium in Pregnant and Lactating Females From Archaeological Samples. *American Journal of Physical Anthropology* 80, 173-85.

Boessneck, J. and von den Driesch, A. 1978. The significance of measuring animal bones from archaeological sites. In R.H. Meadow and M.A. Zeder (eds.) *Approaches to Faunal Analysis in the Middle East*, 25-39. Peabody Museum Bulletin 2. Cambridge, Harvard University.

Bogaard, A. 2004. *Neolithic farming in central Europe, an archaeobotanical study of crop husbandry practices*. London and New York, Routledge.

Bogaard, A. and Jones, G. 2007. Neolithic farming in Britain and central Europe: contrast or continuity? A.Whittle and V.Cummings (eds.) *Going over: The Mesolithic-Neolithic transition in north-west Europe*, 357-75. Oxford, Oxford University Press.

Bogucki, P. 1984. Ceramic sieves of the Linear Pottery Culture and their economic implications. *Oxford Journal of Archaeology* 3(1), 15-30.

Bogucki, P. 1989. The exploitation of domestic animals in Neolithic central Europe. In P. Crabtree, D. Campana and K. Ryan (eds.) *Early animal domestication and its cultural context*, 118-34. Pennsylvania, The Museum Applied Science Center for Archaeology (MASCA).

Bogucki, P. 1993. Animal traction and household economies in Neolithic Europe. *Antiquity* 67, 492-503.

Bökönyi, S. 1969. Archaeological problems and methods of recognizing animal domestication. In P.J. Ucko and G.W. Dimbleby (eds.) *The domestication and exploitation of plants and animals*, 219-29. Chicago, Aldine Publishing Company.

Bollongino, R. and Burger, J. 2007. Neolithic cattle domestication as seen from ancient DNA. In A. Whittle and V. Cummings (eds.) *Going over: The Mesolithic-Neolithic transition in north-west Europe*, 165-87. Oxford, Oxford University Press.

Bonsall, C. Macklin, M.G. Anderson, D.E. Payton, R.W. 2002. Climate change and the adoption of agriculture in north-west Europe. *European Journal of Archaeology* 5(1), 9-23.

Boyle, K.V. 2006. Neolithic wild game animals in western Europe: the question of hunting. In D. Serjeantson and D. Field (eds.) *Animals in the Neolithic of Britain and Europe*, 10-23. Oxford, Oxbow.

Bradley, R. 1972. Prehistorians and pastoralists in Neolithic and Bronze Age England. *World Archaeology* 4(2), 192-204.

Bradley, R. 1978. *The prehistoric settlement of Britain*. London, Henley, Boston, Routledge and Kegan Paul.

Bradley, R. 1998. *The significance of monuments*. London and New York, Routledge.

Bratton, S.P. 1975. The effect of the European wild boar, *Sus scrofa*, on gray beech forest in the Great Smoky Mountains. *Ecology* 56(6), 1356-66.

Britton, K. Grimes, V. Dau, J. and Richards, M.P. 2009. Reconstructing faunal migrations using intra-tooth sampling and strontium and oxygen isotope analyses: a case study of modern caribou (*Rangifer tarandus granti*). *Journal of Archaeological Science* 36, 1163-72.

Brown, P. 1978. *Highland peoples of New Guinea*. Cambridge, Cambridge University Press.

Bruford, M.W. Bradley, D.G. Luikart, G. 2003. DNA markers reveal the complexity of livestock domestication. *Nature* 4, 900-10.

Budd, P. Montgomery, J. Barreiro, B. Thomas, R.G. 2000. Differential diagenesis of strontium in archaeological human dental tissues. *Applied Geochemistry* 15, 687-94.

Bull, G. and Payne, S. 1982. Tooth eruption and epiphysial fusion in pigs and wild boar. In B. Wilson, C. Grigson, S. Payne (eds.) *Ageing and Sexing Animal Bones from Archaeological Sites*, 55–71. British Archaeological Reports, British Series 109.

Carter, H.H. 1976. Fauna of an area of Mesolithic occupation in the Kennet valley considered in relation to contemporary eating habits. *Berkshire Archaeological Journal* 68, 1-3.

Carter, R. 1997. Age estimation of the roe deer (*Capreolus capreolus*) mandibles from the Mesolithic at Star Carr, Yorkshire, based on radiographs of mandibular tooth development. *Journal of Zoology, London* 241, 495-502.

Carter R. 1998. Reassessment of seasonality at the early Mesolithic site of Star Carr, Yorkshire based on radiographs of mandibular tooth development in red deer (*Cervus elaphus*). *Journal of Archaeological Science* 25, 851-6.

Carter, R. 2001. New evidence for seasonal human presence at the Early Mesolithic site of Thatcham, Berkshire, England. *Journal of Archaeological Science* 28, 1055-60.

Chaix, L. and Arbogast, R-M. 1999. Holocene aurochs from western Europe. In G-C. Weniger (ed.) *Archäologie und biologie des Auerochsen*, 35-46. Mettmann, Neanderthal Museum, Wissenschafliche Schriften 1.

Chamberlain, A.T. 1996. More dating evidence for human remains in British caves. *Antiquity* 70, 950-53.

Chatterton, R. 2006. Ritual. In C. Conneller and G.Warren (eds.) Mesolithic Britain and Ireland: new approaches, 101-20. Stroud, Tempus.

Chenery, C. Chisham, C. 2006. Thames and Solent research framework the Upper Palaeolithic and Mesolithic of Berkshire. Unpublished report, Wessex Archaeology.

Churchill, D.M. 1962. The stratigraphy of the Mesolithic sites III and V at Thatcham, Berkshire, England. *Proceedings of the Prehistoric Society* 28, 362-70.

Clark, J.G.D. 1954. *Excavations at Star Carr. Cambridge*. Cambridge, Cambridge University Press.

Clason, A.T. 1967. *Animal and man in Holland's past.* Groningen, Wolters.

Cleal, R. and MacSween, A. 1999. *Grooved Ware in Britain and Ireland.* Oxford, Oxbow.

Cleal, R. 2004. The dating and diversity of the earliest ceramics of Wessex and south west England. In R. Cleal and J. Pollard (eds.) *Monuments and material culture. Papers in honour of an Avebury archaeologist: Isobel Smith*, 164-92. East Knoyle, Hobnob press.

Clutton-Brock, J. 1978. Bones for the Zoologist. In R.H. Meadow and M.A. Zeder (eds.) *Approaches to Faunal Analysis in the Middle East,* 49-51. Peabody Museum Bulletin 2. Cambridge, Harvard University.

Clutton-Brock, J. 1981. *Domesticated animals from early times*. London, Heinemann.

Clutton-Brock, J. 1999. *A Natural History of Domesticated Mammals.* Cambridge, Cambridge University Press.

Clutton-Brock J. and Noe-Nygaard N. 1990. New osteological and C-isotope evidence on Mesolithic dogs: companions to hunters and fishers at Star Carr, Seamer Carr and Kongemose. *Journal of Archaeological Science* 17, 643-53.

Coard, R. 2000. Animal bones. In M. Bell, A. Caseldine and H. Neumann (eds.) *Prehistoric Intertidal archaeology in the Welsh Severn Estuary*. Council for British Archaeology Research Report. 120.

Colyer, R.J. 1976. *The Welsh cattle drovers*. Cardiff, University of Wales Press.

Cohen, M.N. 1977. *The food crisis in prehistory*. New Haven and London, Yale University Press.

Conneller, C. 2006. Death. In C. Conneller and G.Warren (eds.) Mesolithic Britain and Ireland: new approaches, 139-64. Stroud, Tempus.

Copley, M.S. Berstan, R. Duss, S.N. Docherty, G. Mukherjee, A.J. Straker, V. Payne, S. and Evershed, R.P. 2003. Direct chemical evidence for widespread dairying in prehistoric Britain. *Proceedings of the National Academy of Sciences USA* 100(4), 1524-9.

Copley, M.S. Berstan, R. Straker, V. Payne, S. and Evershed, R.P. 2005. Dairying in antiquity. II. Evidence from absorbed lipid residues dating to the British Bronze Age. *Journal of Archaeological Science* 32, 505-21.

Cotton, J. Elsden, N. Pipe, A. and Rayner, L. 2006. Taming the wild: a final Neolithic/earlier Bronze Age aurochs deposit from west London. In D. Serjeantson and D.Field (eds.) *Animals in the Neolithic of Britain and Europe*, 149-67. Oxford, Oxbow.

Craig, O.E. Chapman, J. Heron, C. Willis, L.H. Bartosiewicz, L. Taylor, G. Whittle, A. and Collins, M. 2005. Did the first farmers of central and eastern Europe produce dairy foods? *Antiquity* 79, 882-84.

Cymbron, T. Freeman, A.R. Malheiro, M.I. Vigne, J-D. and Bradley, D.G. 2005. Microsatellite diversity suggests different histories for Mediterranean and northern European cattle populations. *Proceedings of the Royal Society B* 272, 1837-43.

Darvill, T. 1987. *Prehistoric Britain.* London, Routledge.

Davidson, I. 1989. Escaped domestic animals and the introduction of agriculture to Spain. In J.Clutton-Brock (ed.) *The Walking Larder: Patterns of Domestication, Pastoralism and Predation.* 59-71. London, Allen and Unwin.

Davis, S.J.M. 1981. The effects of temperature change and domestication on the body size of Late Pleistocene to Holocene mammals of Israel. *Paleobiology* 7(1), 101-14.

Davis, S.J.M. 1987. *The archaeology of animals.* London, Batsford.

Davis, S.J.M. 1991. When and why did prehistoric people domesticate animals? Some evidence from Israel and Cyprus. In O.Bar-Yosef and F.R.Valla (eds.) *The Natufian culture in the Levant*, 381-90. Ann Arbour, International Monographs in Prehistory, archaeological series 1.

Davis, S.J.M. 1992. *A rapid method for recording information about mammal bones from archaeological sites.* London, Historic Buildings and Monuments

Commission, Ancient Monuments Laboratory report 19/92.

Davis, S.J.M. 1996. Measurements of a group of adult female Shetland sheep from a single flock: a baseline for zooarchaeologists. *Journal of Archaeological Science* 23, 593-612.

Davis, S.J.M. 2000. The effect of castration and age on the development of the Shetland sheep skeleton and a metric comparison between bones of males, females and castrates. Journal of Archaeological Science 27(5), 373-90.

Davis, S.J.M. 2005. Why domesticate food animals? Some zoo-archaeological evidence from the Levant. *Journal of Archaeological Science* 32(9), 1408-16.

Davis, S.J.M. and Payne, S. 1993. A barrow full of cattle skulls. *Antiquity* 67, 12-22.

Day, S.P. and Mellars, P. 1994. 'Absolute' dating of Mesolithic human activity at Star Carr, Yorkshire: new palaeoecological studies and identification of 9600BP radiocarbon 'plateau'. *Proceedings of the Prehistoric Society* 60, 417-22.

Degerbøl, M. 1963. Prehistoric cattle in Denmark and adjacent areas. In A.E.Mourant and F.E.Zeuner (eds.) *Man and cattle*, 69-79. Royal Anthropological Institute Occasional Paper 18. London, Royal Anthropological Institute.

Degerbøl, M. and Fredskild, B. 1970. The urus (*Bos primigenius* bojanus) and Neolithic domesticated cattle (*Bos taurus domesticus* Linné) in Denmark. Copenhagen, Kommissionaer: Munksgaard.

Delcroix, I. Mauget, R. Signoret, J.P. 1990. Existence of synchronization of reproduction at the level of the social group of the European wild boar (*Sus scrofa*). *Journal of reproduction and fertility* 89, 613-7.

Dennell, R. 1983. *European economic prehistory, a new approach.* London and New York, Academic Press.

Dobney, K. Ervynck A. Albarella, U. and Rowley-Conwy, P. 2004. The chronology and frequency of a stress marker (linear enamel hypoplasia) in recent and archaeological populations of *Sus scrofa*, and the effects of early domestication. *Journal of Zoology* 264,197-208.

Dobney, K. and Ervynck, A. 1998. A Protocol for Recording Linear Enamel Hypoplasia on Archaeological Pig Teeth. *International Journal of Osteoarchaeology* 8, 263-73.

Dobney, K. Ervynck, A. Albarella, U. Rowley-Conwy, P. 2007. The transition from wild boar to domestic pig in Eurasia, illustrated by a tooth developmental defect and biometrical data. In U. Albarella, K. Dobney, A. Ervynck and P. Rowley-Conwy (eds.) *Pigs and Humans; 10,000 years of interaction*,57-82. Oxford, Oxford University Press.

Domańska, L. 1998. The initial stage of food-production in the Polish lowlands – the Dęby 29 site. In M. Zvelebil, R. Dennell and L. Domańska (eds.) *Harvesting the Sea, Farming the Forest. The Emergence of Neolithic Societies in the Baltic Region*, 129-34. Sheffield, Sheffield Academic Press.

Done, G. 1980. The animal bone. In D. Longley (ed.) *Runnymede Bridge 1976: excavations on the site of a Late Bronze Age settlement*, 74-9. Guildford, Surrey Archaeological Society research volume 6.

Done, G. 1991. The animal bone. In S. Needham (ed.) *Excavation and Salvage at Runnymede Bridge*, 1978, 327-42. London, British Museum Press.

Driesch, A. von den.1976. *A guide to the measurement of animal bones from archaeological sites.* Cambridge Mass., Harvard University, Peabody Museum Bulletin 1.

Dudd, S.N. Evershed, R.P. Gobson, A.M. 1999. Evidence for varying patterns of exploitation of animal products in different prehistoric pottery traditions based on lipids preserved in surface and absorbed residues. *Journal of Archaeological Science* 26, 1473-82.

Ducos, P. 1978. "Domestication" defined and methodological approaches to its recognition in faunal assemblages. In R.H. Meadow and M.A. Zeder (eds.) *Approaches to Faunal Analysis in the Middle East,* 53-6. Peabody Museum Bulletin 2. Cambridge, Harvard University.

Ducos, P. 1989. Defining domestication: a clarification. In J. Clutton-Brock (ed.) *The Walking Larder: Patterns of Domestication, Pastoralism and Predation.* 28-30. London, Allen and Unwin.

Ducos, P. and Kolska Horwitz, L.R. 1998. The influence of climate on artiodactyl size during the Late Pleistocene-Early Holocene of the Levant. *Paléorient* 23(23-2), 229-47.

Dwyer, P.D. 1996. Boars, barrows and breeders: the reproductive status of domestic pig populations in mainland New Guinea. *Journal of Anthropological Research* 52, 481-500.

Dyer, C. 1996. Seasonal settlement in Medieval Gloucestershire: sheepcotes. In H.S.A. Fox (ed.) *Seasonal Settlement*, 25-33. Vaughan Paper No.39. Leicester, University of Leicester.

Edwards, A. 1994. *The Late Neolithic faunal assemblage from West Kennet palisade enclosure.* Unpublished MSc thesis, University of Sheffield.

Edwards, C. Bollongino, R. Scheu, A. Chamberlain, A. Tresset, A. Vigne, J-D. Baird, J.F. Larson, G. Ho, S.Y.W. Heupink, T.H. Shapiro, B. Freeman, A.R. Thomas, M.G. Arbogast, R-M. Arndt, B. Bartosiewicz, L. Benecke, N. Budja, M. Chaix, L. Choyke, A. Coqueugniot, E. Döhle, H-J. Göldner, H. Hartz, S. Helmer, D. Herzig, B. Hongo, H. Mashkour, M. Özdogan, M. Pucher, E. Roth, G. Schade-Lindig, S. Schmölcke, U. Schulting, R.J. Stephan, E. Uerpmann, H-P. Vörös, I. Voytek, B. Bradley, D.G. and Burger, J. 2007. Mitochondrial DNA analysis shows a Near Eastern Neolithic origin for domestic cattle and no indication of domestication of European aurochs. *Proceedings of the Royal Society B* 247, 1377-85.

Edwards, A. and Horne, M. 1994. Animal bone. In A.W.R.Whittle (ed.) *Sacred Mound, Holy Rings: Silbury Hill and the West Kennet Palisade Enclosures – A later Neolithic complex in North Wiltshire*, 117-29. Oxford, Oxbow.

Ellis, C.J. Allen, M.J. Gardiner, J. Harding, P. Ingrem, C. Powell, A. Scaife, R.G. Gale, R. and Heathcote, J. 2003. An Early Mesolithic seasonal hunting site in the Kennet Valley, Southern England. *Proceedings of the Prehistoric Society* 69, 108-35.

Ervynck, A. Dobney, K, Hongo, H. and Meadow, R. 2001. Born Free? New Evidence for the status of *Sus scrofa* At Neolithic Çayönü Tepesi (South Eastern Anatolia, Turkey). *Paleorient* 27(2), 47-73.

Evans, J. Chenery, C.A. Fitzpatrick, A.P. 2006a. Bronze Age childhood migration of individuals near Stonehenge, revealed by strontium and oxygen isotope tooth enamel analysis. *Archaeometry* 48(2), 309-21.

Evans, J. Montgomery, J. And Wildman, G. 2009. Isotope domain mapping of $^{87}Sr/^{86}Sr$ biosphere variation on the Isle of Skye, Scotland. *Journal of the Geological Society* 166(4), 617-31.

Evans, J.A. Montgomery, J. Wildman, G. and Boulton, N. 2010. Spatial variations in biosphere $^{87}Sr/^{86}Sr$ in Britain. *Journal of the Geological Society* 167, 1-4.

Evans, J.A. Stoodley, N. And Chenery, C. 2006b. A strontium and oxygen isotope assessment of a possible fourth century immigrant population in Hampshire cemetery, southern England. *Journal of Archaeological Science* 33, 265-72.

Evans, J.A. Tatham, S. Chenery, S.R. Chenery, C.A. 2007. Anglo-Saxon animal husbandry techniques revealed through isotope and chemical variations in cattle teeth. *Applied Geochemistry* 22, 1994-2005.

Evershed, R.P. Payne, S. Sherratt, A.G. Copley, M.S. Coolidge, J. Urem-Kotsou, D. Kotsakis, K. Özdoğan, M. Özdoğan, A.E. Nieuwenhuyse, O. Akkermans, P.M.M.G. Bailey, D. Andeescu, R-R. Campbell, S. Farid, S. Hodder, I. Yalman, N. Özbaşaran, M. Biçakci, E. Garfinkel, Y. Levy, T. and Burton, M.M. 2008. Earliest date for milk use in the Near East and southeastern Europe linked to cattle herding. *Nature* 455, 528-31.

Fernández-Llario, P. and Carranza, J. 2000. Reproductive performance of the wild boar in a Mediterranean ecosystem under drought conditions. *Ethology ecology and evolution* 12, 335-43.

Fernández-Llario, P. and Mateos-Quesada, P. 2005. Influence of rainfall on the breeding biology of wild boar (*Sus scrofa*) in a Mediterranean ecosystem. *Folia Zoologica* 54(3), 240-8.

Field, N.H. Matthews, C.L. Smith, I.F. and Ewbank, J.M. 1964. New Neolithic sites in Dorset and Bedfordshire, with a note on the distribution of Neolithic storage pits in Britain. *Proceedings of the Prehistoric Society* 15, 352-81.

Fonseca, C. Santos, P. Monzón, A. Bento, P. Alves da Silva, A. Alves, A. Silvério, A. Soares, A.M.V.M. Petrucci-Fonseca, F. 2004. Reproduction in the wild boar (*Sus scrofa* LINNAEUS, 1758) populations of Portugal. *Galemys* 16, 53-65.

Fox, H.S.A. (ed.) 1996. *Seasonal settlement.* Vaughan Paper No.39. Leicester, University of Leicester.

Fraser, F.C. and King, J.E. 1954. Faunal remains. In J.G.D.Clark (ed.) *Excavations at Star Carr*, 70-95. Cambridge, Cambridge University Press.

Fricke, H.C. and O'Neil, J.R. 1996. Inter- and intra-tooth variation in the oxygen isotope composition of mammalian tooth enamel phosphate: implications for palaeoclimatological and palaeobiological research. *Palaeogeography, Palaeoclimatology, Palaeoecology* 126, 91-9.

Friis-Hansen, J. 1990. Mesolithic cutting arrows: functional analysis of arrows used in the hunting of large game. *Antiquity* 64, 494-504.

Froom, F.R. 1972. A Mesolithic site at Wawcott, Kintbury. *BerkshireArchaeological Journal* 66, 23-44.

Froom, F.R. 1976. *Wawcott III: A stratified Mesolithic succession.* British Archaeological Reports 27.

Geddes, D.S. 1983. Neolithic transhumance in the Mediterranean Pyrenees. *World Archaeology* 15(1), 51-66.

Geddes, D. 1985. Mesolithic domestic sheep in west Mediterranean Europe. *Journal of Archaeological Science* 12, 25-48.

Geisser, H. Reyer, H-U. 2005. The influence of food and temperature on population density of wild boar *Sus scrofa* Thurgau (Switzerland). *Journal of Zoology* 267, 89-96.

Gentry A, Clutton-Brock J and Groves C P. 2004. The naming of wild animal species and their domestic derivatives. *Journal of Archaeological Science* 31. 645-51.

Giuffra, E. Kijas, J.M.H. Amarger, V. Carlborg, O. Jeon, J.T. and Andersson, L. 2000. The origin of the domestic pig: independent domestication and subsequent introgression. *Genetics* 154(4), 1785-91.

Glass, M. 1991. *Animal production systems in Neolithic central Europe*. Oxford, British Archaeological Reports, International Series 572.

Götherström, A. Anderung, C. Hellborg, L. Elburg, R. Smith, C. Bradley, D.G. Ellegren, H. 2005. Cattle domestication in the Near East was followed by hybridization with aurochs bulls in Europe. *Proceedings of the Royal Society B* 272, 1-6.

Goulding, M. 2003. *Wild boar in Britain*. Wiltshire, Whittet Books.

Goulding, M. Kitchener, A.C. and Yalden, D. 2008. Family Suidae (Pigs). In S. Harris and D. Yalden (eds.) *Mammals of the British Isles*, 561-564. Southampton, The Mammal Society.

Grant, A. 1978. Variation in dental attrition in mammals and its relevance to age estimation. In D. R. Brothwell, K. D. Thomas, J. Clutton-Brock (eds.) *Research problems in zooarchaeology*, 103-6. London, Institute of Archaeology Occasional Paper.

Grant, A. 1982. The use of tooth wear as a guide to the age of domestic ungulates. In B. Wilson, C. Grigson and S. Payne (eds.) *Ageing and sexing animal bones from archaeological sites*, 91-108. Oxford, British Archaeological Reports, British Series 109.

Graves, H.B. 1984. Behaviour and ecology of wild and feral swine (*Sus scrofa*). *Journal of Animal Science* 58, 482-92.

Greenfield, H.J. 1988. The origin of milk and wool production in the Old World. *Current Archaeology* 29(4), 573-92.

Genov, P.V. 1999. A review of the cranial characteristics of the wild boar (*Sus scrofa* Linnaeus 1758), with systematic conclusions. *Mammal Review* 29(4), 205-38.

Grigson, C. 1965. Measurements of bones, horncores, antlers and teeth. In I.F. Smith (ed.) *Windmill Hill and Avebury, excavations by Alexander Keiller 1925-1939*, 145-67. Oxford, Clarendon press.

Grigson, C. 1969. The uses and limitations of differences in absolute size in the distinction between the bones of aurochs (*Bos primigenius*) and domestic cattle (*Bos taurus*). In P.J. Ucko and G.W. Dimbleby (eds.) *The domestication and exploitation of plants and animals,* 277-94. Chicago, Aldine Publishing Company.

Grigson, C. 1976. The animal bones from Neolithic pit 6. In C.L. Matthews (ed.) *Occupation sites on a Chiltern ridge. Part 1: Neolithic, Bronze Age and Early Iron Age*, 11-8. British Archaeological Report 29.

Grigson, C. 1982a. Sex and Age Determination of Some Bones and Teeth of Domestic Cattle: A Review of the Literature. In B. Wilson, C. Grigson and S. Payne (eds.) *Ageing and Sexing Animal Bones from Archaeological Sites,* 7-23. Oxford, British Archaeological Reports, British Series 109.

Grigson, C. 1982b. Porridge and pannage: pig husbandry in Neolithic England. In M.Bell and S.Limbrey (eds.) *Aspects of woodland ecology*, 291-313. British Archaeological Reports International Series 146.

Grigson, C. 1989. Size and sex: Evidence for the domestication of cattle in the Near East. In A. Milles, D. Williams and N. Gardner (eds.) *The beginnings of Agriculture*, 77-109. Oxford, British Archaeological Reports, International Series 496.

Grigson, C. 1999. The mammalian remains. In A.Whittle, J.Pollard and C.Grigson (eds.) *The harmony of symbols: The Windmill Hill causewayed enclosure*, 164-252. Oxford, Oxbow.

Groves, C. 1981. *Ancestors for the pigs: taxonomy and phylogeny of the genus* Sus. Technical bulletin no.3. Canberra, Australian National University.

Groves, C. 2007. Current views on taxonomy and zoogeography of the genus *Sus*. In U. Albarella, K. Dobney, A. Ervynck and P. Rowley-Conwy (eds.) *Pigs and Humans; 10,000 years of interaction*, 15-29. Oxford, Oxford University Press.

Guintard, C. 1999. On the size of the ure-ox or aurochs (*Bos primigenius* BOJANUS, 1827). In G-C. Weniger (ed.) *Archäologie und biologie des auerochsen*, 7-21. Mettmann, Neanderthal Museum, Wissenschafliche Schriften 1.

Haber, A. and Dayan, T. 2004. Analyzing the process of domestication: Hagoshrim as a case study. *Journal of Archaeological Science* 31, 1587-601.
Haber, A., Dayan, T. and Getzov, N. 2005. Pig exploitation at Hagoshrim; a prehistoric site in the southern Levant. In J.-D. Vigne, J. Peters and D. Helmer (eds.) *First Steps of Animal Domestication; New archaeozoological approaches,* 80-5. Oxford, Oxbow Books.

Hadjikoumis, A. 2010. *The origins and evolution of pig domestication in prehistoric Spain.* Unpublished thesis, University of Sheffield.

Hall, S.J.G. 2008. A comparative analysis of the habitat of the extinct aurochs and other prehistoric mammals in Britain. *Ecography* 31(2), 187-90.

Hall, S.J.G. and Moore, C.F. 1986. Feral cattle of Swona, Orkney Islands. *Mammal Review* 16(2), 89-96.

Hall, S.J. and Moore, C.F. 1987. The feral cattle of Swona, Orkney Islands. *Animal Genetic Resources Information* 6, 1-10.

Hamilton, J. Hedges, R.E.M. Robinson, M. 2009. Rooting for pigfruit: pig feeding in Neolithic and Iron Age Britain compared. *Antiquity* 83, 998-1011.

Hammon, A. 2005. Late Romano-British – early medieval socio-economic and cultural change: Analysis of the mammal and bird bone assemblages from the Roman city of *Viroconium Cornoviorum*, Shropshire: Volume 1: Text and Volume 2: Figures. Unpublished PhD thesis, University of Sheffield.

Harcourt, R.A. 1971. The animal bones. In G.J. Wainwright and I.H. Longworth (eds.) *Durrington Walls: Excavations 1966-1968*, 338-50. London, The Society of Antiquaries of London.

Harcourt, R.A. 1979. The animal bones. In Wainwright, G.J. 1979. *Mount Pleasant, Dorset: Excavations 1970-1971,* 214-23. London, the society of antiquaries of London.

Harris, M. 1989. *Cows, pigs, wars and witches: the riddles of culture.* New York, Vintage books.

Healey, F. Heaton, M. Lobb, S.J. 1992. Excavations of a Mesolithic Site at Thatcham, Berkshire. *Proceedings of the Prehistoric Society* 58, 41-76.

Hedges, R.E.M. Saville, A. and O'Connell, T. 2008. Characterizing the diet of individuals at the Neolithic chambered tomb of Hazleton North, Gloucestershire, England, using stable isotopic analysis. *Archaeometry* 50(1), 114-28.

Helmer, D. Gourichon, L. Monchot, H. Peters, J and Saña Segui, M. 2005. Identifying early domestic cattle from pre-pottery Neolithic sites on the Middle Euphrates using sexual dimorphism. In J-D. Vigne, J. Peters and D. Helmer (eds.) *The first steps of animal domestication*, 86-95. Oxford, Oxbow.

Helmer, D. and Vigne, J-D. 2007. Was milk a "secondary product" in the Old World Neolithisation process? Its role in the domestication of cattle, sheep and goats. *Anthropozoologica* 42(2), 9-40.

Herraro, J. García-Serrano, A. Couto, S. Ortuño, V.M. García-González, R. 2006. Diet of wild boar *Sus scrofa* L. and crop damage in an intensive agroecosystem. *European Journal of wildlife research* 52, 245-50.

Herring, P. 1996. Transhumance in Medieval Cornwall. In H.S.A. Fox (ed.) *Seasonal settlement*, 35-44. Vaughan Paper No.39. Leicester, University of Leicester.

Hesse, B. 1982. Slaughter patterns and domestication: The beginnings of pastoralism in western Iran. *Man new series* 17(3), 403-17.

Hey, G. and Barclay, A. 2007. The Thames Valley in the late fifth and early fourth millennium cal BC: the appearance of domestication and the evidence for change. In A. Whittle and V. Cummings (eds.) *Going over: The Mesolithic-Neolithic transition in north-west Europe*, 399-422. Oxford, Oxford University Press.

Hey, G. Mulville, J. and Robinson, M. 2003. Diet and culture in southern Britain: the evidence from Yarnton. In M. Parker Pearson (ed.) *Food, culture and identity in the Neolithic and Early Bronze Age*, 79-88. British Archaeological Report international series 1117.

Higgs, E.S. and Jarman, M.R. 1972. The origins of animal and plant husbandry. In E.S. Higgs (ed.) *Papers in economic prehistory*, 3-12. Cambridge, Cambridge University Press.

Higham, C.F.W. 1969. Towards an economic prehistory of Europe. *Current Anthropology* 10(2/3), 139-50.

Hodder, I. 1990. *The domestication of Europe*. Oxford, Basil Blackwell.

Hongo, H. and Meadow, R.H. 1998. Pig exploitation at Neolithic Çayönü Tepesi (southeastern Anatolia). In S. Nelson (ed.) *Ancestors for the Pigs: Pigs in prehistory*, 77-98. MASCA Research Papers in Science and Archaeology 15.

Hoppe, K.A. Koch, P.L. Furutani, T.T. 2003. Assessing the preservation of biogenic strontium in fossil bones and tooth enamel. *International Journal of Osteoarchaeology* 13, 20-28.

Horne, M. 1994. A study of fauna from West Kennet palisade enclosure 2 – a Late Neolithic monument. University of Sheffield, unpublished MSc. thesis.
Howe, T.D. and Bratton, S.P. 1976. Winter rooting activity of the European wild boar in the Great Smoky Mountains National Park. *Castanea* 41(3), 256-64.

Ingrem, C. and Powell, A. 1999. *Animal bone*. Unpublished report, Wessex Archaeology.

Jarman, M.R. 1976. Early animal husbandry. *Philosophical transactions of the Royal Society of London B* 275, 85-97.

Jarman, M.R. and Wilkinson, P.F. 1972. Criteria of animal domestication. In E.S. Higgs (ed.) *Papers in economic prehistory*, 83-96. Cambridge, Cambridge University Press.

Jewell, P. 1963. Cattle from British archaeological sites. In A.E. Mourant and F.E. Zeuner (eds.) *Man and cattle*, 80-91. London, The Royal Anthropological Institute.

Jope, M. 1965. Faunal remains. In I.F. Smith (ed.) *Windmill Hill and Avebury, excavations by Alexander Keiller 1925-1939*, 142-45. Oxford, Clarendon press.

Kienlin, T.L. and Valde-Nowak, P. 2004. Neolithic transhumance in the Black Forest mountains, sw Germany. *Journal of Field Archaeology* 29(1/2), 29-44.

King, J.E. 1962. Report on animal bones. In J. Wymer Excavations at the Maglemosian sites at Thatcham, Berkshire. *Proceedings of the Prehistoric Society* 13, 355-61.

Kitchener, A.C. Bonsall, C. and Bartosiewicz, L. 2004. Missing mammals from Mesolithic middens: a comparison of the fossil and archaeological records from Scotland. In A. Saville (ed.) *Mesolithic Scotland and its neighbours*, 73-82. Edinburgh, Society of Antiquaries of Scotland.

Krasiński, A.Z. and Raczyński, J. 1967. The reproduction biology of European bison living in reserves and in freedom. *Acta Theriologica* 12(19-35), 407-44.

Krause-Kyora, B. Makarewicz, C. Evin, A. Girdland-Flink, L. Dobney, K. Larson, G. Hartz, S. Schreiber, S. von Carnap-Bornheim, C. von Wurmb-Schwark, N. Nebel, A. 2013. Use of domesticated pigs by Mesolithic hunter-gatherers in northwest Europe. *Nature Communications* 4:2348 doi: 10.1038/ncomms3348.

Kyselý, R. 2008. Aurochs and potential crossbreeding with domestic cattle in Central Europe in the Eneolithic period. *Anthropozoologica* 43(2), 7-37.

Larson, G. Albarella, U. Dobney, K. Rowley-Conwy, P. Schibler, J. Tresset, A. Vigne, J-D. Edwards, C. J. Schlumbaum, A. Dinu, A. Bălăçsescu, A. Dolman, G. Tagliacozzo, A. Manaseryan, N. Miracle, P. Van Wijngaarden-Bakker, L. Masseti, M. Bradley, D.G. Cooper, A. 2007. Ancient DNA, pig domestication, and the spread of the Neolithic into Europe. *Proceedings of the National Academy of Sciences USA* 104(39), 15276-81.

Larson, G. Dobney, K. Albarella, U. Fang, M. Matisoo-Smith, E. Robins, J. Lowden, S. Finlayson, H. Brand, T. Willerslev, E. Rowley-Conwy, P. Andersson, L. and Cooper, A. 2005. Worldwide phylogeography of wild boar reveals multiple centers of domestication. *Science* 307, 1618-21.

Larson, G. Albarella, U. Dobney, K. Rowley-Conwy, P. 2007. Current views on *Sus* phylogeography and pig domestication as seen through modern mtDNA studies. In U. Albarella, K. Dobney, A. Ervynck and P. Rowley-Conwy (eds.) *Pigs and Humans; 10,000 years of interaction*, 30-41. Oxford, Oxford University Press.

Lasota-Moskalewska, A. 1998. Archaeozoological reconstruction of the animal husbandry at Dęby in Kujavia, Poland. In M. Zvelebil, R. Dennell and L. Domańska (eds.) *Harvesting the Sea, Farming the Forest. The Emergence of Neolithic Societies in the Baltic Region*, 135-6. Sheffield, Sheffield Academic Press.

Lasota-Moskalewska, A. Kobryń, H. and Swiezynski, K. 1987. Changes in the size of the domestic and wild pig in the territory of Poland from the Neolithic to the Middle Ages. *Acta Theriologica* 32(5), 51-81.

Legge, A.J. 1975. The fauna of Tell Abu Hureyra: preliminary analysis. In A.M.T. Moore (ed.) The excavation of Tell Abu Hureyra in Syria. *Proceedings of the Prehistoric Society* 41, 74-77.

Legge, A.J. 1981. Aspects of cattle husbandry. In R. Mercer (ed.) *Farming practice in British prehistory*, 169-81. Edinburgh, Edinburgh University Press.

Legge, A.J. 1996. The beginning of caprine domestication in Southwest Asia. In D.R. Harris (ed.) *The origins and spread of agriculture and pastoralism in Eurasia, 238-62*. London, University College London Press.

Legge, T. 2005. Milk use in prehistory: the osteological evidence. In J. Mulville and A. Outram (eds.) *The zooarchaeology of milk and fats*, 8-13. Oxford, Oxbow.

Legge, A.J. 2009. Livestock and Neolithic society at Hambledon Hill. In R. Mercer and F. Healy (eds.) *Hambledon Hill, Dorset, England. Excavation and survey of a Neolithic monument complex and its surrounding landscape, volume 2*, 536-86. Swindon, English Heritage.

Legge, A. and Rowley-Conwy, P. 1988. *Star Carr revisited: a re-analysis of the large mammals*. London, The archaeological laboratory centre for extra-mural studies.

Legge, T. and Rowley-Conwy, P. 1989. Some preliminary results of a re-examination of the Star Carr fauna. In C. Bonsall (ed.) *The Mesolithic in Europe*, 225-30. Edinburgh, John Donald.

Levitan, B. 1990. The non-vertebrate remains. In A. Saville (ed.) *Hazleton North, Gloucestershire, 1979-82. The excavation of a Neolithic long cairn of the Cotswold-Severn group*, 199-214. English Heritage Archaeological report 13. London, Historic Buildings and Monuments Commission for England.

Levitan, B. and Serjeantson, D. 1999. Animal Bone. In A. Barclay and C. Halpin (eds.) *Excavations at Barrow Hills, Radley, Oxfordshire. Volume 1: The Neolithic and Bronze Age monument complex*, 236-41. Oxford, Oxford Archaeological Unit.

Liddle, J. 1998. *A re-evaluation of the faunal remains from West Kennet Long Barrow, Wiltshire*. Unpublished MSc thesis, University of Sheffield.

Lister, A. M. 1996. Dwarfing in island elephants and deer: processes in relation to time and isolation. Symposia of the Zoolological Society of London 69, 277–92.

Lomolino, M.V. 1985. Body size of mammals on islands. The island rule reconsidered. *The American naturalist* 125(2), 310-16.

Louwe-Kooijmans, L.P.L. and Jongste, P.F.B. 2006. Schipluiden, a Neolithic settlement on the Dutch North Sea coast c.3500 cal BC. *Analecta Praehistorica Leidensia*, 37-38.

Lucas, A.T. 1989. *Cattle in ancient Ireland.* Kilkenny, Ireland, Boethius Press.

Lyman, R. L. 1994 *Vertebrate Taphonomy*. Cambridge. Cambridge University Press.

Lynch, A. 1981. *Man and Environment in S.W. Ireland.* Oxford, British Archaeological Reports, British Series 85.

Lynch, F. Aldhouse-Green, S. and Davies, J.L. 2000. *Prehistoric Wales*. Gloucestershire, Sutton Publishing.

Lynch, T. Hamilton, J. and Hedges, R.E.M. 2008. Where the wild things are: aurochs and cattle in England. *Antiquity* 82, 1025-39.

Madgwick, R. Mulville, J. and Evans, J. 2012. Investigating diagenesis and the suitability of porcine enamel for strontium (87Sr/86Sr) isotope analysis. *Journal of Analytical Atomic Spectrometry* 27(5), 733-42.

Maillard, D. and Fournier, P. 2004. Timing and synchrony of births in wild boar (*Sus scrofa* LINNAEUS, 1758) in a Mediterranean habitat: the effect of food availability. *Galemys* 16, 67-74.

Maltby, M. 1990. The exploitation of animals in the Stonehenge environs in the Neolithic and Bronze Age. In J. Richards (ed.) *The Stonehenge environs project*, 247-50. Swindon, English Heritage Archaeological Report 16.

Malcolmson, R. and Mastoris, S. 2001. *The English pig: a history*. London, New York, Hambledon and London.

Maroo, S. and Yalden, D. 2000. The Mesolithic mammal fauna of Great Britain. *Mammal Review* 30(3 and 4), 243-48.

Massei, G. and Genov, P.V. 2004. The environmental impact of wild boar. *Galemys* 16, 135-45.

Matschke, G.H. 1967. Ageing European wild hogs by dentition. *Journal of Wildlife Management* 31(1). 109-113.

Mayer, J.J. and Brisbin, I.L. 1988. Sex identification of *Sus scrofa* based on canine morphology. *Journal of Mammalogy* 69(20), 408-12.

Mayer, J.J. and Brisbin, I.L. 1991. *Wild pigs in the United States*. Athens, London, Georgia University Press.

Mayer, J. J. Novak, J.M. and Brisbin, I.L. 1998 Evaluation of molar size as a basis for distinguishing wild boar from domestic swine: employing the present to decipher the past. In S. Nelson (ed.) *Ancestors for the Pigs: Pigs in prehistory*, 39-53. MASCA Research Papers in Science and Archaeology 15.

McCarthy, M. 1999. Faunal remains. In P. Woodman, E. Anderson and N. Finlay (eds.) *Excavations at Ferriter's Cove, 1983-95*, 85-92. Bray, Wordwell.

McCormick, F. 1992. Early faunal evidence for dairying. *Oxford Journal of Archaeology* 11(2), 201-10.

McCormick, F. 2004. Hunting wild pig in the Late Mesolithic. In H. Roche, E. Grogan, J. Bradley, J. Coles and B. Raftery (ed.) *From megaliths to metal, essays in honour of George Eogan*, 1-5. Oxford, Oxbow.

McFadyen, L. 2007. Mobile spaces of Mesolithic Britain. *Home Cultures* 4, 117-28.

Meadow, R. H. 1989. Osteological evidence for the process of animal domestication. In J. Clutton-Brock (ed.) *The Walking Larder: Patterns of Domestication, Pastoralism and Predation*. 80-90. London, Allen and Unwin.

Meadow, R.H. 1999. The use of size index scaling techniques for research on archaeozoological collections from the Middle East. In C. Becker, H. Manhart, J. Peters and J. Schibler (eds.) *Historie Animalium ex Ossibus*, 285-300. Rahden, Verlag Marie Leidorf.

Meadows, J. Barclay, A. and Bayliss, A. 2007. A short passage of time: The dating of the Hazleton long cairn revisited. *Cambridge Archaeological Journal* 17 (1) supplement, 45-64.

Mellars, P. (ed.) 1998. *Star Carr in context.* Cambridge, McDonald Institute Monographs.

Mercer, R. 1980. *Hambledon Hill a Neolithic Landscape*. Edinburgh, Edinburgh University Press.

Mercer, R. and Healy, F. (eds.) 2009. *Hambledon Hill, Dorset, England. Excavation of a Neolithic monument complex and its surrounding landscape volume 1 and 2*. Swindon, English Heritage.

Milner, N. Craig, O.E. Bailey, G.N. Pedersen, K. and Andersen, S.H. 2004. Something fishy in the Neolithic? A re-evaluation of stable isotope analysis of Mesolithic and Neolithic coastal populations. *Antiquity* 78(299), 9-22.

Milisauskas, S. and Kruk, J. 1991. Utilization of cattle for traction during the later Neolithic in southeastern Poland. *Antiquity* 65, 562-6.

Mona, S. Randi, E. and Tommaseo-Ponzetta, M. 2007. Evolutionary history of the genus *Sus* inferred from cytochrome b sequences. *Molecular phylogenetics and evolution* 45, 757-62.

Montgomery, J. Budd, P. And Evans, J. 2000. Reconstructing the lifetime movements of ancient people: a Neolithic case study from southern England. *European Journal of Archaeology* 3(3), 370-85.

Montgomery, J. Evans, J.E. and Wildman, G. 2006. $^{87}Sr/^{86}Sr$ isotope composition of bottled British mineral waters for environmental and forensic purposes. *Applied Geochemistry* 21(10), 1626-34.

Moran, N.C. and O'Connor, T.P. 1994. Age attribution in domestic sheep by skeletal and dental maturation: a pilot study of available sources. *International Journal of Osteoarchaeology* 4, 267-85.

Moore, A.M.T. 2003. The Abu Hureyra project: investigating the beginning of farming in western Asia. In A.J. Ammerman and P. Biagi (eds.) *The widening harvest, the Neolithic transition in Europe: looking back, looking forward*, 59-74. Boston, Massachusetts, Archaeological Institute of America.

Mukherjee, A.J. Copley, M.S. Berstan, R. Clark, K.A. and Evershed, R.P. 2005. Interpretation of δ13C values of fatty acids in relation to animal husbandry, food processing and consumption in prehistory. In J. Mulville and A. Outram (eds.) *The zooarchaeology of milk and fats*, 77-93. Oxford, Oxbow.

Mulville, J. and Grigson, C. 2007. The animal bones. In D. Benson and A.Whittle (eds.) *Building memories: The Neolithic Cotswold long barrow at Ascott-Under-Wychwood, Oxfordshire*, 237-53. Oxford, Oxbow.

Needham, S. 1991. Radiocarbon dating. In S. Needham (ed.) *Excavation and Salvage at Runnymede Bridge*, 1978, 345-7. London, British Museum Press.

Needham, S. and Spence, T. 1996. *Refuse and disposal at area 16 east, Runnymede*. London, British Museum Press.

Nehlich, O. Montgomery, J. Evans, J. Schade-Lindig, S. Pichler, S.L. Richards, M.P. and Alt, K.W. 2009. Mobility of migration: a case study from the Neolithic settlement of Nieder-Mörlen (Hessen, Germany). *Journal of Archaeological Science* 36, 1791-9.

Nelson, B.K. Deniro, M.J. Schoeninger, MJ. De Paolo, D.J. and Hare, P.E. 1986. Effects of diagenesis on strontium, carbon, nitrogen and oxygen concentration and isotopic composition of bone. *Geochemica et Cosmochimica Acta* 50(9), 1941-9.

Nemeth, D.J. 1998. Privy-pigs in prehistory? A Korean analog for Neolithic Chinese subsistence practices. In S. Nelson (ed.) *Ancestors for the pigs: Pigs in prehistory*, 11-25. The Museum Applied Science Center for Archaeology (MASCA).

Noddle, B. 1989. Cattle and sheep in Britain and northern Europe up to the Atlantic period: a personal viewpoint. In A. Milles, D. Williams and N. Gardner (eds.) *The beginnings of agriculture*, 179-202. Oxford, British archaeological reports international series 496.

Noe-Nygaard, N. 1974. Mesolithic hunting in Denmark illustrated by bone injuries caused by human weapons. *Journal of archaeological science* 1(3), 217-24.

Noe-Nygaard, N. 1983. A new find of Brown Bear (*Ursus arctos*) from Star Carr and other finds in the Late Glacial and Post Glacial of Britain and Denmark. *Journal of Archaeological Science* 10, 317-25.

Noe-Nygaard, N. Price, T.D. and Hede, S.U. 2005. Diet of aurochs and early cattle in southern Scandinavia: evidence from 15N and 13C stable isotopes. *Journal of Archaeological Science* 32, 855-71.

O'Connor, T.P. 1988. Bones from the general accident site, Tanner row. London, York Archaeological Trust.

O'Connor, T.P. 2003. *The Analysis of Urban Animal Bone Assemblages: A Handbook for Archaeologists*. The Archaeology of York Principles and Methods 19/2.

Oliver, W.L.R. Brisbin, I.L. and Takahashi, S. 1993a. The Eurasian wild pig (*Sus scrofa*). In W.L.R. Oliver (ed.) *Pigs, peccaries and hippos: status survey and conservation action plan*, 112-20. Gland, International Union for Conservation of Nature (IUCN).

Oliver, W.L.R. Groves, C.P. Cox, R. and Blouch, R.A. 1993b. Origins of domestication and pig culture. In W.L.R. Oliver (ed.) *Pigs, peccaries and hippos: status survey and conservation action plan*, 171-9. Gland, International Union for Conservation of Nature (IUCN).

Parker, A.G. Goudie, A.S. Anderson, D.E. Robinson, M.A. Bonsall, C. 2002. A review of the mid-Holocene elm decline in the British Isles. *Progress in physical geography* 26(1), 1-45.

Parker Pearson, M. 2003. Food, culture and identity: an introduction and overview. In M. Parker Pearson (ed.) *Food, culture and identity in the Neolithic and Early Bronze Age*. British Archaeological Report international series 1117.

Parker Pearson, M. Pollard, J. Richards, C. Thomas, J. Tilley, C. Welham, K. and Albarella, U. 2006. Materializing Stonehenge. *Journal of Material Culture* 11, 227-61.

Parker Pearson, M. Pollard, J. Tilley, C. Thomas, J. Richards, C and Welham, 2005. *The Stonehenge Riverside Project Interim Report 2005*. Unpublished interim report.

Parker Pearson, M. Pollard, J. Tilley, C. Thomas, J. Richards, C. Welham, K. Albarella, U. Allen, M. Cleal, R. Knüsel, C. Lewis, H. Linford, N. Payne, A. Scaife, R. and Schmidt, A. 2004. The Stonehenge Riverside project. Unpublished interim report.

Payne, S. 1973. Kill-off patterns in sheep and goats: the mandibles from Aşvan Kale. *Journal of Anatolian Studies* 23, 281-303.

Payne, S. and Bull, G. 1988. Components of variation in measurements of pig bones and teeth, and the use of measurements to distinguish wild from domestic pig remains. *Archaeozoologia* II (1.2), 27-66.

Pellegrini, M. Donahue, R.E. Chenery, C. Evans, J. Lee-Thorp, J. Montgomery, J. and Mussi, M. 2008. Faunal migration in late-glacial central Italy: implications for human resource exploitation. *Rapid communications in mass spectrometry* 22, 1714-26.

Piggott, S. 1962. *The West Kennet long barrow* excavations 1955-56. London, Her Majesty's stationery office.

Piggott, S. 1981. Early prehistory. In J. Thirsk (ed.) *The agrarian history of England and Wales, vol.1.1 prehistory*, 3-59. Cambridge, Cambridge University Press.

Pollard, J. 1999. 'These places have their moments': thoughts on settlement practices in the British Neolithic. In J. Brück and M. Goodman (eds.) *Making places in the prehistoric world*, 76-93. London, University College London Press.

Price, T.D. 2000. Europe's first farmers: an introduction. In T.D. Price (ed.) *Europe's first farmers*, 1-18. Cambridge, Cambridge University Press.

Price, T.D. 2003. The arrival of agriculture in Europe as seen from the north. In A.J. Ammerman and P. Biagi (eds.) *The widening harvest, the Neolithic transition in Europe: looking back, looking forward*, 273-94. Boston, Archaeological Institute of America.

Price, T.D. Bentley, A. Lüning, J. Gronenborn, D. and Wahl, J. 2001. Prehistoric human migration in the Linearbandkeramik of central Europe. *Antiquity* 75, 593-603.

Price, T.D. Burton, J.H. Bentley, R.A. 2002. The characterization of biologically available strontium isotope ratios for the study of prehistoric migration. *Archaeometry* 44(1), 117-35.

Price, T.D. and Gestdóttir, H. 2005. The first settlers of Iceland: an isotopic approach to colonisation. *Antiquity* 80, 130-44.

Price, T.D. Grupe, G. Schröter, P. 1994. Reconstruction of migration patterns in the Bell Beaker period by stable strontium isotope analysis. *Applied Geochemistry* 9, 413-7.

Prummel, W. 1987. The faunal remains from the Neolithic site of Hekelingen III. *Helinium* 27(2), 190-258.

Randi, E. Lucchini, V. and Diong, C.H. 1996. Evolutionary genetics of the suiforms as reconstructed using mtDNA sequencing. *Journal of mammalian evolution* 3(2), 163-94.

Rappaport, R.A. 1968. *Pigs for the ancestors: Ritual ecology of a New Guinea People*. New Haven and London, Yale University Press.

Ray, K. And Thomas, J. 2003 In the kinship of cows: the social centrality of cattle in the earlier Neolithic of southern Britain. In M. Parker Pearson (ed.) *Food, culture and identity in the Neolithic and Early Bronze Age*, 37-44. Oxford, British Archaeological Reports, International Series 117.

Reitz, E. J. and Wing E. S. 1999. *Zooarchaeology*. Cambridge, Cambridge University Press.

Reynier, M.J. 1998 Early Mesolithic settlement in England and Wales: Some preliminary observations. In J. Wymer, N. Ashton, F. Healy and P. Pettitt (eds.) Stone Age archaeology: essays in honour of John Wymer. Oxford, Oxbow.

Reynier, M.J. 2000. Thatcham revisited: Spatial and stratigraphic analyses of two subassemblages from site III and its implications for Early Mesolithic typo-chronology in Britain. In R. Young (ed.) *Mesolithic lifeways: Current research from Britain and Ireland*, 33-46. University of Leicester, Leicester Archaeology Monographs 7.

Richards, J. 1990. *The Stonehenge Environs Project*. London, English Heritage.

Richards, M. Harvati, K. Grimes, V. Smith, C. Smith, T. Hublin, J-J, Karkanas, P. and Panagopoulou, E. 2008. Strontium isotope evidence of Neanderthal mobility at the site of Lakonis, Greece using laser-ablation PIMMS. *Journal of Archaeological Science* 35, 1251-6.

Richards, M. P. 2009. Stable isotope values. In R. Mercer and F. Healy (eds.) *Hambledon Hill, Dorset, England. Excavation and survey of a Neolithic monument complex and its surrounding landscape volume 2*, 522-7. Swindon, English Heritage.

Richards, M.P. and Hedges, R.E.M. 1999. A Neolithic revolution? New evidence of diet in the in the British Neolithic. *Antiquity* 73,891-7.

Richards, C. and Thomas, J. 1984. Ritual activity and structured deposition in later Neolithic Wessex. In R. Bradley and J. Gardiner (eds.) *Neolithic Studies*, 189-218. British Archaeological Report 133.

Robinson, M. 2000. Further considerations of Neolithic charred cereals, fruit and nuts. In A.S. Fairbairn (ed.) *Plants in Neolithic Britain and beyond*, 85-90. Oxford, Oxbow.

Rosman, A. and Rubel, P.G. 1989. Stalking the wild pig: hunting and horticulture in Papua New Guinea. In S. Kent (ed.) *Farmers as hunters, the implications of sedentism*, 25-36. Cambridge, Cambridge University Press.

Rowley-Conwy, P. 1995. Wild or domestic? On the Evidence for the earliest domestic cattle and pigs in south Scandinavia and Iberia. *International Journal of Osteoarchaeology* 5, 115-26.

Rowley-Conwy, P. 1998. Faunal remains and antler artefacts. In P. Mellars (ed.) *Star Carr in context*, 99-107. Cambridge, McDonald Institute papers.

Rowley-Conwy, P. 2000. Milking caprines, hunting pigs: the Neolithic economy of Arene Candide in its west Mediterranean context. In P.Rowley-Conwy (ed.) *Animal bones, human societies*, 124-32.

Rowley-Conwy, P. 2001. Determination of season of death in European Wild Boar (*Sus scrofa ferus*): a preliminary study. In A. Millard (ed.) *Archaeological Sciences '97. Proceedings of the Conference Held at the University of Durham 2nd – 4th September 1997*, 133-9. Oxford, British Archaeological Reports, International Series 939.

Rowley-Conwy, P. 2003. Early domestic animals in Europe: imported or locally domesticated? In A.J. Ammerman and P. Biagi (eds.) *The widening harvest, the Neolithic transition in Europe: looking back, looking forward*, 99-117. Boston, Archaeological Institute of America.

Rowley-Conwy, P. 2004. How the west was lost, a reconsideration of agricultural origins in Britain, Ireland and southern Scandinavia. *Current Anthropology* 45, 83-113.

Rowley-Conwy P. Albarella U & Dobney K. 2010. Wild boar and domestic pigs: approaches and methods. In Zooarchaeology Volume 1. Collection of Papers from International Conference of Zooarchaeology in Zhengzhou, China, 2007, pp.70-115. Bejing: Cultural Relics Press.

Rowley-Conwy, P. and Dobney, K. 2007. Wild boar and domestic pigs in Mesolithic and Neolithic southern Scandinavia. In U. Albarella, K. Dobney, A. Ervynck and P. Rowley-Conwy (eds.) *Pigs and humans; 10,000 years of interaction*, 131-55. Oxford, Oxford University Press.

Ryan, K. (2005) Facilitating milk let-down in traditional cattle herding systems: East Africa and Beyond. In J. Mulville and A.K. Outram (eds.) *The Zooarchaeology of Fats, Oils, Milk and Dairying*, 96-107. Proceedings of the 9th ICAZ Conference of the International Council of Archaeozoology.

Salzman, P.C. 2004. *Pastoralists: equality, hierarchy and the state*. Colorado, Westview Press.

Santos, P. Fernández-Lario, P. Fonseca, C. Monzón, A. Bento, P. Soares, A.M.V.M.

Mateos-Quesada, P. and Petrucci-Fonseca, F. 2006. Habitat and reproductive phenology of wild boar (*Sus scrofa*) in western Iberian peninsula. *European Journal of Wildlife Research* 52(3), 207-12.

Saville, A. 1990. *Hazleton North, Gloucestershire, 1979-82: The excavation of a Neolithic long cairn of the Cotswold-Severn group*. Swindon, English Heritage.

Schadla-Hall, T. 1990.. The Vale of Pickering in the early Mesolithic in context. In: C. Bonsall (ed.) *The Mesolithic in Europe*. John Donald Publishers, Edinburgh. 218-24.

Stallibrass, S. 1995 Review of the vertebrate remains. In: J P. Huntley and S. Stallibrass (eds.) *Plant and vertebrate remains from archaeological sites in northern England: data reviews and future directions.* Research Report No.4. Architectural and Archaeological Society of Durham and Northumberland. Durham. 84-198.

Scheu, A. Hartz, S. Schmölcke, U. Tresset, A. Burger, J. Bollongino, R. 2008. Ancient DNA provides no evidence for independent domestication of cattle in Mesolithic Rosenhof, northern Germany. *Journal of Archaeological Science* 35, 1257-64.

Schley, L. and Roper, T.J. 2003. Diet of wild boar *Sus scrofa* in western Europe, with particular reference to consumption of agricultural crops. *Mammal Review* 33(1), 43-56.

Schulting, R.J. and Richards, M.P. 2002. Finding the coastal Mesolithic in southwest Britain. *Antiquity* 76, 1011-25.

Schweissing, M.M. and Grupe, G. 2003. Tracing migration events in man and cattle by stable isotope analysis of appositionally grown mineralized tissue. *International Journal of Osteoarchaeology* 13(1-2), 96-103.

Sealy, J. Armstrong, R. and Schrire, C. 1991. Beyond lifetime averages: tracing life histories through isotopic analysis of different calcified tissues from archaeological human skeletons. Antiquity 69(263), 290-300.

Serjeantson, D. 2006. Food or Feast at Neolithic Runnymede. In D. Serjeantson and D. Field (eds.) Animals in the Neolithic of Britain and Europe, 113-34. Oxford, Oxbow.

Serjeantson, D.1991. 'Rid grasse of bones': a taphonomic study of the bones from midden deposits at the Neolithic and Bronze Age site of Runnymede, Surrey, England. *International Journal of Osteoarchaeology* 1, 73-89.

Serjeantson, D. 1996. The animal bones. In S. Needham and T. Spence (eds.) *Refuse and disposal at area 16 east, Runnymede*, 194-223. London, British Museum Press.

Serjeantson, D. 2006. Food or Feast at Neolithic Runnymede? In D. Serjeantson and D. Field (eds.) *Animals in the Neolithic of Britain and Europe.* Oxford, Oxbow.

Sharples, N.M. (ed.) 1991. *Maiden Castle, excavations and field survey 1985-6*. London, English Heritage Archaeological Reports 19.

Sheridan, A. 2000. Achnacreebeag and its French connections: vive the "auld alliance". In J.C. Henderson (ed.) *The prehistory and early history of Atlantic Europe*, 1-16. Oxford, British Archaeological Report International Series 861.

Sherratt, A. 1981. Plough and pastoralism: aspects of the secondary products revolution. In I. Hodder, G. Isaac and N. Hammond (eds.) *Pattern of the past, studies in honour of David Clarke*, 261-305. Cambridge, Cambridge University Press.

Sherratt, A. 1983. The secondary exploitation of animals in the old world. *World Archaeology* 15(1), 90-104.

Sherratt, A. 1995. Instruments of conversion? The role of megaliths in the Mesolithic/Neolithic transition in north west Europe. *Oxford Journal of Archaeology* 14, 245-60.

Sherratt, A. 1997. Climatic cycles and behavioural revolutions: the emergence of modern humans and the beginning of farming. *Antiquity* 71, 271-87.

Sillen, A. and Sealy, J.C. 1995. Diagenesis of Strontium in Fossil Bone: A Reconsideration of Nelson *et al.* (1986). *Journal of Archaeological Science* 22, 313-20.

Silver, I.A. 1970. The ageing of domestic animals. In D.R. Brothwell and E.S. Higgs (eds.) *Science in Archaeology: a survey of progress and research*, 283-302. New York, Praeger Publishing.

Simmons, I.G. Dimbleby, G.W. and Grigson, C. 1981. The Mesolithic. In I. Simmons and M. Tooley (eds.) *The environment in British prehistory,* 82-124. London, Duckworth.

Simmons, I.G. 1996. *The environmental impact of later Mesolithic cultures.* Edinburgh, Edinburgh University Press.

Singer, F.J. Otto, D.K. Tipton, A.R. Hable, C.P. 1981. Home ranges, movements and habitat use of European wild boar in Tennessee. *Journal of wildlife management* 45(2), 343-53.

Smith, I.F. 1958. Excavations at Windmill Hill, Avebury, Wilts, 1957-8. *Wiltshire archaeology and natural history magazine* 57, 149-62.

Smith, I.F. 1959. *Windmill Hill and Avebury : a short account of the excavations, 1925-1939.* London, Clover Press.

Smith, I.F. (ed.) 1965. *Windmill Hill and Avebury, excavations by Alexander Keiller 1925-1939.* Oxford, Clarendon press.

Smith, I.F. 1976. Pottery discussion. In C.L. Matthews (ed.) *Occupation sites on a Chiltern ridge. Part 1: Neolithic, Bronze Age and Early Iron Age*, 9-11. British Archaeological Report 29.

Smith, C. 1992. *Late stone age hunters of the British Isles*. London, Routledge.

Street, M. 1999. Remains of aurochs (*Bos primigenius*) from the Early Mesolithic site Bedburg-Königshoven (Rhineland, Germany). In G-C. Weniger (ed.) *Archäologie und biologie des Auerochsen*, 173-94. Mettmann, Neanderthal Museum, Wissenschafliche Schriften 1.

Sykes, N. White, J. Hayes, T.E. and Palmer, M.R. 2006. Tracking animals using strontium isotopes in teeth: the role of fallow deer (*Dama dama*) in Roman Britain. *Antiquity* 80, 948-59.

Tani, Y. 2005. Early techniques as forerunner of milking practices. In J. Mulville and A.K. Outram (eds.) *The Zooarchaeology of Fats, Oils, Milk and Dairying*, 14-12. Oxford, Oxbow.

Teichert, M. 1993. Size and utilisation of the most important domesticated animals in central Europe from the beginning of domestication until the late Middle Ages. In A. Clason, S.Payne and H-P. Uerpmann (eds.) *Skeletons in her cupboard, festschrift for Juliet Clutton-Brock*, 235-8. Oxford, Oxbow Monograph 34.

Thomas, J. 1988. The social significance of Cotswold-Severn Burial practices. *Man new series* 23(3), 540-59.

Thomas, J. 1991. *Understanding the Neolithic.* London, Routledge.

Thomas, J. 1996. The cultural context of the first use of domesticates in north-west Europe. In D. Harris (ed.) *The origins of agriculture and pastoralism in Eurasia*, 357-94. London, Routledge.

Thomas, J. 1997. The materiality of the Mesolithic-Neolithic transition in Britain. *Analecta Prehistorica Leidensia* 29, 57-64.

Thomas, J. 2008. The Mesolithic-Neolithic transition in Britain. In J.Pollard (ed.) *Prehistoric Britain*, 58-89. Oxford, Blackwell.

Thorpe, I.J. 1996. *The origins of agriculture in Europe*. London and New York, Routledge.

Thomas, J.S. and Whittle, A. 1986. Anatomy of a tomb: West Kennet revisited. *Oxford Journal of Archaeology* 5, 129-56.

Timofeev, V.I. 1998. The beginning of the Neolithic in the eastern Baltic. In M. Zvelebil, R. Dennell and L. Domanska (eds.) *Harvesting the Sea, Farming the Forest*, 225-36. Sheffield, Sheffield Academic Press.

Towers, J. Montgomery, J. Evans, J. Jay, M. and Parker Pearson, M. 2010. An investigation of the origins of cattle and aurochs deposited in the Early Bronze Age barrows at Gayhurst and Irthlingborough. *Journal of Archaeological Science* 37, 508-515.

Tresset, A. 1997. L'approvisionnement carné Cerny dans le context Neolithique du Bassin Parisien. In C. Constantin, D. Mordant and D. Simonin (eds.) *La culture de Cerny: Nouvelle économie, nouvelle société au Néolithique*, 299-314. Mémoires du muse de préhistoire d'lle-de-France 6.

Tresset, A. 2000. Early husbandry in Atlantic areas. Animal introductions, diffusions of techniques and native acculturation at the north-west fringe of Europe. In J. Henderson (ed.) *The prehistory and early history of Atlantic Europe*, 17-32. Oxford, British Archaeological Reports, International Series 861.

Tresset, A. 2003. French Connections II: of cows and men. In I. Armit, E. Murphy, E. Nelis, D. Simpson (eds.) *Neolithic Settlement in Ireland and Western Britain.* Oxford, Oxbow Books.

Tresset, A. and Vigne, J-D. 2007. Substitution of species, techniques and symbols at the Mesolithic-Neolithic transition in Western Europe. In A.Whittle and V.Cummings (eds.) *Going over: The Mesolithic-Neolithic transition in north-west Europe*, 189-210. Oxford, Oxford University Press.

Troy, C.S. MacHugh, D.E. Bailey, J.F. Magee, D.A. Loftus, R.T. Cunningham, P. Chamberlain, A.T. Sykes, B.C. Bradley, D.G. 2001. Genetic evidence for Near Eastern origins of European cattle. *Nature* 410, 1088-91.

Uchiyama, J. Clutton-Brock, J. and Rowley-Conwy, P. 2000. Animal bones from the excavations at Seamer Carr, Yorkshire, 1977-86. Unpublished report.

Uerpmann, H-P. 1978. Metrical analysis of faunal remains from the Middle East. In R.H. Meadow and M.A. Zeder (eds.) *Approaches to faunal analysis in the Middle East*, 41-5. Peabody Museum Bulletin 2. Cambridge Mass., Harvard University.

Uerpmann, H-P. 1987. *The ancient distribution of ungulate mammals in the Middle East.* Wiesbaden, Dr. Ludwig Reichert Verlag.

van der Merwe, N.J. and Medina, E. 1991. The canopy effect, carbon isotope ratios and foodwebs in Amazonia. *Journal of Archaeological Science* 18, 249-59.

Vera, F.W.M. 2000. Grazing ecology and forest history. Wallingford, CABI publication.

Vigne, J-D. Helmer, D. and Peters, J. 2005. New archaeozoological approaches to trace the first steps of animal domestication. In J-D. Vigne, J. Peters and D. Helmer (eds.) *The first steps of animal domestication*, 1-16. Oxford, Oxbow.

Viner, S. 2011. Cattle and pig husbandry in the British Neolithic. In A. Hadjikoumis, E. Robinson and S. Viner (eds.) *The Dynamics of Neolithisation in Europe: Studies in honour of Andrew Sherratt*, 313-52. Oxford, Oxbow books.

Viner, S. Evans, J. Albarella, U. and Parker Pearson, M. 2010. Cattle mobility in prehistoric Britain: strontium isotope analysis of cattle teeth from Durrington Walls (Wiltshire, Britain). *Journal of Archaeological Science* 37, 2812-20.

van Vuure, C. 2005. *Retracing the aurochs*. Sofia, Pensoft.

Waddell, J. 1998. *The prehistoric archaeology of Ireland.* Galway, Galway University Press.

Waddington, C. 1996. Putting rock art to use; a model of Early Neolithic transhumance in north Northumberland. *Northern Archaeology* 13/14, 147-77.

Wainwright, G.J. 1979. *Mount Pleasant, Dorset: Excavations 1970-1971.* London, the Society of Antiquaries of London.

Wainwright, G.J. Evans, J.G. and Longworth, I.H. 1971. The excavation of a Late Neolithic enclosure at Marden, Wiltshire. *The Antiquaries Journal* 51(2), 177-239.

Wainwright, G.J. and Longworth, I.H. 1971. *Durrington Walls: Excavations 1966-1968.* London, The Society of Antiquaries of London.

Walker, M.J. 1983. Laying a mega-myth: dolmens and drovers in prehistoric Spain. *World Archaeology* 15(1), 37-50.

Welinder, S. 1998. Pre-neolithic farming in the Scandinavian peninsula. In M. Zvelebil, R. Dennell and L. Domańska (eds.) *Harvesting the sea, farming the forest. The emergence of Neolithic societies in the Baltic region*, 165-74. Sheffield, Sheffield Academic Press.

White, R. and Barker, P. 1998. Wroxeter: Life and death of a Roman city. Stroud, Tempus.

Whittle, AW.R. 1977. *The earlier Neolithic of s. England and its continental background.* British Archaeological Report 35.

Whittle, A. 1996. *Europe in the Neolithic, the creation of new worlds.* Cambridge, Cambridge University Press.

Whittle, A. 1997. Moving on and moving around: Neolithic settlement mobility. In P. Topping (ed.) *Neolithic landscapes*, 15-22. Neolithic studies group seminar papers 2. Oxbow monograph 86. Oxford, Oxbow.

Wickham-Jones, C. 2005. Summer walkers? – mobility and the Mesolithic. In N. Milner and P. Woodman (eds.) *Mesolithic studies at the beginning of the 21st century*, 30-41. Oxford, Oxbow.

van Wijngaarden-Bakker, L.H. 1974. The animal remains from the Beaker settlement at Newgrange, co. Meath: first report. *Proceedings of the Royal Irish Academy* 74, 11-112.

van Wijngaarden-Bakker, L.H. 1986. The animal remains from the Beaker settlement at Newgrange, co. Meath: final report. *Proceedings of the Royal Irish Academy* 86c, 17-111.

van Wijngaarden-Bakker, L.H. 1990. Faunal remains and the Irish Mesolithic. In C. Bonsall (ed.) *The Mesolithic in Europe*, 125-33. Edinburgh, John Donald Publishers.

Wilcox, J.T. and Van Vuren, D.H. 2009. Wild pigs as predators in oak woodlands of California. *Journal of Mammalogy* 90(1), 114-8.

Williams, E. 1989. Dating the introduction of food production into Britain and Ireland. *Antiquity* 63, 510-21.

Wilson, C.J. 2003. Distribution and status of feral wild boar *Sus scrofa* in Dorset, southern England. Mammal review 33(3), 302-7.

Woodman, P. McCarthy, M. and Monaghan, N. 1997. The Irish quaternary fauna Project. *Quaternary science reviews* 16, 129-59.

Woodman, P.C. Anderson, E. and Finlay, N. 1999. *Excavations at Ferriter's Cove, 1983-5*. Bray, Wordwell.

Woodman, P. 2000. Getting back to basics: transitions to farming in Ireland and Britain. In T.D. Price (ed.) *Europe's first farmers*, 219-59. Cambridge, Cambridge University Press.

Wymer, J.J. 1958. Excavations at the Mesolithic Site at Thatcham, Berkshire- 1958. *Transactions of the Newbury District Field Club* 10, 31-48.

Wymer, J.J. 1959. Excavations at the Mesolithic site at Thatcham, Berks – 1958. Interim report. *Berkshire Archaeological Journal* 57, 1-24.

Wymer, J.J. 1962. Excavations at the Maglemosian sites at Thatcham, Berkshire,
England. *Proceedings of the Prehistoric Society* 13, 329-54.

Yalden, D. 1999. *The History of British Mammals*. London, T. and A.D. Poyser.

Yalden, D. and Kitchener, A.C. 2008. History of the fauna. In S. Harris and D. Yalden (eds.) *Mammals of the British Isles*, 17-31. Southampton, The Mammal Society.

Zaliznyak, L. 1998. The ethnographic record, and structural changes in the prehistoric hunter-gatherer economy of boreal Europe. In M. Zvelebil, R. Dennell and L. Domańska (eds.) *Harvesting the Sea, Farming the Forest. The Emergence of Neolithic Societies in the Baltic Region*, 45-51. Sheffield, Sheffield Academic Press.

Zeder, M.A. 2006. Archaeological approaches to documenting animal domestication. In M.A. Zeder, D.G. Bradley, E. Emshwiller and B.D. Smith (eds.) *Documenting domestication: new genetic and archaeological paradigms*, 171-80.

Zeder, M.A. and Hesse, B. 2000. The initial domestication of goats (*Capra hircus*) in the Zagros mountains 10,000 years ago. *Science* 287(5461), 2254-7.

Zeiler, J.T. 2006. Mammals. In L.P. Louwe Kouijmans and P.F.B. Jongste (eds.) *Schipluiden: a Neolithic settlement on the Dutch north sea coast c. 3500 cal BC*, 375-420. Analecta Praehistorica Leidensia 37/38. Leiden, Leiden University Press.

Zeuner, F. 1963a. *A history of domesticated animals*. London, Hutchinson.

Zeuner, F. 1963b. The history of the domestication of cattle. In A.E. Mourant and F.E. Zeuner (eds.) *Man and cattle*, 9-15. London, Royal Anthropological Institute.

Zvelebil, M. 1995. Hunting, gathering, or husbandry? Management of food resources by the Late Mesolithic communities of temperate Europe. In D.V. Campana (ed.) *Before farming: hunter-gatherer society and subsistence*, 79-104. MASCA research papers in science and archaeology volume 12. Philadelphia, University of Pennsylvania.

Zvelebil, M. 1996. Agricultural Frontiers, Neolithic Origins, and the Transition to Farming in the Baltic Basin. In M. Zvelebil, R. Dennell and L. Domanska (eds.) *Harvesting the Sea, Farming the Forest*, 9-27. Sheffield, Sheffield Academic Press.

www.ingramcontent.com/pod-product-compliance
Lightning Source LLC
LaVergne TN
LVHW071208210926
848482LV00015B/752